SOTHIS PRESS

Sothis Press aims to completely reimagine the public domain book with an emphasis on universal access to captivatingly designed print editions. We believe that technology will enable every single book ever published to be perpetually accessible in beautiful print form. Sothis is beauty in ubiquity.

www.sothis.press
1 Waterhouse Square, London, EC1N 2ST

Theory and Calculation of Alternating Current Phenomena

Charles Proteus Steinmetz

SOTHIS PRESS

ISBN 978-1-0365-1254-5

This edition is published by
Sothis Press © 2023. The company
is registered in England and Wales.
The Company Number is 14900389.

THEORY AND CALCULATION

OF

ALTERNATING CURRENT PHENOMENA

BY

CHARLES PROTEUS STEINMETZ

WITH THE ASSISTANCE OF

ERNST J. BERG

THIRD EDITION, REVISED AND ENLARGED

NEW YORK
McGRAW PUBLISHING COMPANY
114 LIBERTY STREET
1900

DEDICATED

TO THE

MEMORY OF MY FATHER,

CARL HEINRICH STEINMETZ.

PREFACE TO THE THIRD EDITION.

In preparing the third edition, great improvements have been made, and a considerable part of the work entirely rewritten, with the addition of much new material. A number of new chapters have been added, as those on vector representation of double frequency quantities as power and torque, and on symbolic representation of general alternating waves. Many chapters have been more or less completely rewritten and enlarged, as those on the topographical method, on distributed capacity and inductance, on frequency converters and induction machines, etc., and the size of the volume thereby greatly increased.

The denotations have been carried through systematically, by distinguishing between complex vectors and absolute values throughout the text; and the typographical errors which had passed into the first and second editions, have been eliminated with the utmost care.

To those gentlemen who so materially assisted me by drawing my attention to errors in the previous editions, I herewith extend my best thanks, and shall be obliged for any further assistance in this direction. Great credit is due to the publishers, who have gone to very considerable expense in bringing out the third edition in its present form, and carrying out all my requests regarding changes and additions. Many thanks are due to Mr. Townsend Wolcott for his valuable and able assistance in preparing and editing the third edition.

CHARLES PROTEUS STEINMETZ.

CAMP MOHAWK, VIELE'S CREEK,
July, 1900.

PREFACE TO FIRST EDITION.

THE following volume is intended as an exposition of the methods which I have found useful in the theoretical investigation and calculation of the manifold phenomena taking place in alternating-current circuits, and of their application to alternating-current apparatus.

While the book is not intended as first instruction for a beginner, but presupposes some knowledge of electrical engineering, I have endeavored to make it as elementary as possible, and have therefore only used common algebra and trigonometry, practically excluding calculus, except in §§ 106 to 115 and Appendix II.; and even §§ 106 to 115 have been paralleled by the elementary approximation of the same phenomenon in §§ 102 to 105.

All the methods used in the book have been introduced and explicitly discussed, with instances of their application, the first part of the book being devoted to this. In the investigation of alternating-current phenomena and apparatus, one method only has usually been employed, though the other available methods are sufficiently explained to show their application.

A considerable part of the book is necessarily devoted to the application of complex imaginary quantities, as the method which I found most useful in dealing with alternating-current phenomena; and in this regard the book may be considered as an expansion and extension of my paper on the application of complex imaginary quantities to electrical engineering, read before the International Electrical Congress at Chicago, 1893. The complex imaginary quantity is gradually introduced, with full explanations, the algebraic operations with complex quantities being discussed in Appendix I., so as not to require from the reader any previous knowledge of the algebra of the complex imaginary plane.

While those phenomena which are characteristic to polyphase systems, as the resultant action of the phases, the effects of unbalancing, the transformation of polyphase systems, etc., have been discussed separately in the last chapters, many of the investigations in the previous parts of the book apply to polyphase systems as well as single-phase circuits, as the chapters on induction motors, generators, synchronous motors, etc.

A part of the book is original investigation, either published here for the first time, or collected from previous publications and more fully explained. Other parts have been published before by other investigators, either in the same, or more frequently in a different form.

I have, however, omitted altogether literary references, for the reason that incomplete references would be worse than none, while complete references would entail the expenditure of much more time than is at my disposal, without offering sufficient compensation; since I believe that the reader who wants information on some phenomenon or apparatus is more interested in the information than in knowing who first investigated the phenomenon.

Special attention has been given to supply a complete and extensive index for easy reference, and to render the book as free from errors as possible. Nevertheless, it probably contains some errors, typographical and otherwise; and I will be obliged to any reader who on discovering an error or an apparent error will notify me.

I take pleasure here in expressing my thanks to Messrs. W. D. Weaver, A. E. Kennelly, and Townsend Wolcott, for the interest they have taken in the book while in the course of publication, as well as for the valuable assistance given by them in correcting and standardizing the notation to conform with the international system, and numerous valuable suggestions regarding desirable improvements.

Thanks are due also to the publishers, who have spared no effort or expense to make the book as creditable as possible mechanically.

CHARLES PROTEUS STEINMETZ.

January, 1897.

CONTENTS.

Appendix II. **Oscillating Currents.—**

THEORY AND CALCULATION

OF

ALTERNATING-CURRENT PHENOMENA.

CHAPTER I.

INTRODUCTION.

1. In the practical applications of electrical energy, we meet with two different classes of phenomena, due respectively to the continuous current and to the alternating current.

The continuous-current phenomena have been brought within the realm of exact analytical calculation by a few fundamental laws :—

1.) Ohm's law : $i = e / r$, where r, the resistance, is a constant of the circuit.

2.) Joule's law : $P = i^2 r$, where P is the rate at which energy is expended by the current, i, in the resistance, r.

3.) The power equation : $P_0 = ei$, where P_0 is the power expended in the circuit of E.M.F., e, and current, i.

4.) Kirchhoff's laws :

a.) The sum of all the E.M.Fs. in a closed circuit = 0, if the E.M.F. consumed by the resistance, ir, is also considered as a counter E.M.F., and all the E.M.Fs. are taken in their proper direction.

b.) The sum of all the currents flowing towards a distributing point = 0.

In alternating-current circuits, that is, in circuits conveying currents which rapidly and periodically change their direction, these laws cease to hold. Energy is expended, not only in the conductor through its ohmic resistance, but also outside of it; energy is stored up and returned, so that large currents may flow, impressed by high E.M.Fs., without representing any considerable amount of expended energy, but merely a surging to and fro of energy; the ohmic resistance ceases to be the determining factor of current strength; currents may divide into components, each of which is larger than the undivided current, etc.

2. In place of the above-mentioned fundamental laws of continuous currents, we find in alternating-current circuits the following:

Ohm's law assumes the form, $i = e / z$, where z, the apparent resistance, or *impedance*, is no longer a constant of the circuit, but depends upon the frequency of the currents; and in circuits containing iron, etc., also upon the E.M.F.

Impedance, z, is, in the system of absolute units, of the same dimensions as resistance (that is, of the dimension LT^{-1} = velocity), and is expressed in ohms.

It consists of two components, the resistance, r, and the reactance, x, or —

$$z = \sqrt{r^2 + x^2}.$$

The resistance, r, in circuits where energy is expended only in heating the conductor, is the same as the ohmic resistance of continuous-current circuits. In circuits, however, where energy is also expended outside of the conductor by magnetic hysteresis, mutual inductance, dielectric hysteresis, etc., r is larger than the true ohmic resistance of the conductor, since it refers to the total expenditure of energy. It may be called then the *effective resistance.* It is no longer a constant of the circuit.

The reactance, x, does not represent the expenditure of power, as does the effective resistance, r, but merely the surging to and fro of energy. It is not a constant of the circuit, but depends upon the frequency, and frequently, as in circuits containing iron, or in electrolytic conductors, upon the E.M.F. also. Hence, while the effective resistance, r, refers to the energy component of E.M.F., or the E.M.F. in phase with the current, the reactance, x, refers to the wattless component of E.M.F., or the E.M.F. in quadrature with the current.

3. The principal sources of reactance are electro-magnetism and capacity.

ELECTRO—MAGNETISM.

An electric current, i, flowing through a circuit, produces a magnetic flux surrounding the conductor in lines of magnetic force (or more correctly, lines of magnetic induction), of closed, circular, or other form, which alternate with the alternations of the current, and thereby induce an E.M.F. in the conductor. Since the magnetic flux is in phase with the current, and the induced E.M.F. 90°, or a quarter period, behind the flux, this *E.M.F. of self-inductance* lags 90°, or a quarter period, behind the current; that is, is in quadrature therewith, and therefore wattless.

If now Φ = the magnetic flux produced by, and interlinked with, the current i (where those lines of magnetic force, which are interlinked n-fold, or pass around n turns of the conductor, are counted n times), the ratio, Φ / i, is denoted by L, and called *self-inductance*, or the *coefficient of self-induction* of the circuit. It is numerically equal, in absolute units, to the interlinkages of the circuit with the magnetic flux produced by unit current, and is, in the system of absolute units, of the dimension of length. Instead of the self-inductance, L, sometimes its ratio with the ohmic resistance, r, is used, and is called the *Time-Constant* of the circuit:

$$T = \frac{L}{r}.$$

If a conductor surrounds with n turns a magnetic circuit of reluctance, $\mathcal{R}$, the current, i, in the conductor represents the M.M.F. of ni ampere-turns, and hence produces a magnetic flux of $ni / \mathcal{R}$ lines of magnetic force, surrounding each n turns of the conductor, and thereby giving $\Phi = n^2 i / \mathcal{R}$ interlinkages between the magnetic and electric circuits. Hence the inductance is $L = \Phi / i = n^2 / \mathcal{R}$.

The fundamental law of electro-magnetic induction is, that the E.M.F. induced in a conductor by a varying magnetic field is the rate of cutting of the conductor through the magnetic field.

Hence, if i is the current, and L is the inductance of a circuit, the magnetic flux interlinked with a circuit of current, i, is Li, and $4 NLi$ is consequently the average rate of cutting; that is, the number of lines of force cut by the conductor per second, where $N =$ frequency, or number of complete periods (double reversals) of the current per second.

Since the maximum rate of cutting bears to the average rate the same ratio as the quadrant to the radius of a circle (a sinusoidal variation supposed), that is the ratio $\pi / 2 \div 1$, the maximum rate of cutting is $2 \pi N$, and, consequently, the maximum value of E.M.F. induced in a circuit of maximum current strength, i, and inductance, L, is,

$$e = 2 \pi NLi.$$

Since the maximum values of sine waves are proportional (by factor $\sqrt{2}$) to the effective values (square root of mean squares), if $i =$ effective value of alternating current, $e = 2 \pi NLi$ is the effective value of E.M.F. of self-inductance, and the ratio, $e / i = 2 \pi NL$, is the magnetic reactance:

$$x_m = 2 \pi NL.$$

Thus, if $r =$ resistance, $x_m =$ reactance, $z =$ impedance,–

the E.M.F. consumed by resistance is: $e_1 = ir$;
the E.M.F. consumed by reactance is: $e_2 = ix_m$;

and, since both E.M.Fs. are in quadrature to each other, the total E.M.F. is—

$$e = \sqrt{e_1^2 + e_2^2} = i\sqrt{r^2 + x_m^2} = iz;$$

that is, the impedance, z, takes in alternating-current circuits the place of the resistance, r, in continuous-current circuits.

CAPACITY.

4. If upon a condenser of capacity, C, an E.M.F., e, is impressed, the condenser receives the electrostatic charge, Ce.

If the E.M.F., e, alternates with the frequency, N, the average rate of charge and discharge is $4N$, and $2\pi N$ the maximum rate of charge and discharge, sinusoidal waves supposed, hence, $i = 2\pi NCe$ the current passing into the condenser, which is in quadrature to the E.M.F., and leading.

It is then:— $$x_c = \frac{e}{i} = \frac{1}{2\pi NC},$$

the "*capacity reactance*," or "*condensance*."

Polarization in electrolytic conductors acts to a certain extent like capacity.

The capacity reactance is inversely proportional to the frequency, and represents the leading out-of-phase wave; the magnetic reactance is directly proportional to the frequency, and represents the lagging out-of-phase wave. Hence both are of opposite sign with regard to each other, and the total reactance of the circuit is their difference, $x = x_m - x_c$.

The total resistance of a circuit is equal to the sum of all the resistances connected in series; the total reactance of a circuit is equal to the algebraic sum of all the reactances connected in series; the total impedance of a circuit, however, is not equal to the sum of all the individual impedances, but in general less, and is the resultant of the total resistance and the total reactance. Hence it is not permissible directly to add impedances, as it is with resistances or reactances.

A further discussion of these quantities will be found in the later chapters.

5. In Joule's law, $P = i^2 r$, r is not the true ohmic resistance any more, but the "effective resistance;" that is, the ratio of the energy component of E.M.F. to the current. Since in alternating-current circuits, besides by the ohmic resistance of the conductor, energy is expended, partly outside, partly even inside, of the conductor, by magnetic hysteresis, mutual inductance, dielectric hysteresis, etc., the effective resistance, r, is in general larger than the true resistance of the conductor, sometimes many times larger, as in transformers at open secondary circuit, and is not a constant of the circuit any more. It is more fully discussed in Chapter VII.

In alternating-current circuits, the power equation contains a third term, which, in sine waves, is the cosine of the difference of phase between E.M.F. and current:—

$$P_0 = ei \cos \phi.$$

Consequently, even if e and i are both large, P_0 may be very small, if $\cos \phi$ is small, that is, ϕ near 90°.

Kirchhoff's laws become meaningless in their original form, since these laws consider the E.M.Fs. and currents as directional quantities, counted positive in the one, negative in the opposite direction, while the alternating current has no definite direction of its own.

6. The alternating waves may have widely different shapes; some of the more frequent ones are shown in a later chapter.

The simplest form, however, is the sine wave, shown in Fig. 1, or, at least, a wave very near sine shape, which may be represented analytically by:—

$$i = I \sin \frac{2\pi}{T}(t - t_1) = I \sin 2\pi N (t - t_1);$$

where I is the maximum value of the wave, or its *amplitude;* T is the time of one complete cyclic repetition, or the *period* of the wave, or $N = 1/T$ is the *frequency* or number of complete periods per second; and t_1 is the time, where the wave is zero, or the *epoch* of the wave, generally called the *phase.**

Obviously, "phase" or "epoch" attains a practical meaning only when several waves of different phases are considered, as "difference of phase." When dealing with one wave only, we may count the time from the moment

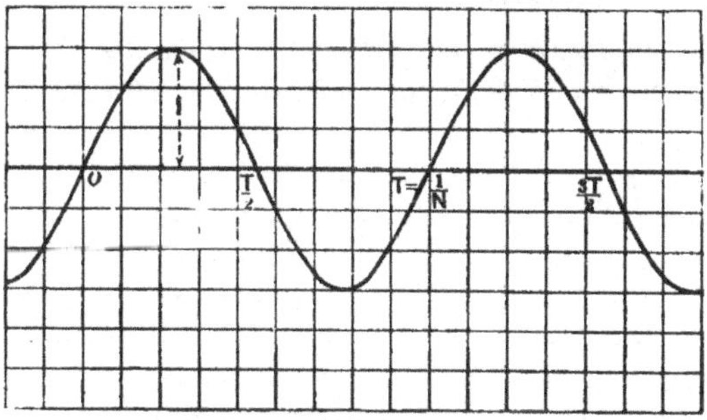

Fig. 1. Sine Wave.

where the wave is zero, or from the moment of its maximum, and then represent it by:—

$$i = I \sin 2\pi Nt;$$

or,

$$i = I \cos 2\pi Nt.$$

Since it is a univalent function of time, that is, can at a given instant have one value only, by Fourier's theorem, any alternating wave, no matter what its shape may be, can be represented by a series of sine functions of different frequencies and different phases, in the form:—

$$i = I_1 \sin 2\pi N(t - t_1) + I_2 \sin 4\pi N(t - t_2) + I_3 \sin 6\pi N(t - t_3) + \ldots$$

* "Epoch" is the time where a periodic function reaches a certain value, for instance, zero; and "phase" is the angular position, with respect to a datum position, of a periodic function at a given time. Both are in alternate-current phenomena only different ways of expressing the same thing.

where $I_1, I_2, I_3, \ldots$ are the maximum values of the different components of the wave, $t_1, t_2, t_3 \ldots$ the times, where the respective components pass the zero value.

The first term, $I_1 \sin 2\pi N(t - t_1)$, is called the *fundamental wave*, or the *first harmonic;* the further terms are called the *higher harmonics*, or "overtones," in analogy to the overtones of sound waves. $I_n \sin 2n\pi N(t - t_n)$ is the nth harmonic.

By resolving the sine functions of the time differences, $t - t_1, t - t_2 \ldots$, we reduce the general expression of the wave to the form:

$$i = A_1 \sin 2\pi Nt + A_2 \sin 4\pi Nt + A_3 \sin 6\pi Nt + \ldots$$
$$+ B_1 \cos 2\pi Nt + B_2 \cos 4\pi Nt + B_3 \cos 6\pi Nt + \ldots$$

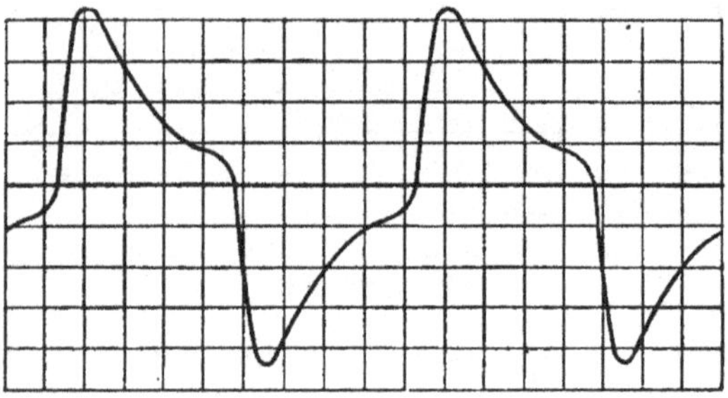

Fig. 2. Wave without Even Harmonics.

The two half-waves of each period, the *positive wave* and the *negative wave* (counting in a definite direction in the circuit), are almost always identical. Hence the even higher harmonics, which cause a difference in the shape of the two half-waves, disappear, and only the odd harmonics exist, except in very special cases.

Hence the general alternating-current wave is expressed by:

$$i = I_1 \sin 2\pi N(t - t_1) + I_3 \sin 6\pi N(t - t_3)$$
$$+ I_5 \sin 10\pi N(t - t_5) + \ldots$$

or,

$$i = A_1 \sin 2\pi Nt + A_3 \sin 6\pi Nt + A_5 \sin 10\pi Nt + \ldots$$
$$+ B_1 \cos 2\pi Nt + B_3 \cos 6\pi Nt + B_5 \cos 10\pi Nt + \ldots$$

Such a wave is shown in Fig. 2, while Fig. 3 shows a wave whose half-waves are different. Figs. 2 and 3 represent the secondary currents of a Ruhmkorff coil, whose secondary coil is closed by a high external resistance: Fig. 3 is the coil operated in the usual way, by make and break of the primary battery current; Fig. 2 is the coil fed with reversed currents by a commutator from a battery.

7. Self-inductance, or electro-magnetic momentum, which is always present in alternating-current circuits, — to a large extent in generators, transformers, etc., — tends to

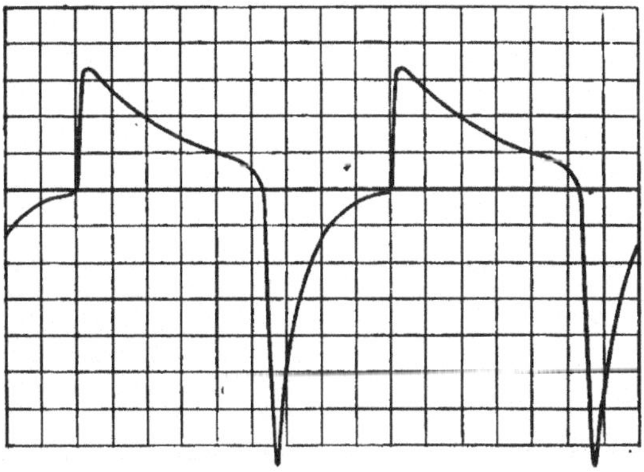

Fig. 3. Wave with Even Harmonics.

suppress the higher harmonics of a complex harmonic wave more than the fundamental harmonic, since the self-inductive reactance is proportional to the frequency, and is thus greater with the higher harmonics, and thereby causes a general tendency towards simple sine shape, which has the effect, that, in general, the alternating currents in our light and power circuits are sufficiently near sine waves to make the assumption of sine shape permissible.

Hence, in the calculation of alternating-current phenomena, we can safely assume the alternating wave as a sine wave, without making any serious error; and it will be sufficient to keep the distorticn from sine shape in mind as a possible disturbing factor, which generally, however, is in practice negligible — perhaps with the only exception of low-resistance circuits containing large magnetic reactance, and large condensance in series with each other, so as to produce resonance effects of these higher harmonics.

CHAPTER II

INSTANTANEOUS VALUES AND INTEGRAL VALUES.

8. In a periodically varying function, as an alternating current, we have to distinguish between the *instantaneous value,* which varies constantly as function of the time, and the *integral value,* which characterizes the wave as a whole.

As such integral value, almost exclusively the *effective*

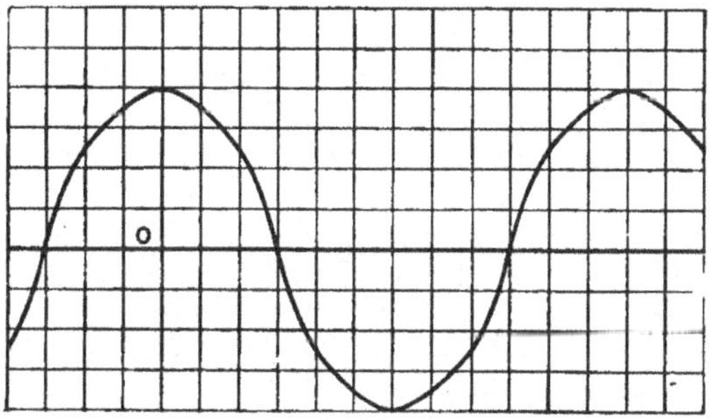

Fig. 4. Alternating Wave.

value is used, that is, the square root of the mean squares; and wherever the intensity of an electric wave is mentioned without further reference, the effective value is understood.

The *maximum value* of the wave is of practical interest only in few cases, and may, besides, be different for the two half-waves, as in Fig. 3.

As *arithmetic mean,* or *average value,* of a wave as in Figs. 4 and 5, the arithmetical average of all the instantaneous values during one complete period is understood.

This arithmetic mean is either = 0, as in Fig. 4, or it differs from 0, as in Fig. 5. In the first case, the wave is called an *alternating wave,* in the latter a *pulsating wave.*

Thus, an alternating wave is a wave whose positive values give the same sum total as the negative values ; that is, whose two half-waves have in rectangular coördinates the same area, as shown in Fig. 4.

A pulsating wave is a wave in which one of the half waves preponderates, as in Fig. 5.

By electromagnetic induction, pulsating waves are produced only by commutating and unipolar machines (or by the superposition of alternating upon direct currents, etc.).

All inductive apparatus without commutation give exclusively alternating waves, because, no matter what conditions may exist in the circuit, any line of magnetic force, which during a complete period is cut by the circuit, and thereby induces an E.M.F., must during the same period be cut again in the opposite direction, and thereby induce the same total amount of E.M.F. (Obviously, this does not apply to circuits consisting of different parts movable with regard to each other, as in unipolar machines.)

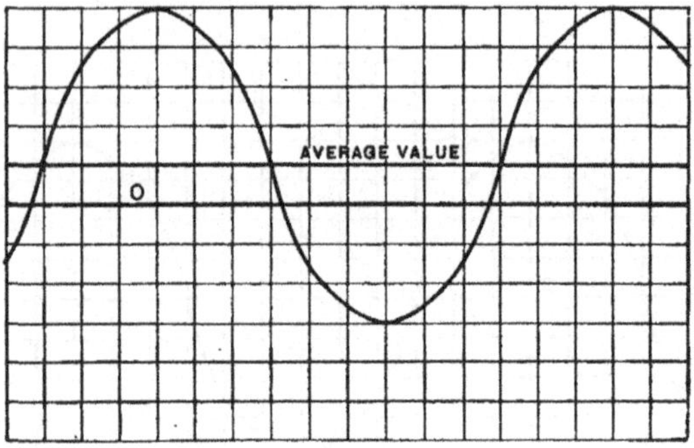

Fig. 5. Pulsating Wave.

In the following we shall almost exclusively consider the alternating wave, that is the wave whose true arithmetic mean value = 0.

Frequently, by mean value of an alternating wave, the average of one half-wave only is denoted, or rather the average of all instantaneous values without regard to their sign. This *mean value* is of no practical importance, and is, besides, in many cases indefinite.

9. In a sine wave, the relation of the mean to the maximum value is found in the following way:—

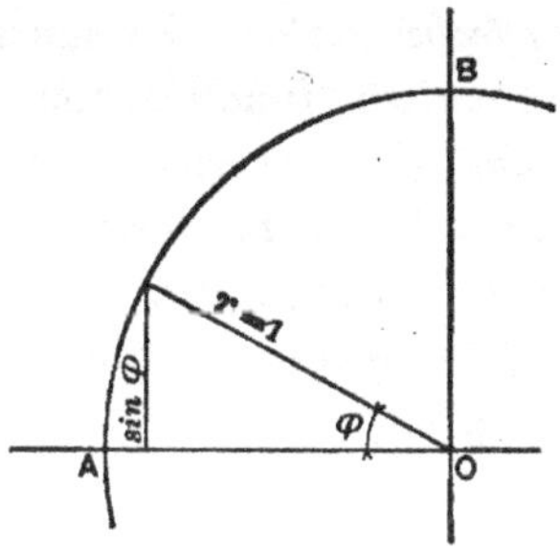

Fig. 6.

Let, in Fig. 6, AOB represent a quadrant of a circle with radius 1.

Then, while the angle ϕ traverses the arc $\pi / 2$ from A to B, the sine varies from 0 to $OB = 1$. Hence the average variation of the sine bears to that of the corresponding arc the ratio $1 \div \pi / 2$, or $2 / \pi \div 1$. The maximum variation of the sine takes place about its zero value, where the sine is equal to the arc. Hence the maximum variation of the sine is equal to the variation of the corresponding arc, and consequently the maximum variation of the sine bears to its average variation the same ratio as the average variation of the arc to that of the sine; that is, $1 \div 2 / \pi$, and since the variations of a sine-function are sinusoidal also, we have,

$$\text{Mean value of sine wave} \div \text{maximum value} = \frac{2}{\pi} \div 1$$
$$= .63663.$$

The quantities, "current," "E.M.F.," "magnetism," etc., are in reality mathematical fictions only, as the components of the entities, "energy," "power," etc.; that is, they have no independent existence, but appear only as squares or products.

Consequently, the only integral value of an alternating wave which is of practical importance, as directly connected with the mechanical system of units, is that value which represents the same *power* or *effect* as the periodical wave. This is called the *effective value.* Its square is equal to the mean square of the periodic function, that is:—

The effective value of an alternating wave, or the value representing the same effect as the periodically varying wave, is the square root of the mean square.

In a sine wave, its relation to the maximum value is found in the following way:

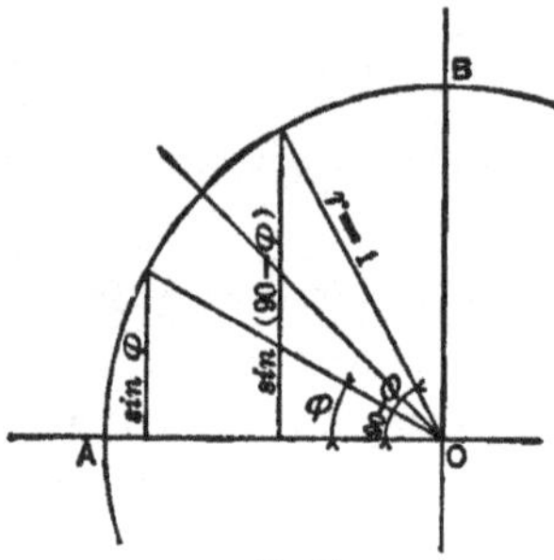

Fig. 7.

Let, in Fig. 7, *AOB* represent a quadrant of a circle with radius 1.

Then, since the sines of any angle ϕ and its complementary angle, $90° - \phi$, fulfill the condition,—

$$\sin^2 \phi + \sin^2 (90 - \phi) = 1,$$

the sines in the quadrant, *AOB*, can be grouped into pairs, so that the sum of the squares of any pair $= 1$; or, in other words, the mean square of the sine $= 1/2$, and the square root of the mean square, or the effective value of the sine, $= 1/\sqrt{2}$. That is:

The effective value of a sine function bears to its maximum value the ratio, —

$$\frac{1}{\sqrt{2}} \div 1 = .70711.$$

Hence, we have for the sine curve the following relations:

Max.	Eff.	Aritm. Mean.	
		Half Period.	Whole Period.
1	$\frac{1}{\sqrt{2}}$	$\frac{2}{\pi}$	0
1	.7071	.63663	0
1.4142	1	.90034	0
1.5708	1.1107	1	0

10. Coming now to the general alternating wave,

$$i = A_1 \sin 2\pi Nt + A_2 \sin 4\pi Nt + A_3 \sin 6\pi Nt + \ldots$$
$$+ B_1 \cos 2\pi Nt + B_2 \cos 4\pi Nt + B_3 \cos 6\pi Nt + \ldots,$$

we find, by squaring this expression and canceling all the products which give 0 as mean square, the *effective value,*—

$$I = \sqrt{\tfrac{1}{2}\left(A_1^2 + A_2^2 + A_3^2 + \cdots + B_1^2 + B_2^2 + B_3^2 \cdots\right)}.$$

The *mean value* does not give a simple expression, and is of no general interest.

CHAPTER III.

LAW OF ELECTRO-MAGNETIC INDUCTION.

11. If an electric conductor moves relatively to a magnetic field, an E.M.F. is induced in the conductor which is proportional to the intensity of the magnetic field, to the length of the conductor, and to the speed of its motion perpendicular to the magnetic field and the direction of the conductor; or, in other words, proportional to the number of lines of magnetic force cut per second by the conductor.

As a practical unit of E.M.F., the *volt* is defined as the E.M.F. induced in a conductor, which cuts $10^8 = 100,000,000$ lines of magnetic force per second.

If the conductor is closed upon itself, the induced E.M.F. produces a current.

A closed conductor may be called a turn or a convolution. In such a turn, the number of lines of magnetic force cut per second is the increase or decrease of the number of lines inclosed by the turn, or n times as large with n turns.

Hence the E.M.F. in volts induced in n turns, or convolutions, is n times the increase or decrease, per second, of the flux inclosed by the turns, times 10^{-8}.

If the change of the flux inclosed by the turn, or by n turns, does not take place uniformly, the product of the number of turns, times change of flux per second, gives the average E.M.F.

If the magnetic flux, Φ, alternates relatively to a number of turns, n — that is, when the turns either revolve through the flux, or the flux passes in and out of the turns, the total flux is cut four times during each complete period or cycle twice passing into, and twice out of, the turns.

Hence, if N = number of complete cycles per second, or the frequency of the flux Φ, the average E.M.F. induced in n turns is,

$$E_{avg.} = 4\, n\, \Phi\, N\, 10^{-8} \text{ volts.}$$

This is the fundamental equation of electrical engineering, and applies to continuous-current, as well as to alternating-current, apparatus.

12. In continuous-current machines and in many alternators, the turns revolve through a constant magnetic field; in other alternators and in induction motors, the magnetic field revolves; in transformers, the field alternates with respect to the stationary turns.

Thus, in the continuous-current machine, if n = number of turns in series from brush to brush, Φ = flux inclosed per turn, and N = frequency, the E.M.F. induced in the machine is $E = 4n\Phi N 10^{-8}$ volts, independent of the number of poles, of series or multiple connection of the armature, whether of the ring, drum, or other type.

In an alternator or transformer, if n is the number of turns in series, Φ the maximum flux inclosed per turn, and N the frequency, this formula gives,

$$E_{avg.} = 4\, n\, \Phi\, N\, 10^{-8} \text{ volts.}$$

Since the maximum E.M.F. is given by, —

$$E_{max.} = \frac{\pi}{2} E_{avg}$$

we have

$$E_{max.} = 2\, \pi\, n\, \Phi\, N\, 10^{-8} \text{ volts.}$$

And since the effective E.M.F. is given by, —

$$E_{eff.} = \frac{E_{max.}}{\sqrt{2}}$$

we have

$$E_{eff.} = \sqrt{2}\, \pi\, n\, \Phi\, N\, 10^{-8}$$
$$= 4.44\, n\, \Phi\, N\, 10^{-8} \text{ volts,}$$

which is the fundamental formula of alternating-current induction by sine waves.

13. If, in a circuit of n turns, the magnetic flux, Φ, inclosed by the circuit is produced by the current flowing in the circuit, the ratio —

$$\frac{\text{flux} \times \text{number of turns} \times 10^{-8}}{\text{current}}$$

is called the *inductance*, L, of the circuit, in henrys.

The product of the number of turns, n, into the maximum flux, Φ, produced by a current of I amperes effective, or $I\sqrt{2}$ amperes maximum, is therefore —

$$n\Phi = LI\sqrt{2}\,10^8;$$

and consequently the effective E.M.F. of self-inductance is:

$$E = \sqrt{2}\,\pi n \Phi N 10^{-8}$$
$$= 2\,\pi NLI \text{ volts.}$$

The product, $x = 2\,\pi NL$, is of the dimension of resistance, and is called the *reactance* of the circuit; and the E.M.F. of self-inductance of the circuit, or the reactance voltage, is

$$E = Ix,$$

and lags 90° behind the current, since the current is in phase with the magnetic flux produced by the current, and the E.M.F. lags 90° behind the magnetic flux. The E.M.F. lags 90° behind the magnetic flux, as it is proportional to the change in flux; thus it is zero when the magnetism is at its maximum value, and a maximum when the flux passes through zero, where it changes quickest.

CHAPTER IV.

GRAPHIC REPRESENTATION.

14. While alternating waves can be, and frequently are, represented graphically in rectangular coördinates, with the time as abscissæ, and the instantaneous values of the wave as ordinates, the best insight with regard to the mutual relation of different alternate waves is given by their representation in polar coördinates, with the time as an angle or the amplitude, — one complete period being represented by one revolution, — and the instantaneous values as radii vectores.

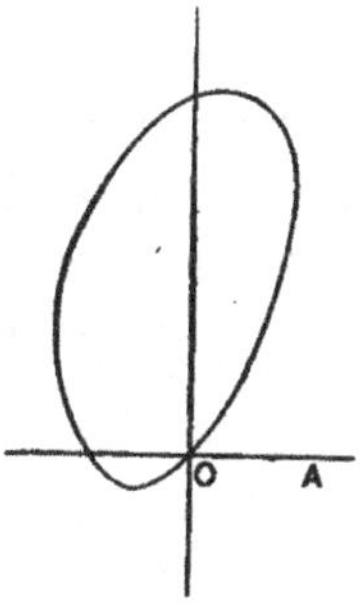

Fig. 8.

Thus the two waves of Figs. 2 and 3 are represented in polar coördinates in Figs. 8 and 9 as closed characteristic curves, which, by their intersection with the radius vector, give the instantaneous value of the wave, corresponding to the time represented by the amplitude of the radius vector.

These instantaneous values are positive if in the direction of the radius vector, and negative if in opposition. Hence the two half-waves in Fig. 2 are represented by the same polar characteristic curve, which is traversed by the point of intersection of the radius vector twice per period, — once in the direction of the vector, giving the positive half-wave,

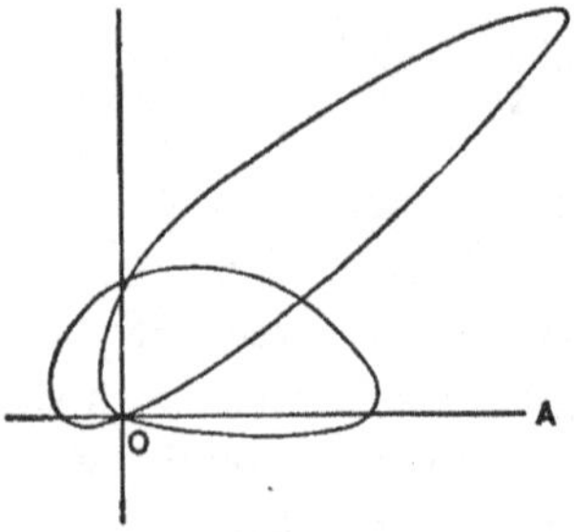

Fig. 9.

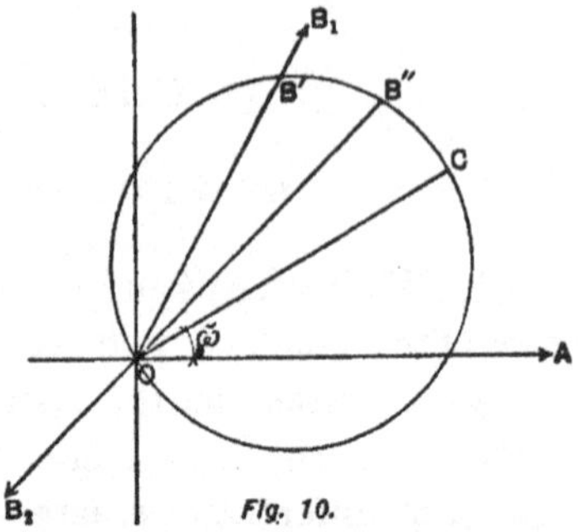

Fig. 10.

and once in opposition to the vector, giving the negative half-wave. In Figs. 3 and 9, where the two half-waves are different, they give different polar characteristics.

15. The sine wave, Fig. 1, is represented in polar coördinates by one circle, as shown in Fig. 10. The diameter of the characteristic curve of the sine wave, $I = \overline{OC}$, represents the *intensity* of the wave; and the amplitude of the diameter, $\overline{OC}$, $\angle \hat{\omega} = AOC$, is the *phase* of the wave, which, therefore, is represented analytically by the function: —

$$i = I \cos (\phi - \hat{\omega}),$$

where $\phi = 2 \pi t / T$ is the instantaneous value of the amplitude corresponding to the instantaneous value, i, of the wave.

The instantaneous values are cut out on the movable radius vector by its intersection with the characteristic circle. Thus, for instance, at the amplitude $AOB_1 = \phi_1 = 2 \pi t_1 / T$ (Fig. 10), the instantaneous value is $\overline{OB'}$; at the amplitude $AOB_2 = \phi_2 = 2 \pi t_2 / T$, the instantaneous value is $\overline{OB''}$, and negative, since in opposition to the radius vector OB_2.

The characteristic circle of the alternating sine wave is determined by the length of its diameter — the intensity of the wave; and by the amplitude of the diameter — the phase of the wave.

Hence, wherever the integral value of the wave is considered alone, and not the instantaneous values, the characteristic circle may be omitted altogether, and the wave represented in intensity and in phase by the diameter of the characteristic circle.

Thus, in polar coördinates, the alternate wave is represented in intensity and phase by the length and direction of a vector, $\overline{OC}$, Fig. 10, and its analytical expression would then be $c = \overline{OC} \cos(\phi - \omega)$.

Instead of the maximum value of the wave, the *effective* value, or square root of mean square, may be used as the vector, which is more convenient; and the maximum value is then $\sqrt{2}$ times the vector $\overline{OC}$, so that the instantaneous values, when taken from the diagram, have to be increased by the factor $\sqrt{2}$.

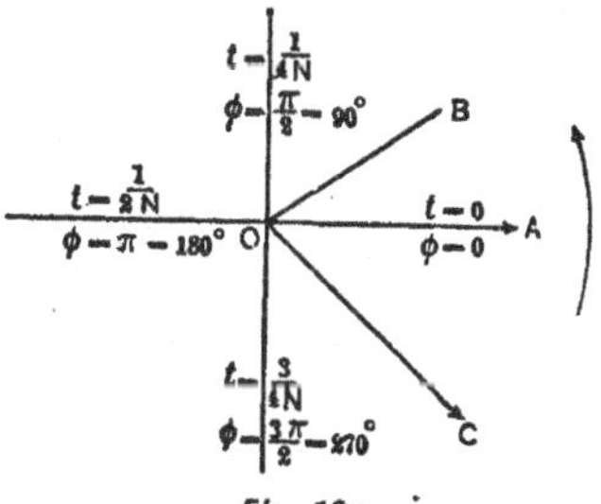

Fig. 10a.

Thus the wave,

$$b = B \cos 2\pi N (t - t_1)$$
$$= B \cos (\phi - \hat{\omega}_1)$$

is in Fig. 10*a* represented by vector $\overline{OB} = \frac{B}{\sqrt{2}}$, of phase $AOB = \hat{\omega}_1$; and the wave,

$$c = C \cos 2\pi N (t + t_2)$$
$$= C \cos (\phi + \hat{\omega}_2)$$

is in Fig. 10*a* represented by vector $\overline{OC} = \frac{C}{\sqrt{2}}$, of phase $AOC = -\hat{\omega}_2$.

The former is said to *lag* by angle $\hat{\omega}_1$, the latter to *lead* by angle $\hat{\omega}_2$, with regard to the zero position.

The wave b lags by angle $(\hat{\omega}_1 + \hat{\omega}_2)$ behind wave c, or c leads b by angle $(\hat{\omega}_1 + \hat{\omega}_2)$.

16. To combine different sine waves, their graphical representations, or vectors, are combined by the parallelogram law.

If, for instance, two sine waves, $\overline{OB}$ and $\overline{OC}$ (Fig. 11), are superposed, — as, for instance, two E.M.F's. acting in the same circuit, — their resultant wave is represented by

$\overline{OD}$, the diagonal of a parallelogram with $\overline{OB}$ and $\overline{OC}$ as sides.

For at any time, *t*, represented by angle $\phi = AOX$, the instantaneous values of the three waves, $\overline{OB}$, $\overline{OC}$, $\overline{OD}$, are their projections upon $\overline{OX}$, and the sum of the projections of $\overline{OB}$ and $\overline{OC}$ is equal to the projection of $\overline{OD}$; that is, the instantaneous values of the wave $\overline{OD}$ are equal to the sum of the instantaneous values of waves $\overline{OB}$ and $\overline{OC}$.

From the foregoing considerations we have the conclusions:

The sine wave is represented graphically in polar coördinates by a vector, which by its length, $\overline{OC}$, denotes the intensity, and by its amplitude, AOC, the phase, of the sine wave.

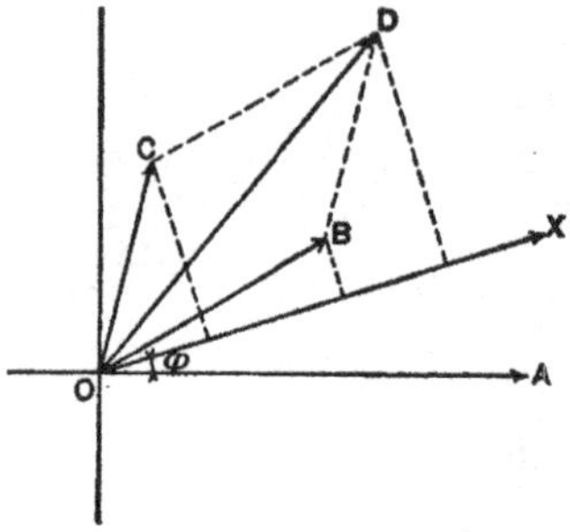

Fig. 11.

Sine waves are combined or resolved graphically, in polar coördinates, by the law of parallelogram or the polygon of sine waves.

Kirchhoff's laws now assume, for alternating sine waves, the form:—

a.) The resultant of all the E.M.Fs. in a closed circuit, as found by the parallelogram of sine waves, is zero if the counter E.M.Fs. of resistance and of reactance are included.

b.) The resultant of all the currents flowing towards a

distributing point, as found by the parallelogram of sine waves, is zero.

The energy equation expressed graphically is as follows:

The power of an alternating-current circuit is represented in polar coördinates by the product of the current, I, into the projection of the E.M.F., E, upon the current, or by the E.M.F., E, into the projection of the current, I, upon the E.M.F., or by $IE \cos (I,E)$.

17. Suppose, as an instance, that over a line having the resistance, r, and the reactance, $x = 2\pi NL$, — where $N =$ frequency and $L =$ inductance, — a current of I amperes be sent into a non-inductive circuit at an E.M.F. of E

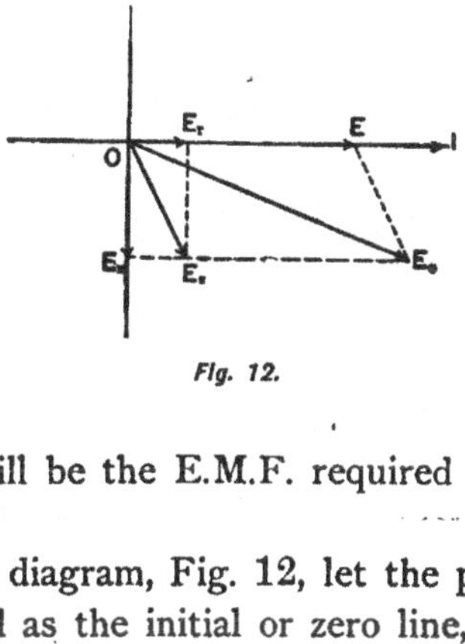

Fig. 12.

volts. What will be the E.M.F. required at the generator end of the line?

In the polar diagram, Fig. 12, let the phase of the current be assumed as the initial or zero line, $\overline{OI}$. Since the receiving circuit is non-inductive, the current is in phase with its E.M.F. Hence the E.M.F., E, at the end of the line, impressed upon the receiving circuit, is represented by a vector, $\overline{OE}$. To overcome the resistance, r, of the line, an E.M.F., Ir, is required in phase with the current, represented by $\overline{OE_r}$ in the diagram. The self-inductance of the line induces an E.M.F. which is proportional to the current I and reactance x, and lags a quarter of a period, or 90°, behind the current. To overcome this counter E.M.F.

of self-induction, an E.M.F. of the value Ix is required, in phase 90° ahead of the current, hence represented by vector $\overline{OE}_x$. Thus resistance consumes E.M.F. in phase, and reactance an E.M.F. 90° ahead of the current. The E.M.F. of the generator, E_0, has to give the three E.M.Fs., E, E_r, and E_x, hence it is determined as their resultant. Combining by the parallelogram law, $\overline{OE}_r$ and $\overline{OE}_x$, give $\overline{OE}_z$, the E.M.F. required to overcome the impedance of the line, and similarly $\overline{OE}_z$ and $\overline{OE}$ give $\overline{OE}_0$, the E.M.F. required at the generator side of the line, to yield the E.M.F. E at the receiving end of the line. Algebraically, we get from Fig. 12—

$$E_0 = \sqrt{(E + Ir)^2 + (Ix)^2};$$

or,

$$E = \sqrt{E_0^2 - (Ix)^2} - Ir.$$

In this instance we have considered the E.M.F. consumed by the resistance (in phase with the current) and the E.M.F. consumed by the reactance (90° ahead of the current) as parts, or components, of the impressed E.M.F., E_0, and have derived E_0 by combining E_r, E_x, and E.

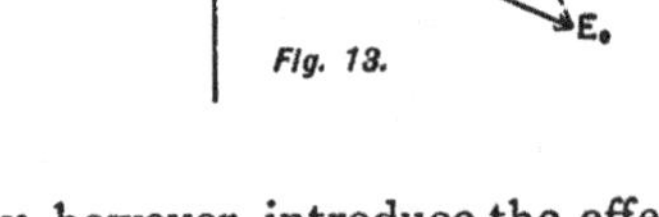

Fig. 13.

18. We may, however, introduce the effect of the inductance directly as an E.M.F., E_x', the counter E.M.F. of self-induction $= Ix$, and lagging 90° behind the current; and the E.M.F. consumed by the resistance as a counter E.M.F., $E_r' = Ir$, but in opposition to the current, as is done in Fig. 13; and combine the three E.M.Fs. E_0, E_r', E_x', to form a resultant E.M.F., E, which is left at the end of the line.

E_r' and E_x' combine to form E_z', the counter E.M.F. of impedance; and since E_z' and E_0 must combine to form E, E_0 is found as the side of a parallelogram, OE_0EE_z', whose other side, $\overline{OE_z'}$, and diagonal, $\overline{OE}$, are given.

Or we may say (Fig. 14), that to overcome the counter E.M.F. of impedance, $\overline{OE_z'}$, of the line, the component, $\overline{OE_z}$, of the impressed E.M.F. is required which, with the other component $\overline{OE}$, must give the impressed E.M.F., $\overline{OE_0}$.

As shown, we can represent the E.M.Fs. produced in a circuit in two ways — either as counter E.M.Fs., which combine with the impressed E.M.F., or as parts, or components,

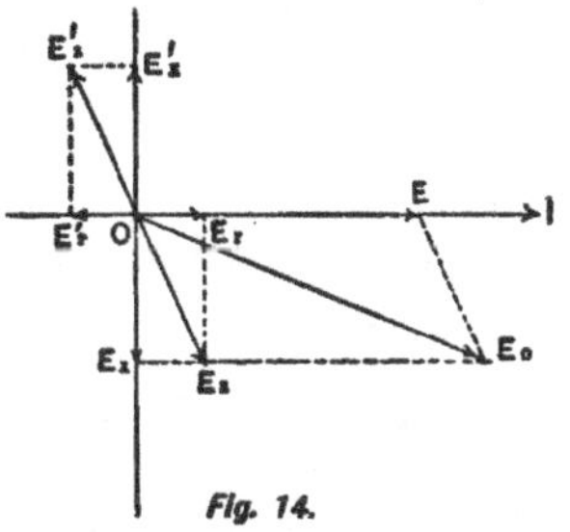

Fig. 14.

of the impressed E.M.F., in the latter case being of opposite phase. According to the nature of the problem, either the one or the other way may be preferable.

As an example, the E.M.F. consumed by the resistance is Ir, and in phase with the current; the counter E.M.F. of resistance is in opposition to the current. The E.M.F. consumed by the reactance is Ix, and 90° ahead of the current, while the counter E.M.F. of reactance is 90° behind the current; so that, if, in Fig. 15, $\overline{OI}$, is the current, —

$\overline{OE_r}$ = E.M.F. consumed by resistance,
$\overline{OE_r'}$ = counter E.M.F. of resistance,
$\overline{OE_x}$ = E.M.F. consumed by inductance,
$\overline{OE_x'}$ = counter E.M.F. of inductance,
$\overline{OE_z}$ = E.M.F. consumed by impedance,
$\overline{OE_z'}$ = counter E.M.F. of impedance.

Obviously, these counter E.M.Fs. are different from, for instance, the counter E.M.F. of a synchronous motor, in so far as they have no independent existence, but exist only through, and as long as, the current flows. In this respect they are analogous to the opposing force of friction in mechanics.

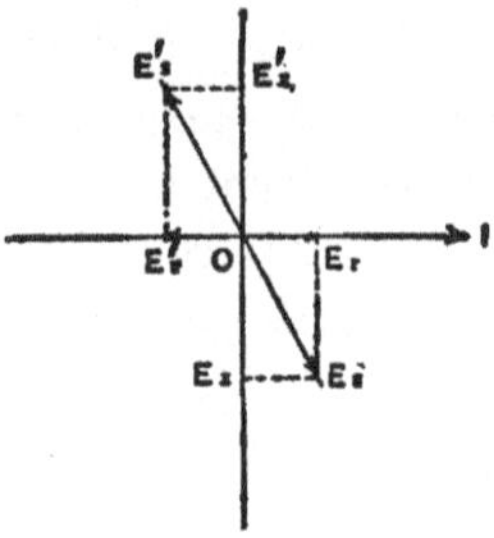

Fig. 15.

19. Coming back to the equation found for the E.M.F. at the generator end of the line, —

$$E_0 = \sqrt{(E + Ir)^2 + (Ix)^2},$$

we find, as the drop of potential in the line —

$$\Delta E = E_0 - E = \sqrt{(E + Ir)^2 + (Ix)^2} - E.$$

This is different from, and less than, the E.M.F. of impedance —

$$E_z = Iz = I\sqrt{r^2 + x^2}.$$

Hence it is wrong to calculate the drop of potential in a circuit by multiplying the current by the impedance; and the drop of potential in the line depends, with a given current fed over the line into a non-inductive circuit, not only upon the constants of the line, r and x, but also upon the E.M.F., E, at end of line, as can readily be seen from the diagrams.

20. If the receiver circuit is inductive, that is, if the current, I, lags behind the E.M.F., E, by an angle $\hat{\omega}$, and we choose again as the zero line, the current $\overline{OI}$ (Fig. 16), the E.M.F., $\overline{OE}$ is ahead of the current by angle $\hat{\omega}$. The

E.M.F. consumed by the resistance, Ir, is in phase with the current, and represented by $\overline{OE_r}$; the E.M.F. consumed by the reactance, Ix, is 90° ahead of the current, and represented by $\overline{OE_x}$. Combining $\overline{OE}$, $\overline{OE_r}$, and $\overline{OE_x}$, we get $\overline{OE_0}$, the E.M.F. required at the generator end of the line. Comparing Fig. 16 with Fig. 13, we see that in the former $\overline{OE_0}$ is larger; or conversely, if E_0 is the same, E will be less with an inductive load. In other words, the drop of potential in an inductive line is greater, if the receiving circuit is inductive, than if it is non-inductive. From Fig. 16,—

$$E_0 = \sqrt{(E \cos \hat{\omega} + Ir)^2 + (E \sin \hat{\omega} + Ix)^2}.$$

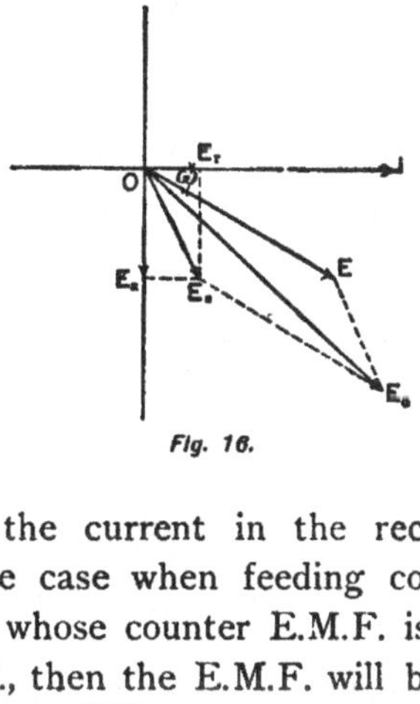

Fig. 16.

If, however, the current in the receiving circuit is leading, as is the case when feeding condensers or synchronous motors whose counter E.M.F. is larger than the impressed E.M.F., then the E.M.F. will be represented, in Fig. 17, by a vector, $\overline{OE}$, lagging behind the current, $\overline{OI}$, by the angle of lead $\hat{\omega}'$; and in this case we get, by combining $\overline{OE}$ with $\overline{OE_r}$, in phase with the current, and $\overline{OE_x}$, 90° ahead of the current, the generator E.M.F., $\overline{OE_0}$, which in this case is not only less than in Fig. 16 and in Fig. 13, but may be even less than E; that is, the potential rises in the line. In other words, in a circuit with leading current, the self-induction of the line raises the potential, so that the drop of potential is less than with a non-inductive load, or may even be negative, and the voltage at the generator lower than at the other end of the line.

These diagrams, Figs. 13 to 17, can be considered polar diagrams of an alternating-current generator of an E.M.F., E_0, a resistance E.M.F., $E_r = Ir$, a reactance E.M.F., $E_x = Ix$, and a difference of potential, E, at the alternator terminals; and we see, in this case, that with an inductive load the potential difference at the alternator terminals will be lower than with a non-inductive load, and that with a non-inductive load it will be lower than when feeding into

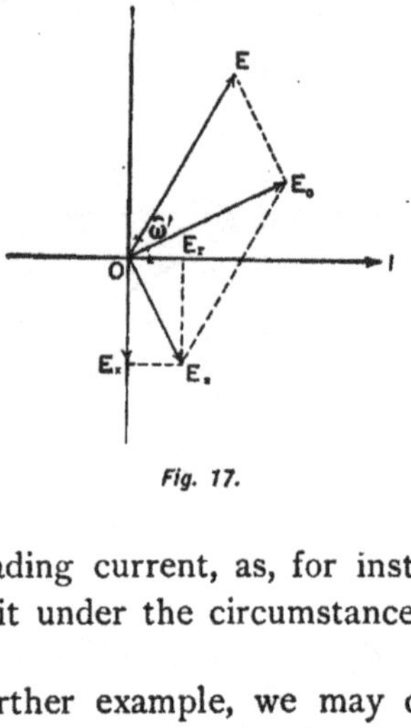

Fig. 17.

a circuit with leading current, as, for instance, a synchro nous motor circuit under the circumstances stated above.

21. As a further example, we may consider the dia gram of an alternating-current transformer, feeding through its secondary circuit an inductive load.

For simplicity, we may neglect here the magnetic hysteresis, the effect of which will be fully treated in a separate chapter on this subject.

Let the time be counted from the moment when the magnetic flux is zero. The phase of the flux, that is, the amplitude of its maximum value, is 90° in this case, and, consequently, the phase of the induced E.M.F., is 180°, since the induced E.M.F. lags 90° behind the inducing flux. Thus the secondary induced E.M.F., E_1, will be represented by a vector, $\overline{OE_1}$, in Fig. 18, at the phase 180°. The secondary current, I_1, lags behind the E.M.F., E_1, by an angle $\hat{\omega}_1$, which is determined by the resistance and inductance of the secondary circuit; that is, by the load in the secondary circuit, and is represented in the diagram by the vector, OF_1, of phase $180 + \hat{\omega}_1$.

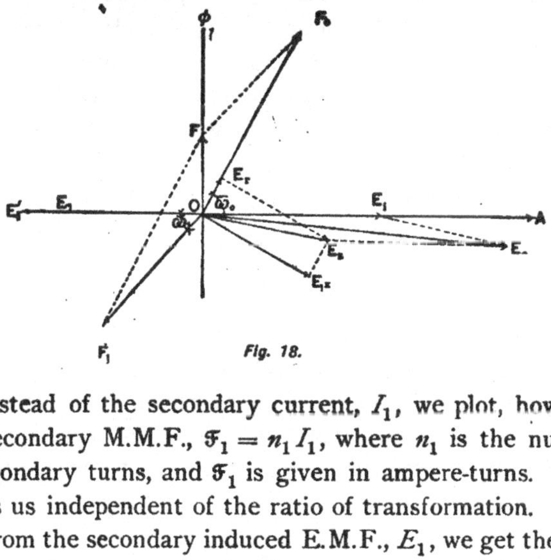

Fig. 18.

Instead of the secondary current, I_1, we plot, however, the secondary M.M.F., $\mathfrak{F}_1 = n_1 I_1$, where n_1 is the number of secondary turns, and $\mathfrak{F}_1$ is given in ampere-turns. This makes us independent of the ratio of transformation.

From the secondary induced E.M.F., E_1, we get the flux, Φ, required to induce this E.M.F., from the equation —

$$E_1 = \sqrt{2}\,\pi n_1 N \Phi 10^{-8};$$

where —

E_1 = secondary induced E.M.F., in effective volts,
N = frequency, in cycles per second,
n_1 = number of secondary turns,
Φ = maximum value of magnetic flux, in webers.

The derivation of this equation has been given in a preceding chapter.

This magnetic flux, Φ, is represented by a vector, $\overline{O\Phi}$, at the phase 90°, and to induce it an M.M.F., $\mathfrak{F}$ is required, which is determined by the magnetic characteristic of the iron, and the section and length of the magnetic circuit of the transformer; it is in phase with the flux Φ, and represented by the vector $\overline{OF}$, in effective ampere-turns.

The effect of hysteresis, neglected at present, is to shift $\overline{OF}$ ahead of $\overline{O\Phi}$, by an angle α, the angle of hysteretic lead. (See Chapter on Hysteresis.)

This M.M.F., $\mathfrak{F}$, is the resultant of the secondary M.M.F., $\mathfrak{F}_1$, and the primary M.M.F., $\mathfrak{F}_0$; or graphically, $\overline{OF}$ is the diagonal of a parallelogram with $\overline{OF_1}$ and $\overline{OF_0}$ as sides. $\overline{OF_1}$ and $\overline{OF}$ being known, we find $\overline{OF_0}$, the primary ampere-turns, and therefrom, and the number of primary turns, n_0, the primary current, $I_0 = \mathfrak{F}_0 / n_0$, which corresponds to the secondary current, I_1.

To overcome the resistance, r_0, of the primary coil, an E.M.F., $E_r = I_0 r_0$, is required, in phase with the current, I_0, and represented by the vector, $\overline{OE_r}$.

To overcome the reactance, $x_0 = 2 \pi n_0 L_0$, of the primary coil, an E.M.F. $E_x = I_0 x_0$ is required, 90° ahead of the current I_0, and represented by vector, $\overline{OE_x}$.

The resultant magnetic flux, Φ, which in the secondary coil induces the E.M.F., E_1, induces in the primary coil an E.M.F. proportional to E_1 by the ratio of turns n_0 / n_1, and in phase with E_1, or, —

$$E_i' = \frac{n_0}{n_1} E_1,$$

which is represented by the vector $\overline{OE_i'}$. To overcome this counter E.M.F., E_i', a primary E.M.F., E_i, is required, equal but opposite to E_i', and represented by the vector, $\overline{OE_i}$.

The primary impressed E.M.F., E_0, must thus consist of the three components, $\overline{OE_i}$, $\overline{OE_r}$, and $\overline{OE_x}$, and is, therefore, their resultant $\overline{OE_0}$, while the difference of phase in the primary circuit is found to be $\hat{\omega}_0 = E_0 O F_0$.

22. Thus, in Figs 18 to 20, the diagram of a transformer is drawn for the same secondary E.M.F., E_1, secondary current, I_1 and therefore secondary M.M.F., $\mathfrak{F}_1$, but with different conditions of secondary displacement: —

In Fig. 18, the secondary current, I_1, lags 60° behind the secondary E.M.F., E_1.

In Fig. 19, the secondary current, I_1, is in phase with the secondary E.M.F., E_1.

In Fig. 20, the secondary current, I_1, leads by 60° the secondary E.M.F., E_1.

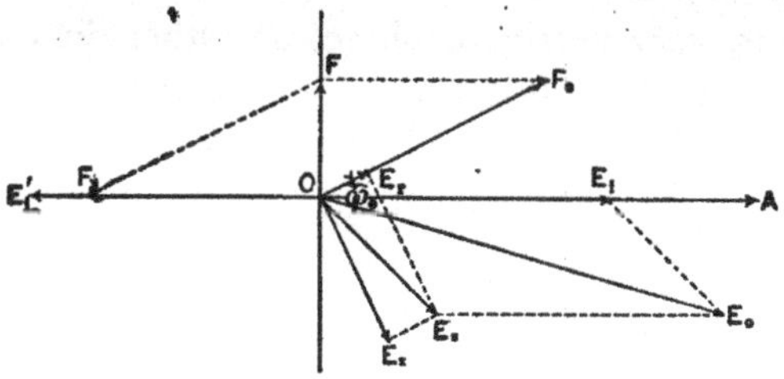

Fig. 19.

These diagrams show that lag in the secondary circuit increases and lead decreases, the primary current and primary E.M.F. required to produce in the secondary circuit the same E.M.F. and current; or conversely, at a given primary

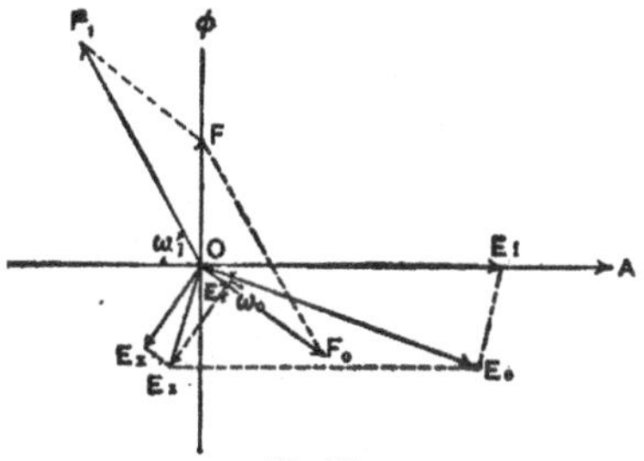

Fig. 20.

impressed E.M.F., E_0, the secondary E.M.F., E_1, will be smaller with an inductive, and larger with a condenser (leading current) load, than with a non-inductive load.

At the same time we see that a difference of phase existing in the secondary circuit of a transformer reappears in the primary circuit, somewhat decreased if leading, and slightly increased if lagging. Later we shall see that hysteresis reduces the displacement in the primary circuit, so that, with an excessive lag in the secondary circuit, the lag in the primary circuit may be less than in the secondary.

A conclusion from the foregoing is that the transformer is not suitable for producing currents of displaced phase; since primary and secondary current are, except at very light loads, very nearly in phase, or rather, in opposition, to each other.

CHAPTER V.

SYMBOLIC METHOD.

23. The graphical method of representing alternating-current phenomena by polar coördinates of time affords the best means for deriving a clear insight into the mutual relation of the different alternating sine waves entering into the problem. For numerical calculation, however, the graphical method is generally not well suited, owing to the widely different magnitudes of the alternating sine waves represented in the same diagram, which make an exact diagrammatic determination impossible. For instance, in the transformer diagrams (*cf.* Figs. 18–20), the different magnitudes will have numerical values in practice, somewhat like $E_1 =$ 100 volts, and $I_1 = 75$ amperes, for a non-inductive secondary load, as of incandescent lamps. Thus the only reactance of the secondary circuit is that of the secondary coil, or, $x_1 = .08$ ohms, giving a lag of $\hat{\omega}_1 = 3.6°$. We have also

$$n_1 = 30 \text{ turns.}$$
$$n_0 = 300 \text{ turns.}$$
$$\mathfrak{F}_1 = 2250 \text{ ampere-turns.}$$
$$\mathfrak{F} = 100 \text{ ampere-turns.}$$
$$E_r = 10 \text{ volts.}$$
$$E_x = 60 \text{ volts.}$$
$$E_i = 1000 \text{ volts.}$$

The corresponding diagram is shown in Fig. 21. Obviously, no exact numerical values can be taken from a parallelogram as flat as OF_1FF_0, and from the combination of vectors of the relative magnitudes 1 : 6 : 100.

Hence the importance of the graphical method consists

not so much in its usefulness for practical calculation, as to aid in the simple understanding of the phenomena involved.

24. Sometimes we can calculate the numerical values trigonometrically by means of the diagram. Usually, however, this becomes too complicated, as will be seen by trying

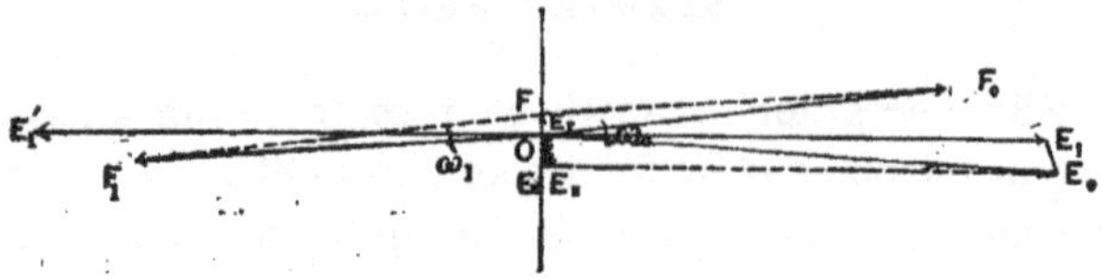

Fig. 21.

to calculate, from the above transformer diagram, the ratio of transformation. The primary M.M.F. is given by the equation:—

$$\mathfrak{F}_0 = \sqrt{\mathfrak{F}^2 + \mathfrak{F}_1^2 + 2\,\mathfrak{F}\mathfrak{F}_1 \sin \omega_1},$$

an expression not well suited as a starting-point for further calculation.

A method is therefore desirable which combines the exactness of analytical calculation with the clearness of the graphical representation.

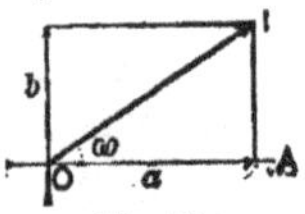

Fig. 22.

25. We have seen that the alternating sine wave is represented in intensity, as well as phase, by a vector, $\overline{OI}$, which is determined analytically by two numerical quantities — the length, $\overline{OI}$, or intensity; and the amplitude, AOI, or phase $\hat{\omega}$, of the wave, I.

Instead of denoting the vector which represents the sine wave in the polar diagram by the polar coördinates,

I and $\hat{\omega}$, we can represent it by its rectangular coördinates, a and b (Fig. 22), where —

$a = I \cos \hat{\omega}$ is the horizontal component,
$b = I \sin \omega$ is the vertical component of the sine wave.

This representation of the sine wave by its rectangular components is very convenient, in so far as it avoids the use of trigonometric functions in the combination or resolution of sine waves.

Since the rectangular components a and b are the horizontal and the vertical projections of the vector representing the sine wave, and the projection of the diagonal of a parallelogram is equal to the sum of the projections of its sides, the combination of sine waves by the parallelogram

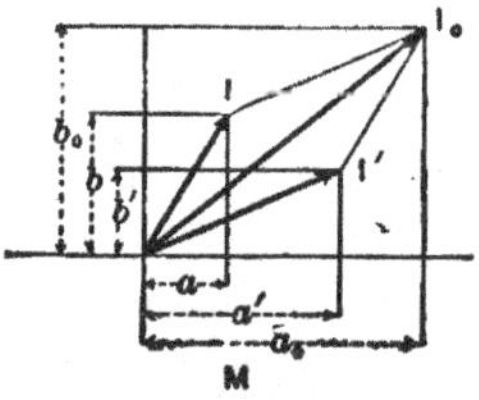

Fig. 23.

law is reduced to the addition, or subtraction, of their rectangular components. That is,

Sine waves are combined, or resolved, by adding, or subtracting, their rectangular components.

For instance, if a and b are the rectangular components of a sine wave, I, and a' and b' the components of another sine wave, I' (Fig. 23), their resultant sine wave, I_0, has the rectangular components $a_0 = (a + a')$, and $b_0 = (b + b')$.

To get from the rectangular components, a and b, of a sine wave, its intensity, i, and phase, $\hat{\omega}$, we may combine a and b by the parallelogram, and derive, —

$$i = \sqrt{a^2 + b^2};$$

$$\tan \hat{\omega} = \frac{b}{a}.$$

Hence we can analytically operate with sine waves, as with forces in mechanics, by resolving them into their rectangular components.

26. To distinguish, however, the horizontal and the vertical components of sine waves, so as not to be confused in lengthier calculation, we may mark, for instance, the vertical components, by a distinguishing index, or the addition of an otherwise meaningless symbol, as the letter j, and thus represent the sine wave by the expression, —

$$I = a + jb,$$

which now has the meaning, that a is the horizontal and b the vertical component of the sine wave I; and that both components are to be combined in the resultant wave of intensity, —

$$i = \sqrt{a^2 + b^2},$$

and of phase, $\tan \hat{\omega} = b / a$.

Similarly, $a - jb$, means a sine wave with a as horizontal, and $-b$ as vertical, components, etc.

Obviously, the plus sign in the symbol, $a + jb$, does not imply simple addition, since it connects heterogeneous quantities — horizontal and vertical components — but implies combination by the parallelogram law.

For the present, j is nothing but a distinguishing index, and otherwise free for definition except that it is not an ordinary number.

27. A wave of equal intensity, and differing in phase from the wave $a + jb$ by 180°, or one-half period, is represented in polar coördinates by a vector of opposite direction, and denoted by the symbolic expression, $-a - jb$. Or —

Multiplying the symbolic expression, $a + jb$, of a sine wave by -1 means reversing the wave, or rotating it through 180°, or one-half period.

A wave of equal intensity, but lagging 90°, or one-quarter period, behind $a + jb$, has (Fig. 24) the horizontal component, $-b$, and the vertical component, a, and is represented symbolically by the expression, $ja - b$.

Multiplying, however, $a + jb$ by j, we get :—

$$ja + j^2 b;$$

therefore, if we define the heretofore meaningless symbol, j, by the condition, —

$$j^2 = -1,$$

we have —

$$j(a + jb) = ja - b;$$

hence :—

Multiplying the symbolic expression, $a + jb$, of a sine wave by j means rotating the wave through 90°, or one-quarter period; that is, retarding the wave through one-quarter period.

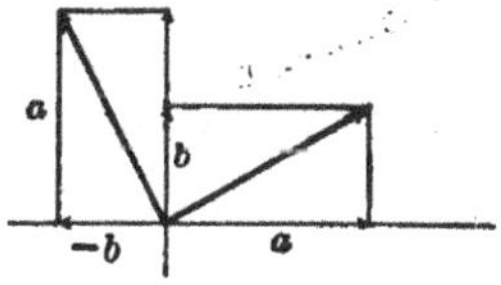

Fig. 24.

Similarly, —

Multiplying by $-j$ means advancing the wave through one-quarter period.

since $\qquad j^2 = -1, \qquad j = \sqrt{-1};$

that is, —

j is the imaginary unit, and the sine wave is represented by a complex imaginary quantity, $a + jb$.

As the imaginary unit j has no numerical meaning in the system of ordinary numbers, this definition of $j = \sqrt{-1}$ does not contradict its original introduction as a distinguishing index. For a more exact definition of this complex imaginary quantity, reference may be made to the text books of mathematics.

28. In the polar diagram of time, the sine wave is represented in intensity as well as phase by one complex quantity —

$$a + jb;$$

where a is the horizontal and b the vertical component of the wave; the intensity is given by —

$$i = \sqrt{a^2 + b^2},$$

the phase by —

$$\tan \hat{\omega} = \frac{b}{a},$$

and

$$a = i \cos \hat{\omega},$$

$$b = i \sin \hat{\omega};$$

hence the wave $a + jb$ can also be expressed by —

$$i (\cos \hat{\omega} + j \sin \hat{\omega}),$$

or, by substituting for $\cos \omega$ and $\sin \hat{\omega}$ their exponential expressions, we obtain —

$$i\epsilon^{j\hat{\omega}}.$$

Since we have seen that sine waves may be combined or resolved by adding or subtracting their rectangular components, consequently: —

Sine waves may be combined or resolved by adding or subtracting their complex algebraic expressions.

For instance, the sine waves, —

$$a + jb$$

and

$$a' + jb',$$

combined give the sine wave —

$$\dot{I} = (a + a') + j(b + b').$$

It will thus be seen that the combination of sine waves is reduced to the elementary algebra of complex quantities.

29. If $\dot{I} = i + ji'$ is a sine wave of alternating current, and r is the resistance, the E.M.F. consumed by the resistance is in phase with the current, and equal to the product of the current and resistance. Or —

$$r\dot{I} = ri + jri'.$$

If L is the inductance, and $x = 2 \pi NL$ the reactance, the E.M.F. produced by the reactance, or the counter

E.M.F. of self-induction, is the product of the current and reactance, and lags 90° behind the current; it is, therefore, represented by the expression —

$$jx\underset{\cdot}{I} = jxi - xi'.$$

The E.M.F. required to overcome the reactance is consequently 90° ahead of the current (or, as usually expressed, the current lags 90° behind the E.M.F.), and represented by the expression —

$$-jx\underset{\cdot}{I} = -jxi + xi'.$$

Hence, the E.M.F. required to overcome the resistance, r, and the reactance, x, is —

$$(r - jx)\,\underset{\cdot}{I};$$

that is —

$Z = r - jx$ *is the expression of the impedance of the circuit, in complex quantities.*

Hence, if $\underset{\cdot}{I} = i + ji'$ is the current, the E.M.F. required to overcome the impedance, $Z = r - jx$, is —

$$\underset{\cdot}{E} = Z\underset{\cdot}{I} = (r - jx)\,(i + ji');$$

hence, since $j^2 = -1$

$$\underset{\cdot}{E} = (ri + xi') + j\,(ri' - xi);$$

or, if $\underset{\cdot}{E} = e + je'$ is the impressed E.M.F., and $Z = r - jx$ the impedance, the current flowing through the circuit is :—

$$\underset{\cdot}{I} = \frac{\dot{E}}{Z} = \frac{e + je'}{r - jx};$$

or, multiplying numerator and denominator by $(r + jx)$ to eliminate the imaginary from the denominator, we have —

$$\underset{\cdot}{I} = \frac{(e + je')\,(r + jx)}{r^2 + x^2} = \frac{er - e'x}{r^2 + x^2} + j\frac{e'r + ex}{r^2 + x^2};$$

or, if $\underset{\cdot}{E} = e + je'$ is the impressed E.M.F., and $\underset{\cdot}{I} = i + ji'$ the current flowing in the circuit, its impedance is —

$$Z = \frac{\dot{E}}{\underset{\cdot}{I}} = \frac{e + je'}{i + ji'} = \frac{(e + je')(i - ji')}{i^2 + i'^2} = \frac{ei + e'i'}{i^2 + i'^2} + j\frac{e'i - ei'}{i^2 + i'^2}.$$

30. If C is the capacity of a condenser in series in a circuit of current $\dot{I} = i + ji'$, the E.M.F. impressed upon the terminals of the condenser is $\dot{E} = \frac{\dot{I}}{2\pi NC}$, 90° behind the current; and may be represented by $\frac{j\dot{I}}{2\pi NC}$, or $jx_1 \dot{I}$, where $x_1 = \frac{1}{2\pi NC}$ is the *capacity reactance* or *condensance* of the condenser.

Capacity reactance is of opposite sign to magnetic reactance; both may be combined in the name reactance.

We therefore have the conclusion that

If r = resistance and L = inductance,
then $x = 2\pi NL$ = magnetic reactance.

If C = capacity, $x_1 = \frac{1}{2\pi NC}$ = capacity reactance, or condensance;

$Z = r - j(x - x_1)$, is the impedance of the circuit.

Ohm's law is then reëstablished as follows:

$$\dot{E} = Z\dot{I}, \quad \dot{I} = \frac{\dot{E}}{Z}, \quad Z = \frac{\dot{E}}{\dot{I}}.$$

The more general form gives not only the intensity of the wave, but also its phase, as expressed in complex quantities.

31. Since the combination of sine waves takes place by the addition of their symbolic expressions, Kirchhoff's laws are now reëstablished in their original form:—

a.) The sum of all the E.M.Fs. acting in a closed circuit equals zero, if they are expressed by complex quantities, and if the resistance and reactance E.M.Fs. are also considered as counter E.M.Fs.

b.) The sum of all the currents flowing towards a distributing point is zero, if the currents are expressed as complex quantities.

If a complex quantity equals zero, the real part as well as the imaginary part must be zero individually, thus if

$$a + jb = 0, \qquad a = 0, b = 0.$$

Resolving the E.M.Fs. and currents in the expression of Kirchhoff's law, we find:—

a.) The sum of the components, in any direction, of all the E.M.Fs. in a closed circuit, equals zero, if the resistance and reactance are considered as counter E.M.Fs.

b.) The sum of the components, in any direction, of all the currents flowing to a distributing point, equals zero.

Joule's Law and the energy equation do not give a simple expression in complex quantities, since the effect or power is a quantity of double the frequency of the current or E.M.F. wave, and therefore requires for its representation as a vector, a transition from single to double frequency, as will be shown in chapter XII.

In what follows, complex vector quantities will always be denoted by dotted capitals when not written out in full; absolute quantities and real quantities by undotted letters.

32. Referring to the instance given in the fourth chapter, of a circuit supplied with an E.M.F., $\dot{E}$, and a current, $\dot{I}$, over an inductive line, we can now represent the impedance of the line by $Z = r - jx$, where r = resistance, x = reactance of the line, and have thus as the E.M.F. at the beginning of the line, or at the generator, the expression —

$$\dot{E}_0 = \dot{E} + Z\dot{I}.$$

Assuming now again the current as the zero line, that is, $\dot{I} = i$, we have in general —

$$\dot{E}_0 = \dot{E} + ir - jix;$$

hence, with non-inductive load, or $\dot{E} = e$,

$$\dot{E}_0 = (e + ir) - jix,$$

$$e_0 = \sqrt{(e + ir)^2 + (ix)^2}, \quad \tan \hat{\omega}_0 = \frac{ix}{e + ir}.$$

In a circuit with lagging current, that is, with leading E.M.F., $\dot{E} = e - je'$, and

$$\dot{E}_0 = e - je' + (r - jx)\,i$$
$$= (e + ir) - j\,(e' + ix),$$

or

$$e_0 = \sqrt{(e + ir)^2 + (e' + ix)^2}, \quad \tan\hat{\omega}_0 = \frac{e' + ix}{e + ir}.$$

In a circuit with leading current, that is, with lagging E.M.F., $\dot{E} = e + je'$, and

$$\dot{E}_0 = (e + je') + (r - jx)\,i$$
$$= (e + ir) + j\,(e' - ix),$$
$$e_0 = \sqrt{(e + ir)^2 + (e' - ix)^2}, \quad \tan\hat{\omega}_0 = \frac{e' - ix}{e + ir};$$

values which easily permit calculation.

CHAPTER VI.

TOPOGRAPHIC METHOD.

33. In the representation of alternating sine waves by vectors in a polar diagram, a certain ambiguity exists, in so far as one and the same quantity — an E.M.F., for instance — can be represented by two vectors of opposite direction, according as to whether the E.M.F. is considered as a part of the impressed E.M.F., or as a counter E.M.F. This is analogous to the distinction between action and reaction in mechanics.

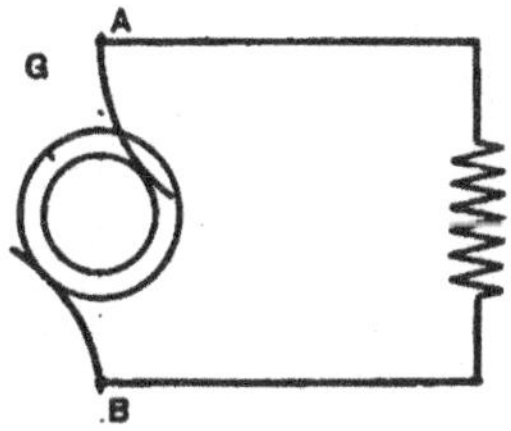

Fig. 25.

Further, it is obvious that if in the circuit of a generator, G (Fig. 25), the current flowing from terminal A over resistance R to terminal B, is represented by a vector $\overline{OI}$ (Fig. 26), or by $\dot{I} = i + ji'$, the same current can be considered as flowing in the opposite direction, from terminal B to terminal A in opposite phase, and therefore represented by a vector $\overline{OI_1}$ (Fig. 26), or by $\dot{I_1} = -i - ji'$.

Or, if the difference of potential from terminal B to terminal A is denoted by the $\dot{E} = e + je'$, the difference of potential from A to B is $\dot{E_1} = -e - je'$.

Hence, in dealing with alternating-current sine waves it is necessary to consider them in their proper direction with regard to the circuit. Especially in more complicated circuits, as interlinked polyphase systems, careful attention has to be paid to this point.

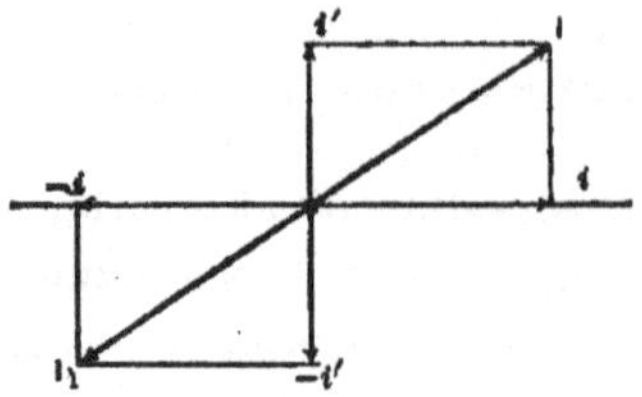

Fig. 26.

34. Let, for instance, in Fig. 27, an interlinked three-phase system be represented diagrammatically, as consisting of three E.M.Fs., of equal intensity, differing in phase by one-third of a period. Let the E.M.Fs. in the direction

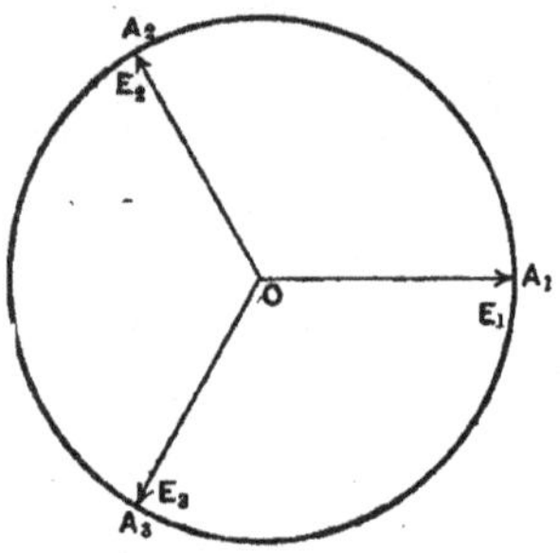

Fig. 27.

from the common connection O of the three branch circuits to the terminals A_1, A_2, A_3, be represented by $\dot{E}_1$, $\dot{E}_2$, $\dot{E}_3$. Then the difference of potential from A_2 to A_1 is $\dot{E}_2 - \dot{E}_1$, since the two E.M.Fs., $\dot{E}_1$ and $\dot{E}_2$, are connected in circuit between the terminals A_1 and A_2, in the direction

$A_1 — O — A_2$; that is, the one, E_2, in the direction OA_2, from the common connection to terminal, the other, E_1, in the opposite direction, A_1O, from the terminal to common connection, and represented by $-E_1$. Conversely, the difference of potential from A_1 to A_2 is $E_1 - E_2$.

It is then convenient to go still a step farther, and drop, in the diagrammatic representation, the vector line altogether; that is, denote the sine wave by a point only, the end of the corresponding vector.

Looking at this from a different point of view, it means that we choose one point of the system — for instance, the common connection O — as a zero point, or point of zero potential, and represent the potentials of all the other points of the circuit by points in the diagram, such that their distances from the zero point gives the intensity; their amplitude the phase of the difference of potential of the respective point with regard to the zero point; and their distance and amplitude with regard to other points of the diagram, their difference of potential from these points in intensity and phase.

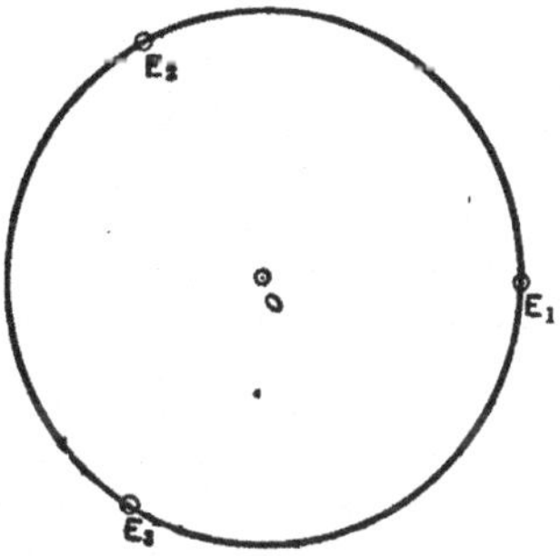

Fig. 28.

Thus, for example, in an interlinked three-phase system with three E.M.Fs. of equal intensity, and differing in phase by one-third of a period, we may choose the common connection of the star-connected generator as the zero point, and represent, in Fig. 28, one of the E.M.Fs., or the potential at one of the three-phase terminals, by point E_1. The potentials at the two other terminals will then be given by the points E_2 and E_3, which have the same distance from O as E_1, and are equidistant from E_1 and from each other.

The difference of potential between any pair of terminals — for instance E_1 and E_2 — is then the distance $\overline{E_2E_1}$, or $\overline{E_1E_2}$, according to the direction considered.

35. If now the three branches $\overline{OE_1}$, $\overline{OE_2}$ and $\overline{OE_3}$, of the three-phase system are loaded equally by three currents equal in intensity and in difference of phase against their

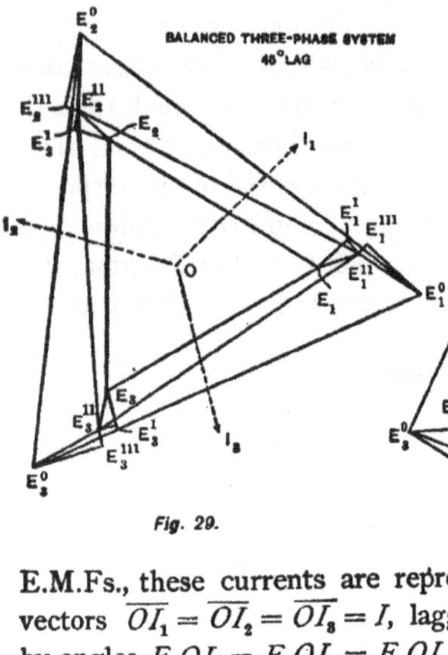

Fig. 29.

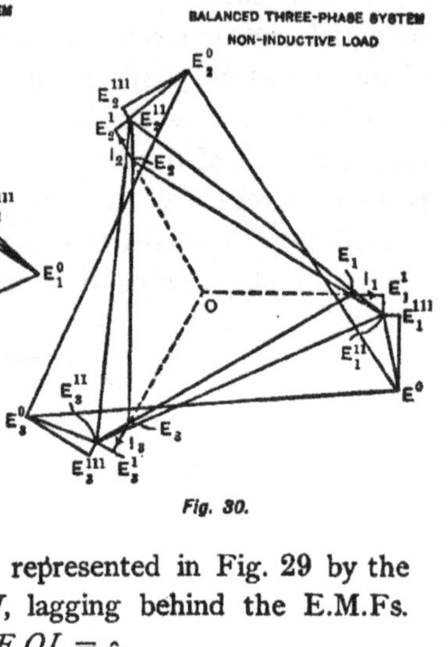

Fig. 30.

E.M.Fs., these currents are represented in Fig. 29 by the vectors $\overline{OI_1} = \overline{OI_2} = \overline{OI_3} = I$, lagging behind the E.M.Fs. by angles $E_1OI_1 = E_2OI_2 = E_3OI_3 = \hat{\omega}$.

Let the three-phase circuit be supplied over a line of impedance $Z_1 = r_1 - jx_1$ from a generator of internal impedance $Z_o = x_o - jx_o$.

In phase $\overline{OE_1}$ the E.M.F. consumed by resistance r_1 is represented by the distance $\overline{E_1E_1^{I}} = Ir_1$, in phase, that is parallel with current $\overline{OI_1}$. The E.M.F. consumed by reactance x_1 is represented by $\overline{E_1^{1}E_1^{II}} = Ix_1$, 90° ahead of current $\overline{OI_1}$. The same applies to the other two phases, and it thus follows that to produce the E.M.F. triangle $E_1E_2E_3$ at the terminals of the consumer's circuit, the E.M.F. triangle $E_1''E_2''E_3''$ is required at the generator terminals.

Repeating the same operation for the internal impedance of the generator we get $\overline{E''E'''} = Ir_0$, and parallel to $\overline{OI_1}$, $\overline{E'''E^0} = Ix_0$, and 90° ahead of $\overline{OI_1}$, and thus as triangle of (nominal) induced E.M.Fs. of the generator $E_1^0E_2^0E_3^0$.

In Fig. 29, the diagram is shown for 45° lag, in Fig. 30 for noninductive load, and in Fig. 31 for 45° lead of the currents with regard to their E.M.Fs.

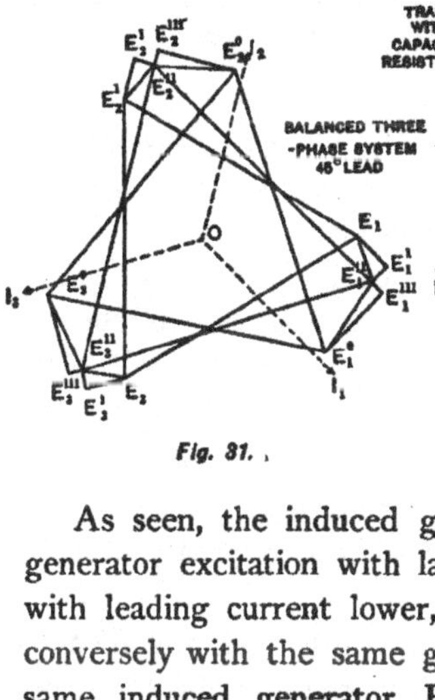

Fig. 31.

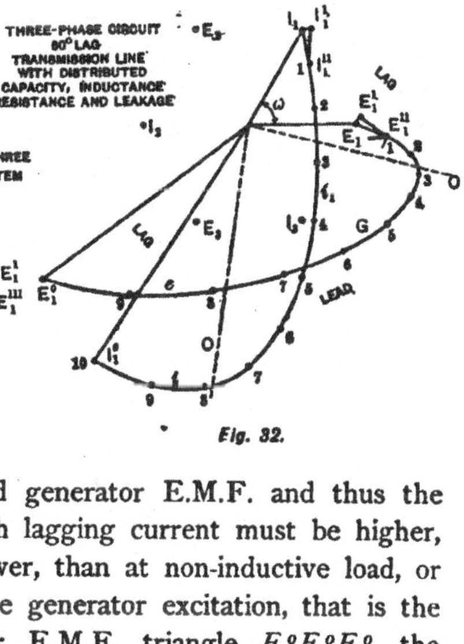

Fig. 32.

As seen, the induced generator E.M.F. and thus the generator excitation with lagging current must be higher, with leading current lower, than at non-inductive load, or conversely with the same generator excitation, that is the same induced generator E.M.F. triangle $E_1^0E_2^0E_3^0$, the E.M.Fs. at the receiver's circuit, E_1, E_2, E_3, fall off more with lagging, less with leading current, than with non-inductive load.

36. As further instance may be considered the case of a single phase alternating current circuit supplied over a cable containing resistance and distributed capacity.

Let in Fig. 33 the potential midway between the two terminals be assumed as zero point 0. The two terminal voltages at the receiver circuit are then represented by the points E and E^1 equidistant from 0 and opposite each other, and the two currents issuing from the terminals are represented by the points I and I^1, equidistant from 0 and opposite each other, and under angle ω with E and E^1 respectively.

Considering first an element of the line or cable next to the receiver circuit. In this an E.M.F. $\overline{EE_1}$ is consumed by the resistance of the line element, in phase with the current $\overline{OI}$, and proportional thereto, and a current $\overline{II_1}$ consumed by the capacity, as charging current of the line element, 90° ahead in phase of the E.M.F. $\overline{OE}$ and proportional thereto, so that at the generator end of this cable element current and E.M.F. are $\overline{OI_1}$ and $\overline{OE_1}$ respectively.

Passing now to the next cable element we have again an E.M.F. $\overline{E_1E_2}$ proportional to and in phase with the current $\overline{OI_1}$ and a current $\overline{I_1I_2}$ proportional to and 90° ahead of the E.M.F. $\overline{OE_1}$, and thus passing from element to element along the cable to the generator, we get curves of E.M.Fs. e and e^1, and curves of currents i and i^1, which can be called the topographical circuit characteristics, and which correspond to each other, point for point, until the generator terminal voltages $\overline{OE_0}$ and $\overline{OE_0^1}$ and the generator currents $\overline{OI_0}$ and $\overline{OI_0^1}$ are reached.

Again, adding $\overline{E_0E''} = I_0r_0$ and parallel OI_0 and $\overline{E''E^0} = I_0x_0$ and 90° ahead of $\overline{OI_0}$, gives the (nominal) induced E.M.F. of the generator $\overline{OE^0}$, where $Z_0 = r_0 - jx_0 =$ internal impedance of the generator.

In Fig. 33 is shown the circuit characteristics for 60° lag, of a cable containing only resistance and capacity.

Obviously by graphical construction the circuit characteristics appear more or less as broken lines, due to the necessity of using finite line elements, while in reality when calculated by the differential method they are smooth curves.

37. As further instance may be considered a three-phase circuit supplied over a long distance transmission line of distributed capacity, self-induction, resistance, and leakage.

Let, in Fig. 33, $\overline{OE_1}, \overline{OE_2}, \overline{OE_3}$ = three-phase E.M.Fs. at receiver circuit, equidistant from each other and $= E$.

Let $\overline{OI_1}, \overline{OI_2}, \overline{OI_3}$ = three-phase currents in the receiver circuit equidistant from each other and $= I$, and making with E the phase angle $\hat{\omega}$.

Considering again as in § 35 the transmission line element by element, we have in every element an E.M.F. $\overline{E_1 E_1{}'}$ consumed by the resistance in phase with the current $\overline{OI_1}$ and proportional thereto, and an E.M.F. $\overline{E_1{}', E_1{}''}$ con-

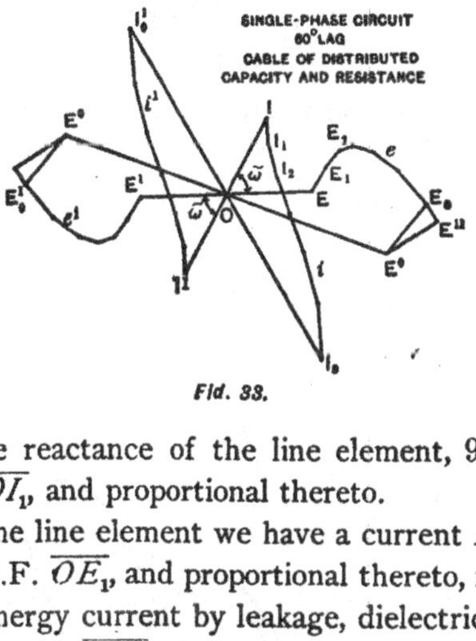

FIG. 33.

sumed by the reactance of the line element, 90° ahead of the current $\overline{OI_1}$, and proportional thereto.

In the same line element we have a current $I_1 I_1{}^1$ in phase with the E.M.F. $\overline{OE_1}$, and proportional thereto, representing the loss of energy current by leakage, dielectric hysteresis, etc., and a current $\overline{I_1{}^1 I_1{}''}$, 90° ahead of the E.M.F. $\overline{OE_1}$, and proportional thereto, the charging current of the line element as condenser, and in this manner passing along the line, element by element, we ultimately reach the generator terminal voltages $E_1{}^o$, $E_2{}^o$, $E_3{}^o$, and generator currents $I_1{}^o$, $I_2{}^o$, $I_3{}^o$, over the topographical characteristics of E.M.F. e_1, e_2, e_3, and of current i_1, i_2, i_3, as shown in Fig. 33.

The circuit characteristics of current i and of E.M.F. e

correspond to each other, point for point, the one giving the current and the other the E.M.F. in the line element.

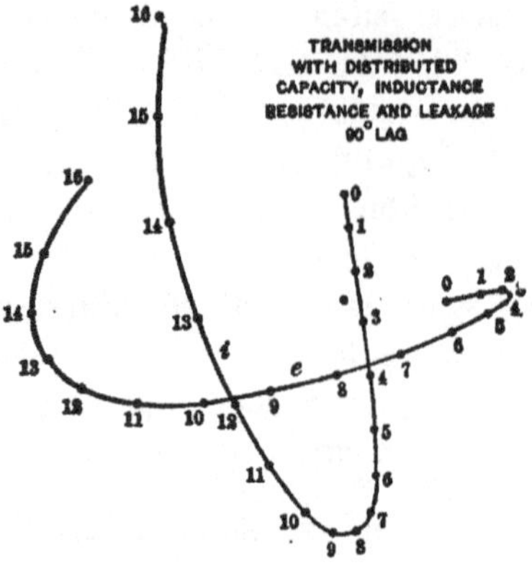

Fig. 34.

Only the circuit characteristics of the first phase are shown as e_1 and i_1. As seen, passing from the receiving end towards the generator end of the line, potential and

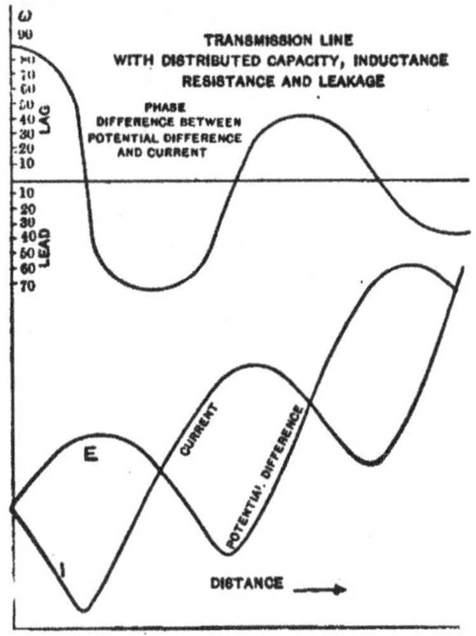

Fig. 35.

current alternately rise and fall, while their phase angle changes periodically between lag and lead.

37. *a.* More markedly this is shown in Fig. 34, the topographic circuit characteristic of one of the lines with 90° lag in the receiver circuit. Corresponding points of the two characteristics e and i are marked by corresponding figures 0 to 16, representing equidistant points of the line. The values of E.M.F., current and their difference of phase are plotted in Fig. 35 in rectangular co-ordinates with the distance as abscissae, counting from the receiving circuit towards the generator. As seen from Fig. 35, E.M.F. and current periodically but alternately rise and fall, a maximum of one approximately coinciding with a minimum of the other and with a point of zero phase displacement.

The phase angle between current and E.M.F. changes from 90° lag to 72° lead, 44° lag, 34° lead, etc., gradually decreasing in the amplitude of its variation.

CHAPTER VII.

ADMITTANCE, CONDUCTANCE, SUSCEPTANCE.

38. If in a continuous-current circuit, a number of resistances, $r_1, r_2, r_3, \ldots$ are connected in series, their joint resistance, R, is the sum of the individual resistances $R = r_1 + r_2 + r_3 + \ldots$

If, however, a number of resistances are connected in multiple or in parallel, their joint resistance, R, cannot be expressed in a simple form, but is represented by the expression: —

$$R = \frac{1}{\frac{1}{r_1} + \frac{1}{r_2} + \frac{1}{r_3} + \ldots}.$$

Hence, in the latter case it is preferable to introduce, instead of the term *resistance*, its reciprocal, or inverse value, the term *conductance*, $g = 1 / r$. If, then, a number of conductances, $g_1, g_2, g_3, \ldots$ are connected in parallel, their joint conductance is the sum of the individual conductances, or $G = g_1 + g_2 + g_3 + \ldots$ When using the term conductance, the joint conductance of a number of series-connected conductances becomes similarly a complicated expression —

$$G = \frac{1}{\frac{1}{g_1} + \frac{1}{g_2} + \frac{1}{g_3} + \ldots}.$$

Hence the term *resistance* is preferable in case of series connection, and the use of the reciprocal term *conductance* in parallel connections; therefore,

The joint resistance of a number of series-connected resistances is equal to the sum of the individual resistances; the

joint conductance of a number of parallel-connected conductances is equal to the sum of the individual conductances.

39. In alternating-current circuits, instead of the term *resistance* we have the term *impedance*, $Z = r - jx$, with its two components, the *resistance*, r, and the *reactance*, x, in the formula of Ohm's law, $\dot{E} = \dot{I}Z$. The resistance, r, gives the component of E.M.F. in phase with the current, or the energy component of the E.M.F., $\dot{I}r$; the reactance, x, gives the component of the E.M.F. in quadrature with the current, or the wattless component of E.M.F., $\dot{I}x$; both combined give the total E.M.F., —

$$Iz = I\sqrt{r^2 + x^2}.$$

Since E.M.Fs. are combined by adding their complex expressions, we have:

The joint impedance of a number of series-connected impedances is the sum of the individual impedances, when expressed in complex quantities.

In graphical representation impedances have not to be added, but are combined in their proper phase by the law of parallelogram in the same manner as the E.M.Fs. corresponding to them.

The term impedance becomes inconvenient, however, when dealing with parallel-connected circuits; or, in other words, when several currents are produced by the same E.M.F., such as in cases where Ohm's law is expressed in the form,

$$\dot{I} = \frac{\dot{E}}{Z}.$$

It is preferable, then, to introduce the reciprocal of impedance, which may be called the *admittance* of the circuit, or

$$Y = \frac{1}{Z}.$$

As the reciprocal of the complex quantity, $Z = r - jx$, the admittance is a complex quantity also, or $Y = g + jb$;

it consists of the component g, which represents the co efficient of current in phase with the E.M.F., or energy current, $g\dot{E}$, in the equation of Ohm's law,—

$$\dot{I} = Y\dot{E} = (g + jb)\,\dot{E},$$

and the component b, which represents the coefficient of current in quadrature with the E.M.F., or wattless component of current, $b\dot{E}$.

g is called the *conductance*, and b the *susceptance*, of the circuit. Hence the conductance, g, is the energy component, and the susceptance, b, the wattless component, of the admittance, $Y = g + jb$, while the numerical value of admittance is —

$$y = \sqrt{g^2 + b^2};$$

the resistance, r, is the energy component, and the reactance, x, the wattless component, of the impedance, $Z = r - jx$, the numerical value of impedance being—

$$z = \sqrt{r^2 + x^2}.$$

40. As shown, the term *admittance* implies resolving the current into two components, in phase and in quadrature with the E.M.F., or the energy current and the wattless current; while the term *impedance* implies resolving the E.M.F. into two components, in phase and in quadrature with the current, or the energy E.M.F. and the wattless E.M.F.

It must be understood, however, that the conductance is not the reciprocal of the resistance, but depends upon the resistance as well as upon the reactance. Only when the reactance $x = 0$, or in continuous-current circuits, is the conductance the reciprocal of resistance.

Again, only in circuits with zero resistance ($r = 0$) is the susceptance the reciprocal of reactance; otherwise, the susceptance depends upon reactance and upon resistance.

The conductance is zero for two values of the resistance:—

1.) If $r = \infty$, or $x = \infty$, since in this case no current passes, and either component of the current $= 0$.

2.) If $r = 0$, since in this case the current which passes through the circuit is in quadrature with the E.M.F., and thus has no energy component.

Similarly, the susceptance, b, is zero for two values of the reactance:—

1.) If $x = \infty$, or $r = \infty$.

2.) If $x = 0$.

From the definition of admittance, $Y = g + jb$, as the reciprocal of the impedance, $Z = r - jx$,

we have $$Y = \frac{1}{Z}, \text{ or, } g + jb = \frac{1}{r - jx};$$

or, multiplying numerator and denominator on the right side by $(r + jx)$

$$g + jb = \frac{r + jx}{(r - jx)(r + jx)};$$

hence, since

$$(r - jx)(r + jx) = r^2 + x^2 = z^2,$$

$$g + jb = \frac{r}{r^2 + x^2} + j\frac{x}{r^2 + x^2} = \frac{r}{z^2} + j\frac{x}{z^2};$$

or

$$g = \frac{r}{r^2 + x^2} = \frac{r}{z^2},$$

$$b = \frac{x}{r^2 + x^2} = \frac{x}{z^2};$$

and conversely

$$r = \frac{g}{g^2 + b^2} = \frac{g}{y^2},$$

$$x = \frac{b}{g^2 + b^2} = \frac{b}{y^2}.$$

By these equations, the conductance and susceptance can be calculated from resistance and reactance, and conversely.

Multiplying the equations for g and r, we get:—

$$gr = \frac{rg}{z^2 y^2};$$

hence, $$z^2 y^2 = (r^2 + x^2)(g^2 + b^2) = 1;$$

and, $$z = \frac{1}{y} = \frac{1}{\sqrt{g^2 + b^2}},$$ the absolute value of impedance;

$$y = \frac{1}{z} = \frac{1}{\sqrt{r^2 + x^2}},$$ the absolute value of admittance.

41. If, in a circuit, the reactance, x, is constant, and the resistance, r, is varied from $r = 0$ to $r = \infty$, the susceptance, b, decreases from $b = 1 / x$ at $r = 0$, to $b = 0$ at $r = \infty$; while the conductance, $g = 0$ at $r = 0$, increases, reaches a maximum for $r = x$, where $g = 1 / 2r$ is equal to the susceptance, or $g = b$, and then decreases again, reaching $g = 0$ at $r = \infty$.

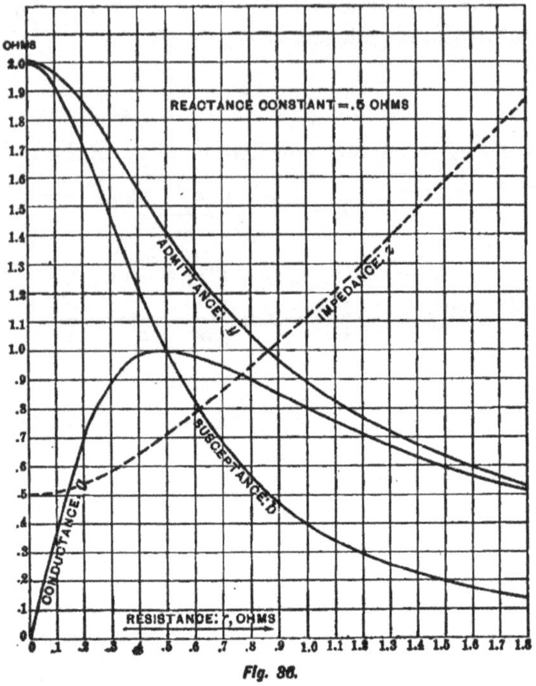

Fig. 36.

In Fig. 36, for constant reactance $x = .5$ ohm, the variation of the conductance, g, and of the susceptance, b, are shown as functions of the varying resistance, r. As shown, the absolute value of admittance, susceptance, and conductance are plotted in full lines, and in dotted line the absolute value of impedance,

$$z = \sqrt{r^2 + x^2} = \frac{1}{y}.$$

Obviously, if the resistance, r, is constant, and the reactance, x, is varied, the values of conductance and susceptance are merely exchanged, the conductance decreasing steadily from $g = 1 / r$ to 0, and the susceptance passing from 0 at $x = 0$ to the maximum, $b = 1 / 2r = g = 1 / 2x$ at $x = r$, and to $b = 0$ at $x = \infty$.

The resistance, r, and the reactance, x, vary as functions of the conductance, g, and the susceptance, b, in the same manner as g and b vary as functions of r and x.

The sign in the complex expression of admittance is always opposite to that of impedance; this is obvious, since if the current lags behind the E.M.F., the E.M.F. leads the current, and conversely.

We can thus express Ohm's law in the two forms —

$$\dot{E} = \dot{I}Z,$$
$$\dot{I} = \dot{E}Y,$$

and therefore —

The joint impedance of a number of series-connected impedances is equal to the sum of the individual impedances; the joint admittance of a number of parallel-connected admittances, if expressed in complex quantities, is equal to the sum of the individual admittances. In diagrammatic representation, combination by the parallelogram law takes the place of addition of the complex quantities.

CHAPTER VIII.

CIRCUITS CONTAINING RESISTANCE, INDUCTANCE, AND CAPACITY.

42. Having, in the foregoing, reëstablished Ohm's law and Kirchhoff's laws as being also the fundamental laws of alternating-current circuits, when expressed in their complex form,

$$\dot{E} = \dot{Z}\dot{I}, \qquad \text{or, } \dot{I} = \dot{Y}\dot{E},$$

and

$$\Sigma \dot{E} = 0 \text{ in a closed circuit,}$$

$$\Sigma \dot{I} = 0 \text{ at a distributing point,}$$

where $\dot{E}$, $\dot{I}$, $\dot{Z}$, $\dot{Y}$, are the expressions of E.M.F., current, impedance, and admittance in complex quantities, — these values representing not only the intensity, but also the phase, of the alternating wave, — we can now — by application of these laws, and in the same manner as with continuous-current circuits, keeping in mind, however, that $\dot{E}$, $\dot{I}$, $\dot{Z}$, $\dot{Y}$, are complex quantities — calculate alternating-current circuits and networks of circuits containing resistance, inductance, and capacity in any combination, without meeting with greater difficulties than when dealing with continuous-current circuits.

It is obviously not possible to discuss with any completeness all the infinite varieties of combinations of resistance, inductance, and capacity which can be imagined, and which may exist, in a system or network of circuits; therefore only some of the more common or more interesting combinations will here be considered.

1.) *Resistance in series with a circuit.*

43. In a constant-potential system with impressed E.M.F.,

$$\dot{E}_0 = e_0 + je_0', \qquad E_0 = \sqrt{e_0^2 + e_0'^2},$$

let the receiving circuit of impedance

$$Z = r - jx, \qquad z = \sqrt{r^2 + x^2},$$

be connected in series with a resistance, r_0.

The total impedance of the circuit is then

$$Z + r_0 = r + r_0 - jx;$$

hence the current is

$$\dot{I} = \frac{\dot{E}_0}{Z + r_0} = \frac{\dot{E}_0}{r + r_0 - jx} = \frac{\dot{E}_0 (r + r_0 + jx)}{(r + r_0)^2 + x^2};$$

and the E.M.F. of the receiving circuit, becomes

$$\dot{E} = \dot{I}Z = \frac{\dot{E}_0 (r - jx)}{r + r_0 - jx} = \frac{\dot{E}_0 \{r(r + r_0) + x^2 - jr_0 x\}}{(r + r_0)^2 + x^2}$$
$$= \frac{\dot{E}_0 \{z^2 + rr_0 - jr_0 x\}}{z^2 + 2rr_0 + r_0^2};$$

or, in absolute values we have the following: —

Impressed E.M.F.,

$$E_0 = \sqrt{e_0^2 + e_0'^2};$$

current,

$$I = \frac{E_0}{\sqrt{(r + r_0)^2 + x^2}} = \frac{E_0}{\sqrt{z^2 + 2rr_0 + r_0^2}};$$

E.M.F. at terminals of receiver circuit,

$$E = E_0 \sqrt{\frac{r^2 + x^2}{(r + r_0)^2 + x^2}} = \frac{E_0 z}{\sqrt{z^2 + 2rr_0 + r_0^2}};$$

difference of phase in receiver circuit, $\tan \hat{\omega} = \frac{x}{r}$;

difference of phase in supply circuit, $\tan \hat{\omega}_0 = \frac{x}{r + r_0}$

since in general,

$$\tan (\text{phase}) = \frac{\text{imaginary component}}{\text{real component}}.$$

a.) If x is negligible with respect to r, as in a non-inductive receiving circuit,

$$I = \frac{E_0}{r + r_0}, \qquad E = E_0 \frac{r}{r + r_0},$$

and the current and E.M.F. at receiver terminals decrease steadily with increasing r_0.

b.) If r is negligible compared with x, as in a wattless receiver circuit,

$$I = \frac{E_0}{\sqrt{r_0^2 + x^2}}, \qquad E = E_0 \frac{x}{\sqrt{r_0^2 + x^2}};$$

or, for small values of r_0,

$$I = \frac{E_0}{x}, \qquad E = E_0;$$

that is, the current and E.M.F. at receiver terminals remain approximately constant for small values of r_0, and then decrease with increasing rapidity.

44. In the general equations, x appears in the expressions for I and E only as x^2, so that I and E assume the same value when x is negative, as when x is positive; or, in other words, series resistance acts upon a circuit with leading current, or in a condenser circuit, in the same way as upon a circuit with lagging current, or an inductive circuit.

For a given impedance, z, of the receiver circuit, the current I, and E.M.F., E, are smaller, as r is larger; that is, the less the difference of phase in the receiver circuit.

As an instance, in Fig. 37 is shown the E.M.F., E, at the receiver circuit, for E_0 = const. = 100 volts, $z = 1$ ohm; hence $I = E$, and —

a.) $r_0 = .2$ ohm (Curve I.)
b.) $r_0 = .8$ ohm (Curve II.)

with values of reactance, $x = \sqrt{z^2 - r^2}$, for abscissæ, from $x = +1.0$ to $x = -1.0$ ohm.

As shown, I and E are smallest for $x = 0$, $r = 1.0$, or for the non-inductive receiver circuit, and largest for $x = \pm 1.0$, $r = 0$, or for the wattless circuit, in which latter a series resistance causes but a very small drop of potential.

Hence the control of a circuit by series resistance depends upon the difference of phase in the circuit.

For $r_0 = .8$, and $x = 0$, $x = +.8$, $x = -.8$, the polar diagrams are shown in Figs. 38 to 40

2.) *Reactance in series with a circuit.*

45. In a constant potential system of impressed E.M.F.,

$$\dot{E}_0 = e_0 + je_0', \qquad E_0 = \sqrt{e_0^2 + e_0'^2},$$

let a reactance, x_0, be connected in series in a receiver circuit of impedance

$$Z = r - jx, \qquad z = \sqrt{r^2 + x^2}.$$

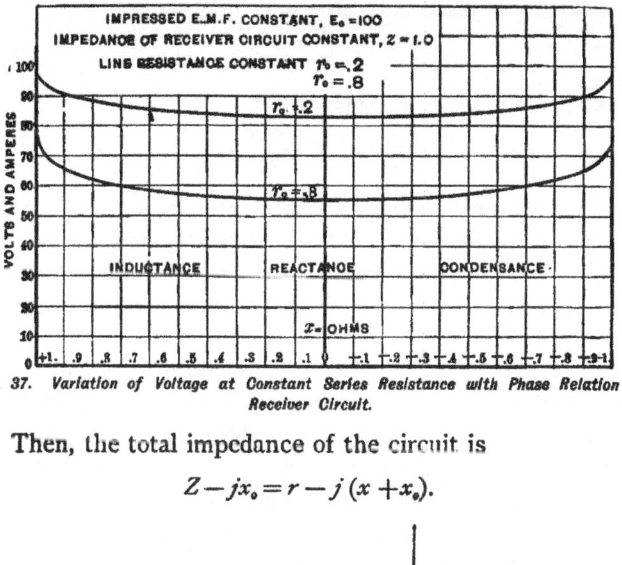

Fig. 37. Variation of Voltage at Constant Series Resistance with Phase Relation of Receiver Circuit.

Then, the total impedance of the circuit is

$$Z - jx_0 = r - j(x + x_0).$$

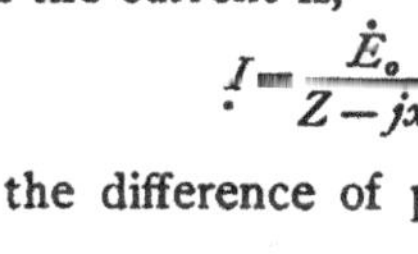

Fig. 38.

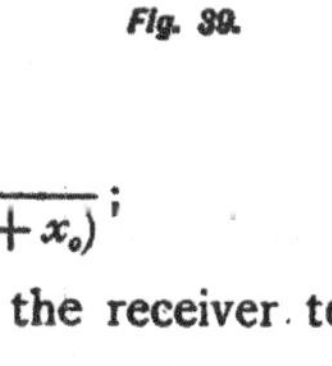

Fig. 39.

and the current is,

$$\dot{I} = \frac{\dot{E}_0}{Z - jx_0} = \frac{\dot{E}_0}{r - j(x + x_0)};$$

while the difference of potential at the receiver terminals is,

$$\dot{E} = \dot{I}Z = \dot{E}_0 \frac{r - jx}{r - j(x + x_0)}.$$

Or, in absolute quantities :—

Current,

$$I = \frac{E_0}{\sqrt{r^2 + (x + x_0)^2}} = \frac{E_0}{\sqrt{z^2 + 2\,xx_0 + x_0^2}};$$

E.M.F. at receiver terminals,

$$E = E_0 \sqrt{\frac{r^2 + x^2}{r^2 + (x + x_0)^2}} = \frac{E_0 z}{\sqrt{z^2 + 2\,xx_0 + x_0^2}};$$

difference of phase in receiver circuit,

$$\tan \hat{\omega} = \frac{x}{r};$$

difference of phase in supply circuit,

$$\tan \hat{\omega}_0 = \frac{x + x_0}{r}.$$

a.) If x is small compared with r, that is, if the receiver circuit is non-inductive, I and E change very little for small values of x_0; but if x is large, that is, if the receiver circuit is of large reactance, I and E change much with a change of x_0.

b.) If x is negative, that is, if the receiver circuit contains condensers, synchronous motors, or other apparatus which produce leading currents — above a certain value of x the denominator in the expression of E, becomes $< z$, or $E > E_0$; that is, the reactance, x_0, raises the potential.

c.) $E = E_0$, or the insertion of a series inductance, x_0, does not affect the potential difference at the receiver terminals, if

$$\sqrt{z^2 + 2\,xx_0 + x_0^2} = z;$$

or,

$$x_0 = -2\,x.$$

That is, if the reactance which is connected in series in the circuit is of opposite sign, but twice as large as the reactance of the receiver circuit, the voltage is not affected, but $E = E_0$, $I = E_0 / z$. If $x_0 < -2\,x$, it raises, if $x_0 > -2\,x$, it lowers, the voltage.

We see, then, that a reactance inserted in series in an alternating-current circuit will lower the voltage at the receiver terminals only when of the same sign as the reactance of the receiver circuit; when of opposite sign, it will lower the voltage if larger, raise the voltage if less, than twice the numerical value of the reactance of the receiver circuit.

d.) If $x = 0$, that is, if the receiver circuit is non-inductive, the E.M.F. at receiver terminals is:

$$E = \frac{E_0 r}{\sqrt{r^2 + x_0^2}} = \frac{E_0}{\sqrt{1 + \left(\frac{x_0}{r}\right)^2}}$$

$$= E_0 \left\{ 1 - \frac{1}{2}\left(\frac{x_0}{r}\right)^2 + \frac{3}{8}\left(\frac{x_0}{r}\right)^4 - + \dots \right\}$$

$\left(\frac{1}{\sqrt{1+x}} = (1+x)^{-\frac{1}{2}} \right.$ expanded by the binomial theorem

$$\left. (1+x)^n = 1 + nx + \frac{n(n-1)}{1 \cdot 2} x^2 + \dots \right).$$

Therefore, if x_0 is small compared with r:—

$$E = E_0 \left(1 - \frac{1}{2}\left(\frac{x_0}{r}\right)^2 \right),$$

$$\frac{E_0 - E}{E_0} = \frac{1}{2}\left(\frac{x_0}{r}\right)^2.$$

That is, the percentage drop of potential by the insertion of reactance in series in a non-inductive circuit is, for small

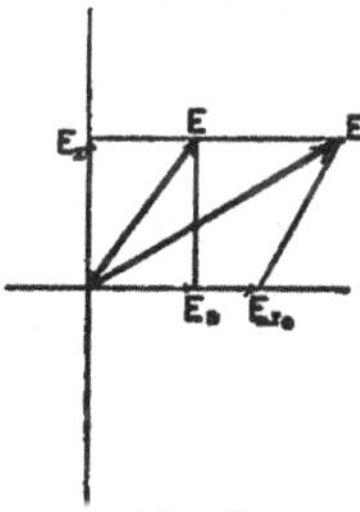

Fig. 40.

values of reactance, independent of the sign, but proportional to the square of the reactance, or the same whether it be inductance or condensance reactance.

46. As an instance, in Fig. 41 the changes of current, I, and of E.M.F. at receiver terminals, E, at constant impressed E.M.F., E_0, are shown for various conditions of a receiver circuit and amounts of reactance inserted in series.

Fig. 41 gives for various values of reactance, x_0 (if positive, inductance — if negative, condensance), the E.M.Fs., E, at receiver terminals, for constant impressed E.M.F.,

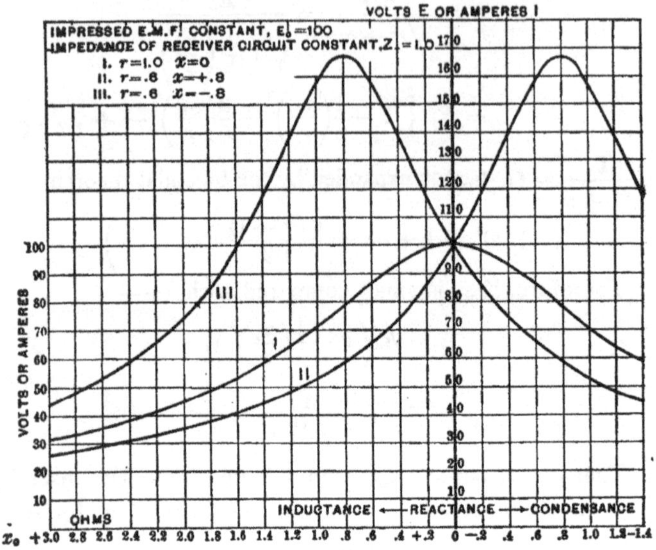

Fig. 41.

$E_0 = 100$ volts, and the following conditions of receiver circuit: —

$$z = 1.0,\ r = 1.0,\ x = 0 \text{ (Curve I.)}$$
$$z = 1.0,\ r = .6,\ x = .8 \text{ (Curve II.)}$$
$$z = 1.0,\ r = .6,\ x = -.8 \text{ (Curve III.)}$$

As seen, curve I is symmetrical, and with increasing x_0 the voltage E remains first almost constant, and then drops off with increasing rapidity.

In the inductive circuit series inductance, or, in a condenser circuit series condensance, causes the voltage to drop off very much faster than in a non-inductive circuit.

Series inductance in a condenser circuit, and series condensance in an inductive circuit, cause a rise of potential. This rise is a maximum for $x_0 = \pm .8$, or, $x_0 = -x$ (the condition of resonance), and the E.M.F. reaches the value, $E = 167$ volts, or, $E = E_0 z / r$. This rise of potential by series reactance continues up to $x_0 = \pm 1.6$, or, $x_0 = -2x$,

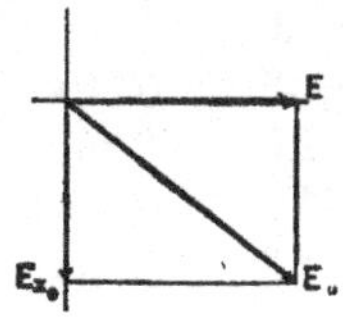

Fig. 42.

where $E = 100$ volts again; and for $x_0 > 1.6$ the voltage drops again.

At $x_0 = \pm .8$, $x = \mp .8$, the total impedance of the circuit is $r - j(x + x_0) = r = .6$, $x + x_0 = 0$, and $\tan \hat{\omega}_0 = 0$; that is, the current and E.M.F. in the supply circuit are in phase with each other, or the circuit is in *electrical resonance.*

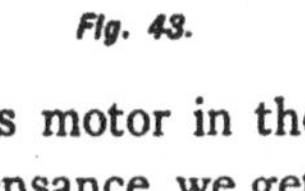

Fig. 43.

Since a synchronous motor in the condition of efficient working acts as a condensance, we get the remarkable result that, in synchronous motor circuits, choking coils, or reactive coils, can be used for raising the voltage.

In Figs. 42 to 44, the polar diagrams are shown for the conditions—

$E_0 = 100$, $x_0 = .6$,	$x = 0$	(Fig. 42) $E = 85.7$
	$x = +.8$	(Fig. 43) $E = 65.7$
	$x = -.8$	(Fig. 44) $E = 158.1$

47. In Fig. 45 the dependence of the potential, E, upon the difference of phase, $\hat{\omega}$, in the receiver circuit is shown for the constant impressed E.M.F., $E_0 = 100$; for the constant receiver impedance, $z = 1.0$ (but of various phase differences $\hat{\omega}$), and for various series reactances, as follows ·

$x_0 = .2$	(Curve I.)
$x_0 = .6$	(Curve II.)
$x_0 = .8$	(Curve III.)
$x_0 = 1.0$	(Curve IV.)
$x_0 = 1.6$	(Curve V.)
$x_0 = 3.2$	(Curve VI.)

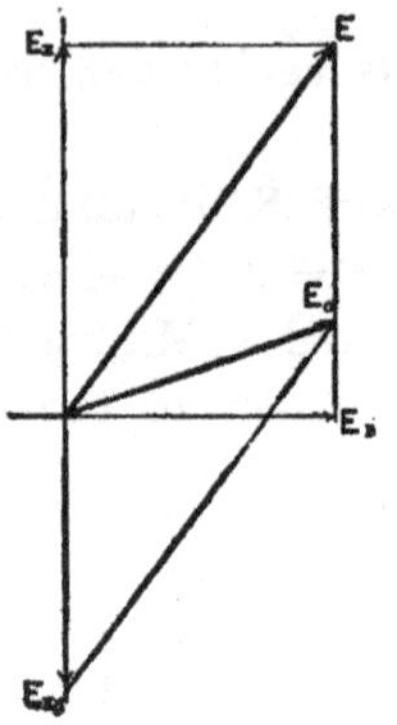

Fig. 44.

Since $z = 1.0$, the current, I, in all these diagrams has the same value as E.

In Figs. 46 and 47, the same curves are plotted as in Fig. 45, but in Fig. 46 with the reactance, x, of the receiver circuit as abscissæ; and in Fig. 47 with the resistance, r, of the receiver circuit as abscissæ.

As shown, the receiver voltage, E, is always lowest when x_0 and x are of the same sign, and highest when they are of opposite sign.

The rise of voltage due to the balance of x_0 and x is a maximum for $x_0 = +1.0$, $x = -1.0$, and $r = 0$, where

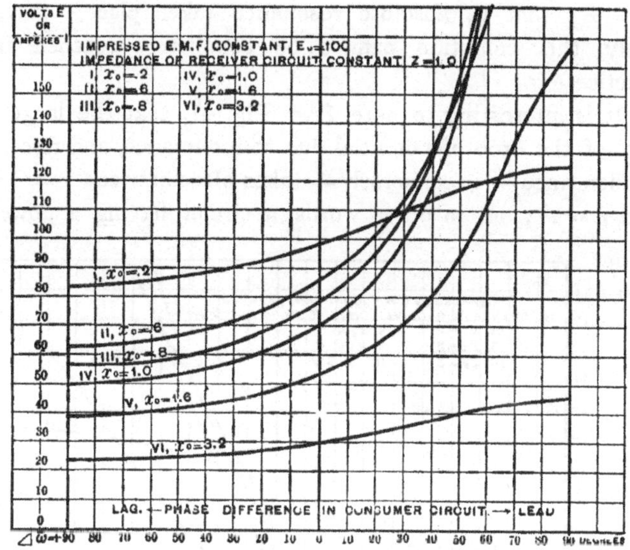

Fig. 45. Variation of Voltage at Constant Series Reactance with Phase Angle of Receiver Circuit.

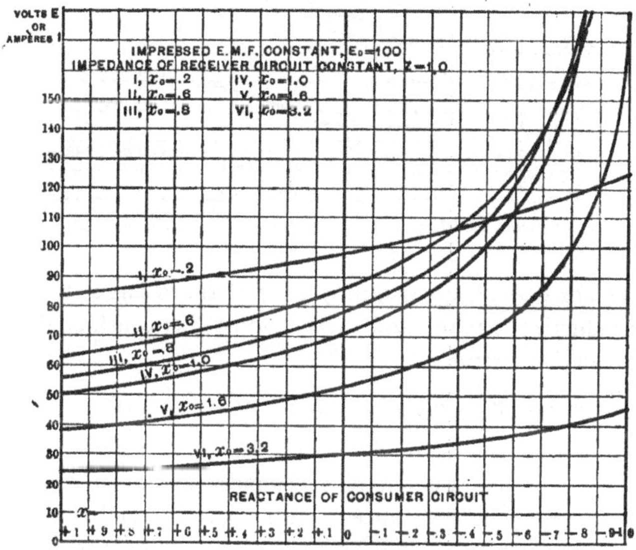

Fig. 46. Variation of Voltage at Constant Series Reactance with Reactance of Receiver Circuit.

$E = \infty$; that is, absolute resonance takes place. Obviously, this condition cannot be completely reached in practice.

It is interesting to note, from Fig. 47, that the largest part of the drop of potential due to inductance, and rise to condensance — or conversely — takes place between $r = 1.0$ and $r = .9$; or, in other words, a circuit having a power

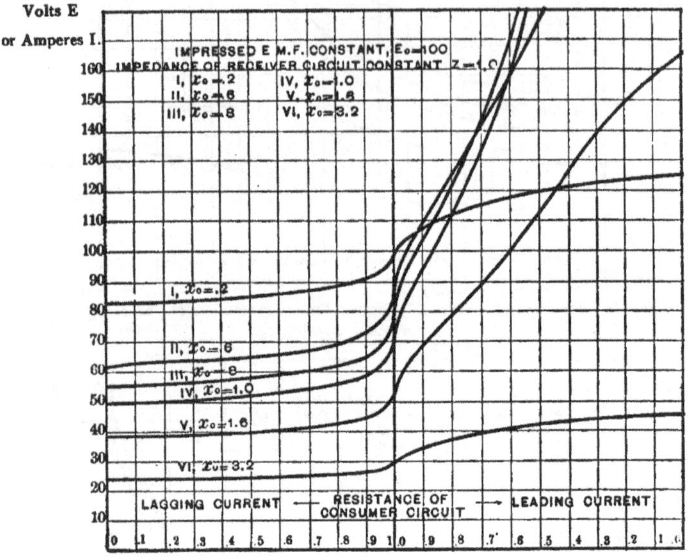

Fig. 47. Variation of Voltage at Constant Series Reactance with Resistance of Receiver Circuit.

factor $\cos \hat{\omega} = .9$, gives a drop several times larger than a non-inductive circuit, and hence must be considered as an inductive circuit.

3.) *Impedance in series with a circuit.*

48. By the use of reactance for controlling electric circuits, a certain amount of resistance is also introduced, due to the ohmic resistance of the conductor and the hysteretic loss, which, as will be seen hereafter, can be represented as an effective resistance.

Hence the impedance of a reactive coil (choking coil) may be written thus:—

$$Z_0 = r_0 - jx_0, \qquad z_0 = \sqrt{r_0^2 + x_0^2},$$

where r_0 is in general small compared with x_0.

From this, if the impressed E.M.F. is

$$\dot{E}_0 = e_0 + je_0', \qquad E_0 = \sqrt{e_0^2 + e_0'^2}$$

and the impedance of the consumer circuit is

$$Z = r - jx \qquad z = \sqrt{r^2 + x^2}$$

we get the current, $\dot{I} = \frac{\dot{E}_0}{Z + Z_0} = \frac{\dot{E}_0}{(r + r_0) - j(x + x_0)}$

and the E.M.F. at receiver terminals,

$$\dot{E} = \dot{E}_0 \frac{Z}{Z + Z_0} = \dot{E}_0 \frac{r - jx}{(r + r_0) - j(x + x_0)}.$$

Or, in absolute quantities,

the current is,

$$I = \frac{E_0}{\sqrt{(r + r_0)^2 + (x + x_0)^2}} = \frac{E_0}{\sqrt{z^2 + z_0^2 + 2(rr_0 + xx_0)}};$$

the E.M.F. at receiver terminals is,

$$E = \frac{E_0 z}{\sqrt{(r + r_0)^2 + (x + x_0)^2}} = \frac{E_0 z}{\sqrt{z^2 + z_0^2 + 2(rr_0 + xx_0)}};$$

the difference of phase in receiver circuit is,

$$\tan \hat{\omega} = \frac{x}{r};$$

and the difference of phase in the supply circuit is,

$$\tan \hat{\omega} = \frac{x + x_0}{r + r_0}.$$

49. In this case, the maximum drop of potential will not take place for either $x = 0$, as for resistance in series, or for $r = 0$, as for reactance in series, but at an intermediate point. The drop of voltage is a maximum; that is, E is a minimum if the denominator of E is a maximum; or, since z, z_0, r_0, x_0 are constant, if $rr_0 + xx_0$ is a maximum, that is, since $x = \sqrt{z^2 - r^2}$, if $rr_0 + x_0\sqrt{z^2 - r^2}$ is a maximum

A function, $f = rr_0 + x_0 \sqrt{z^2 - r^2}$ is a maximum when its differential coefficient equals zero. For, plotting f as curve with r as abscissæ, at the point where f is a maximum or a minimum, this curve is for a short distance horizontal, hence the tangens-function of its tangent equals zero. The tangens-function of the tangent of a curve, however, is the ratio of the change of ordinates to the change of abscissæ, or is the differential coefficient of the function represented by the curve.

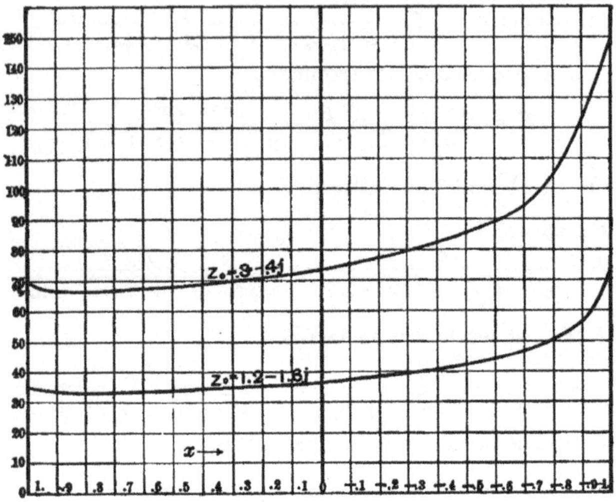

Fig. 48.

Thus we have :—

$$f = rr_0 + x_0 \sqrt{z^2 - r^2} = \text{maximum or minimum, if}$$

$$\frac{df}{dr} = 0.$$

Differentiating, we get :—

$$r_0 + \frac{1}{2} \frac{x_0}{\sqrt{z^2 - r^2}} (-2\, r) = 0\,;$$

or, expanded,—

$$r_0 \sqrt{z^2 - r^2} - x_0 r = r_0 x - x_0 r = 0,$$

or,

$$r \div x = r_0 \div x_0.$$

That is, the drop of potential is a maximum, if the reactance factor, x/r, of the receiver circuit equals the reactance factor, x_0/r_0, of the series impedance.

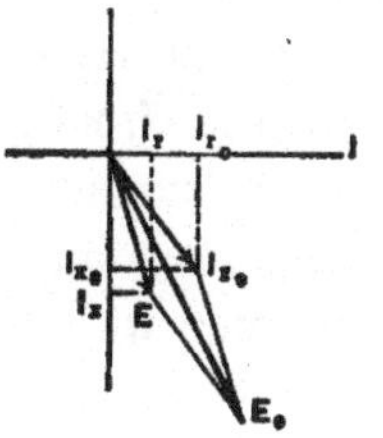

Fig. 49.

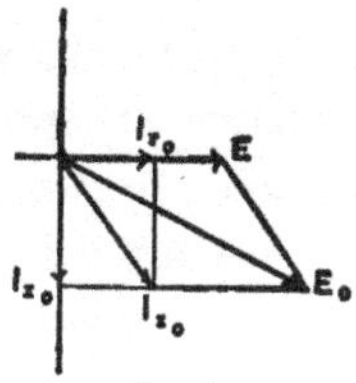

Fig. 50.

50. As an example, Fig. 48 shows the E.M.F., E, at the receiver terminals, at a constant impressed E.M.F., $E_0 = 100$, a constant impedance of the receiver circuit, $z = 1.0$, and constant series impedances,

$$Z_0 = .3 - j.4 \qquad \text{(Curve I.)}$$
$$Z_0 = 1.2 - j1.6 \qquad \text{(Curve II.)}$$

as functions of the reactance, x, of the receiver circuit.

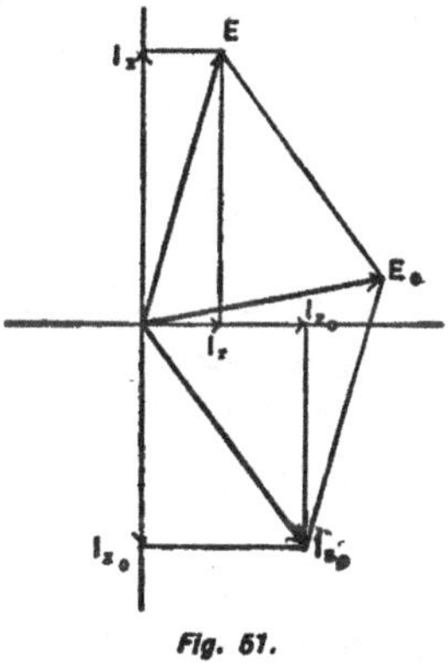

Fig. 51.

Figs. 49 to 51 give the polar diagram for $E_0 = 100$, $x = .95$, $x = 0$, $x = -.95$, and $Z_0 = .3 - j.4$.

4.) *Compensation for Lagging Currents by Shunted Condensance.*

51. We have seen in the latter paragraphs, that in a constant potential alternating-current system, the voltage at the terminals of a receiver circuit can be varied by the use of a variable reactance in series to the circuit, without loss of energy except the unavoidable loss due to the resistance and hysteresis of the reactance; and that, if the series reactance is very large compared with the resistance of the receiver circuit, the current in the receiver circuit becomes more or less independent of the resistance, — that is, of the power consumed in the receiver

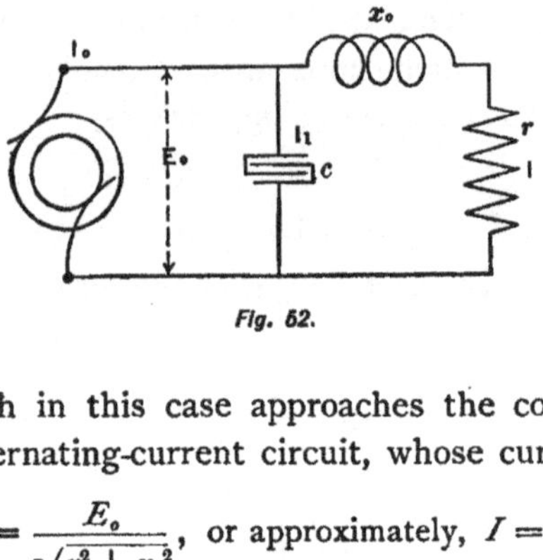

Fig. 52.

circuit, which in this case approaches the conditions of a constant alternating-current circuit, whose current is.

$$I = \frac{E_0}{\sqrt{r^2 + x_0^2}}, \text{ or approximately, } I = \frac{E_0}{x_0}.$$

This potential control, however, causes the current taken from the mains to lag greatly behind the E.M.F., and thereby requires a much larger current than corresponds to the power consumed in the receiver circuit.

Since a condenser draws from the mains a leading current, a condenser shunted across such a circuit with lagging current will compensate for the lag, the leading and the lagging current combining to form a resultant current more or less in phase with the E.M.F., and therefore proportional to the power expended.

In a circuit shown diagrammatically in Fig. 52, let the non-inductive receiver circuit of resistance, r, be connected in series with the inductance, x_0, and the whole shunted by a condenser of condensance, c, entailing but a negligible loss of energy.

Then, if $\dot{E}_0$ = impressed E.M.F., —

the current in receiver circuit is,

$$\dot{I} = \frac{\dot{E}_0}{r - jx_0}, \quad I = \frac{E_0}{\sqrt{r^2 + x_0^2}};$$

the current in condenser circuit is,

$$\dot{I}_1 = \frac{\dot{E}_0}{jc}, \qquad I_1 = \frac{E_0}{c}.$$

and the total current is

$$\dot{I}_0 = \dot{I} + \dot{I}_1 = \dot{E}_0 \left\{ \frac{1}{r - jx_0} + \frac{1}{jc} \right\}$$

$$= \dot{E}_0 \left\{ \frac{r}{r^2 + x_0^2} + j\left(\frac{x_0}{r^2 + x_0^2} - \frac{1}{c} \right) \right\},$$

or, in absolute terms, $I_0 = E_0 \sqrt{\left(\frac{r}{r^2 + x_0^2} \right)^2 + \left(\frac{x_0}{r^2 + x_0^2} - \frac{1}{c} \right)^2};$

while the E.M.F. at receiver terminals is,

$$\dot{E} = \dot{I}r = \dot{E}_0 \frac{r}{r - jx_0}, \qquad E = \frac{E_0 r}{\sqrt{r^2 + x_0^2}}.$$

52. The main current, $\dot{I}_0$, is in phase with the impressed E.M.F., $\dot{E}_0$, or the lagging current is completely balanced, or supplied by, the condensance, if the imaginary term in the expression of $\dot{I}_0$ disappears; that is, if

$$\frac{x_0}{r^2 + x_0^2} - \frac{1}{c} = 0.$$

This gives, expanded: $c = \frac{r^2 + x_0^2}{x_0}$

Hence the capacity required to compensate for the lagging current produced by the insertion of inductance in series to a non-inductive circuit depends upon the resistance and the inductance of the circuit. x_0 being constant, with increasing resistance, r, the condensance has to be increased, or the capacity decreased, to keep the balance.

Substituting
$$c = \frac{r^2 + x_0^2}{x_0},$$

we get, as the equations of the inductive circuit balanced by condensance: —

$$\dot{I} = \frac{\dot{E}_0}{r - jx_0} = \frac{\dot{E}_0 (r + jx_0)}{r^2 + x_0^2}, \qquad I = \frac{E_0}{\sqrt{r^2 + x_0^2}};$$

$$\dot{I}_1 = -\frac{j\dot{E}_0 x_0}{r^2 + x_0^2}, \qquad I_1 = \frac{E_0 x_0}{r^2 + x_0^2};$$

$$\dot{I}_0 = \frac{\dot{E}_0 r}{r^2 + x_0^2}, \qquad I_0 = \frac{E_0 r}{r^2 + x_0^2};$$

$$\dot{E} = \frac{\dot{E}_0 r}{r - jx_0}, \qquad E = \frac{E_0 r}{\sqrt{r^2 + x_0^2}};$$

and for the power expended in the receiver circuit: —

$$I^2 r = \frac{E_0^2 r}{r^2 + x_0^2} = I_0 E_0,$$

that is, the main current is proportional to the expenditure of power.

For $r = 0$ we have $c = x_0$, or the condition of balance.

Complete balance of the lagging component of current by shunted capacity thus requires that the condensance, c, be varied with the resistance, r; that is, with the varying load on the receiver circuit.

In Fig. 53 are shown, for a constant impressed E.M.F., $E_0 = 1000$ volts, and a constant series reactance, $x_0 = 100$ ohms, values for the balanced circuit of,

current in receiver circuit (Curve I.),
current in condenser circuit (Curve II.),
current in main circuit (Curve III.),
E.M.F. at receiver terminals (Curve IV.),

with the resistance, r, of the receiver circuit as abscissæ.

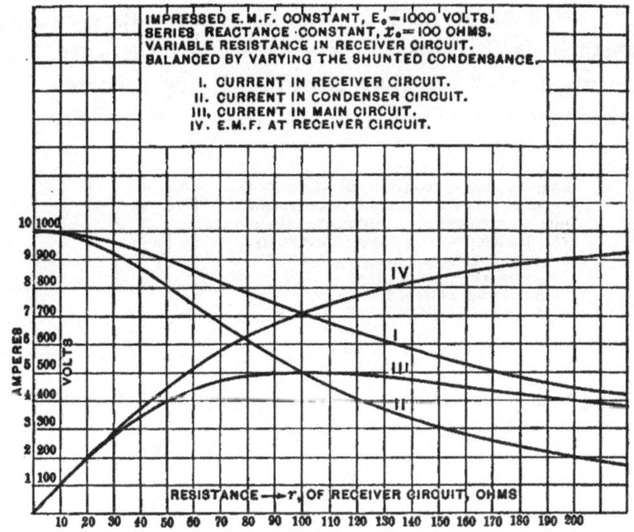

Fig. 53. Compensation of Lagging Currents in Receiving Circuit by Variable Shunted Condensance.

53. If, however, the condensance is left unchanged, $c = x_0$ at the no-load value, the circuit is balanced for $r = 0$, but will be overbalanced for $r > 0$, and the main current will become leading.

We get in this case:—

$$c = x_0;$$

$$\underset{\cdot}{I} = \frac{\dot{E}_0}{r - jx_0}, \qquad I = \frac{E_0}{\sqrt{r^2 + x_0^2}};$$

$$\underset{\cdot}{I_1} = -\frac{j\dot{E}_0}{x_0}, \qquad I_1 = \frac{E_0}{x_0};$$

$$\underset{\cdot}{I_0} = \underset{\cdot}{I} + \underset{\cdot}{I_1} = \frac{\dot{E}_0 r}{x_0(x_0 + jr)}, \quad I_0 = \frac{E_0 r}{x_0\sqrt{r^2 + x_0^2}};$$

$$\underset{\cdot}{E} = \underset{\cdot}{I}r = \frac{\dot{E}_0 r}{r - jx_0}, \qquad E = \frac{E_0 r}{\sqrt{r^2 + x_0^2}}.$$

The difference of phase in the main circuit is,—

$$\tan \omega_0 = -\frac{r}{x_0}, \qquad \text{which is} = 0.$$

when $r = 0$ or at no load, and increases with increasing resistance, as the lead of the current. At the same time, the current in the receiver circuit, I, is approximately constant for small values of r, and then gradually decreases.

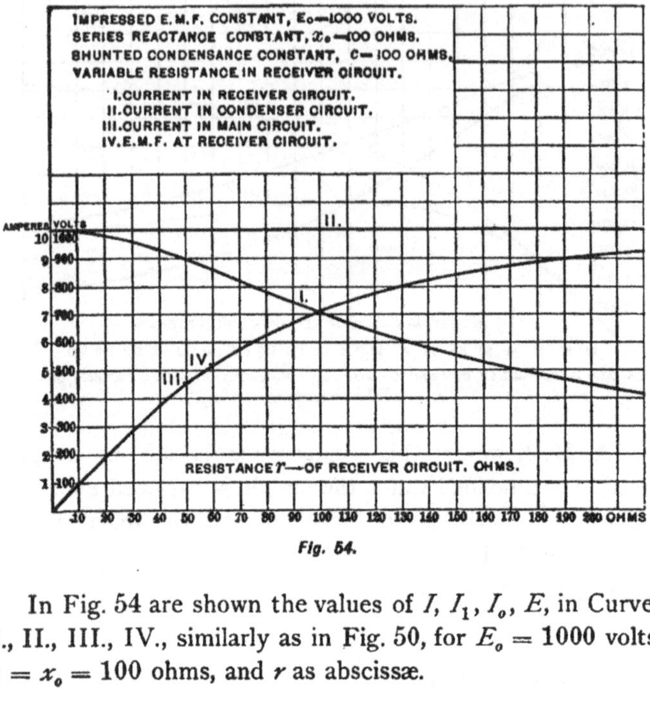

Fig. 54.

In Fig. 54 are shown the values of I, I_1, I_0, E, in Curves I., II., III., IV., similarly as in Fig. 50, for $E_0 = 1000$ volts, $c = x_0 = 100$ ohms, and r as abscissæ.

5.) *Constant Potential — Constant Current Transformation.*

54. In a constant potential circuit containing a large and constant reactance, x_0, and a varying resistance, r, the current is approximately constant, and only gradually drops off with increasing resistance, r, — that is, with increasing load, — but the current lags greatly behind the E.M.F. This lagging current in the receiver circuit can be supplied by a shunted condensance. Leaving, however, the condensance constant, $c = x_0$, so as to balance the lagging current at no load, that is, at $r = 0$, it will overbalance with increasing load, that is, with increasing r, and thus the main current will become leading, while the receiver current decreases if the impressed E.M.F., E_0, is kept constant. Hence, to keep the current in the receiver circuit entirely constant, the impressed E.M.F., E_0, has to be increased with increasing resistance, r; that is, with increasing lead of the main current. Since, as explained before, in a circuit with leading current, a series inductance raises the potential, to maintain the current in the receiver circuit constant under all loads, an inductance, x_2, inserted in the main circuit, as shown in the diagram, Fig. 55, can be used for raising the potential, E_0, with increasing load.

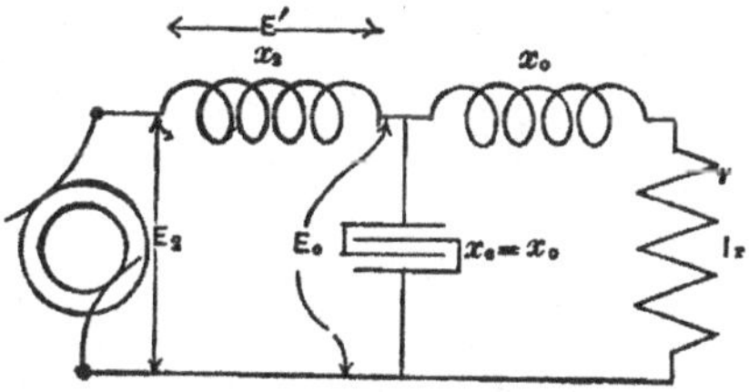

Fig. 55.

Let —

$$\underset{\cdot}{E}_2 = e_2 + je_2', \quad E_2 = \sqrt{e_2^2 + e_2'^2},$$

be the impressed E.M.F. of the generator, or of the mains, and let the condensance be $x_c = x_0$; then —

Current in receiver circuit,

$$\underset{\cdot}{I} = \frac{\dot{E}_0}{r - jx_0};$$

current in condenser circuit,

$$\underset{\cdot}{I}_1 = -\frac{j\dot{E}_0}{x_0}.$$

Hence, the total current in main line is

$$\underset{\cdot}{I}_0 = \underset{\cdot}{I} + \underset{\cdot}{I}_1 = \underset{\cdot}{E}_0 \left\{ \frac{1}{r - jx_0} - \frac{j}{x_0} \right\}$$

$$= \frac{\dot{E}_0 r}{jx_0 (r - jx_0)},$$

and the E.M.F. at receiver terminals,

$$\dot{E} = \dot{I}r = \frac{\dot{E}_0 r}{r - jx_0};$$

E.M.F. at condenser terminals,

$$\dot{E}_0;$$

E.M.F. consumed in main line,

$$\dot{E}' = -j\dot{I}_0 x_2 = -\frac{\dot{E}_0 r x_2}{x_0(r - jx_0)};$$

hence, the E.M.F. at generator is

$$\dot{E}_2 = \dot{E}_0 + \dot{E}' = \dot{E}_0 \left\{ 1 - \frac{r x_2}{x_0(r - jx_0)} \right\};$$

or,

$$= \dot{E}_0 \frac{r(x_0 - x_2) - jx_0^2}{x_0(r - jx_0)}.$$

and conversely the E.M.F. at condenser terminals,

$$\dot{E}_0 = \frac{\dot{E}_2 x_0 (r - jx_0)}{r(x_0 - x_2) - jx_0^2};$$

current in receiver circuit,

$$\dot{I} = \frac{\dot{E}_0}{r - jx_0} = \frac{\dot{E}_2 x_0}{r(x_0 - x_2) - jx_0^2}.$$

This value of $\dot{I}$ contains the resistance, r, only as a factor to the difference, $x_0 - x_2$; hence, if the reactance, x_2, is chosen $= x_0$, r cancels altogether, and we find that if $x_2 = x_0$, the current in the receiver circuit is constant,

$$\dot{I} = j\frac{\dot{E}_2}{x_0},$$

and is independent of the resistance, r; that is, of the load.

Thus, by substituting $x_2 = x_0$, we have,

Impressed E.M.F. at generator,

$$\dot{E}_2 = e_2 + je_2', \qquad E_2 = \sqrt{e_2^2 + e_2'^2} = \text{constant};$$

current in receiver circuit,

$$\dot{I} = j\frac{\dot{E}_2}{x_0}, \qquad I = \frac{E_2}{x_0} = \text{constant};$$

E.M.F. at receiver circuit,

$$\dot{E} = \dot{I}r = j\frac{\dot{E}_2 r}{x_0}, \qquad E = \frac{E_2 r}{x_0}, \text{ or proportional to load } r;$$

E.M.F. at condenser terminals,

$$\dot{E}_0 = \frac{\dot{E}_2(x_0 + jr)}{x_0}$$

$$= \dot{E}_2\left(1 + j\frac{r}{x_0}\right), \quad E_0 = E_2\sqrt{1 + \left(\frac{r}{x_0}\right)^2}, \text{ hence } > E_2;$$

current in condenser circuit,

$$\dot{I}_1 = -j\frac{\dot{E}_2(x_0 + jr)}{x_0^2}, \quad I_1 = \frac{E_2}{x_0}\sqrt{1 + \left(\frac{r}{x_0}\right)^2};$$

main current,

$$\dot{I}_0 = \frac{\dot{E}_0 r}{x_0(x_0 + jr)};$$

$$= \frac{\dot{E}_2 r}{x_0^2}, \qquad I_0 = \frac{E_2 r}{x_0^2}, \quad \left\{\begin{array}{l}\text{proportional to the load,} \\ r, \text{ and in phase with} \\ \text{E.M.F., } E_2.\end{array}\right.$$

The power of the receiver circuit is,

$$IE = \frac{E_2^2 r}{x_0^2};$$

the power of the main circuit,

$$I_0 E_2 = \frac{E_2^2 r}{x_0^2}, \text{ hence the same.}$$

55. This arrangement is entirely reversible; that is,

if E_2 = constant, I = constant; and
if I_0 = constant, E = constant.

In the latter case we have, by expressing all the quantities by I_0:—

Current in main line,

$$I_0 = \text{constant};$$

E.M.F. at receiver circuit,

$$E = I_0 x_0 = \text{constant};$$

current in receiver circuit,

$$I = I_0\frac{x_0}{r}, \text{ proportional to the load } \frac{1}{r};$$

current in condenser circuit,

$$I_1 = I_0\sqrt{1 + \left(\frac{x_0}{r}\right)^2};$$

E.M.F. at condenser terminals,

$$E_c = x_o I_o \sqrt{1 + \left(\frac{x_o}{r}\right)^2};$$

Impressed E.M.F. at generator terminals,

$$E_2 = \frac{x_o^2}{r} I_o, \text{ or proportional to the load } \frac{1}{r}.$$

From the above we have the following deduction:

Connecting two reactances of equal value, x_o, in series to a non-inductive receiver circuit of variable resistance, r, and shunting across the circuit from midway between the inductances by a capacity of condensance, $x_c = x_o$, transforms a constant potential main circuit into a constant current receiver circuit, and, inversely, transforms a constant current main circuit into a constant potential receiver circuit. This combination of inductance and capacity acts as a transformer, and converts from constant potential to constant current and inversely, without introducing a displacement of phase between current and E.M.F.

It is interesting to note here that a short circuit in the receiver circuit acts like a break in the supply circuit, and a break in the receiver circuit acts like a short circuit in the supply circuit.

As an instance, in Fig. 56 are plotted the numerical values of a transformation from constant potential of 1,000 volts to constant current of 10 amperes.

Since $E_2 = 1{,}000$, $I = 10$, we have: $x_o = 100$; hence the constants of the circuit are:—

$E_2 = 1000$ volts;

$I = 10$ amperes;

$E = 10r$, plotted as Curve I., with the resistances, r, as abscissæ;

$E_c = 1000\sqrt{1 + \left(\frac{r}{100}\right)^2}$, plotted as Curve II.;

$I_1 = 10\sqrt{1 + \left(\frac{r}{100}\right)^2}$, plotted as Curve III.

$I_o = .1\, r$, plotted as Curve IV.

56. In practice, the power consumed in the main circuit will be larger than the power delivered to the receiver circuit, due to the unavoidable losses of power in the inductances and condensances.

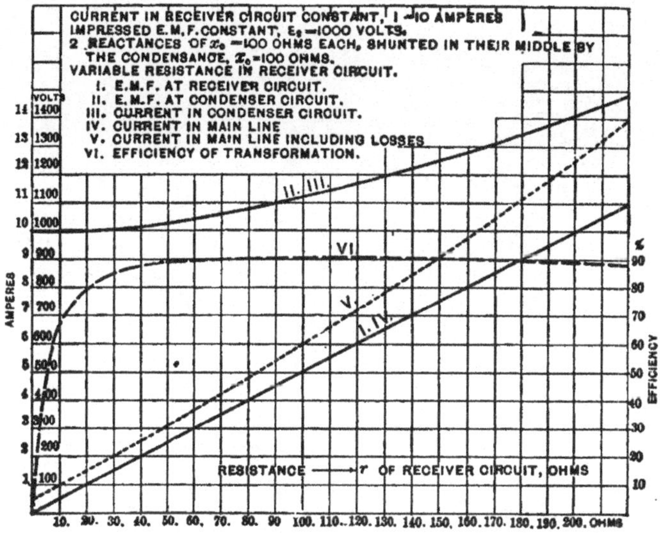

Fig. 56. Constant-Potential — Constant-Current Transformation.

Let —

$r_1 = 2$ ohms = effective resistance of condensance;

$r_o = 3$ ohms = effective resistance of each of the inductances.

We then have: —

Power consumed in condensance, $I_1^2 r_1 = 200 + .02\, r^2$;
power consumed by first inductance, $I^2 r_o = 300$;
power consumed by second inductance, $I_o^2 r_o = .03\, r^2$.

Hence, the total loss of energy is $500 + .05\, r^2$;

output of system, $I^2 r = 100\, r$

input, $500 + 100\, r + .05\, r^2$;

efficiency, $\dfrac{100\, r}{500 + 100\, r + .05\, r^2}$.

It follows that the main current, I_o, increases slightly by the amount necessary to supply the losses of energy in the apparatus.

This curve of current, I_0, including losses in transformation, is shown in dotted lines as Curve V. in Fig. 56 ; and the efficiency is shown in broken line, as Curve VI. As shown, the efficiency is practically constant within a wide range.

CHAPTER IX.

RESISTANCE AND REACTANCE OF TRANSMISSION LINES.

57. In alternating-current circuits, E.M.F. is consumed in the feeders of distributing networks, and in the lines of long-distance transmissions, not only by the resistance, but also by the reactance, of the line. The E.M.F. consumed by the resistance is in phase, while the E.M.F. consumed by the reactance is in quadrature, with the current. Hence their influence upon the E.M.F. at the receiver circuit depends upon the difference of phase between the current and the E.M.F. in that circuit. As discussed before, the drop of potential due to the resistance is a maximum when the receiver current is in phase, a minimum when it is in quadrature, with the E.M.F. The change of potential due to line reactance is small if the current is in phase with the E.M.F., while a drop of potential is produced with a lagging, and a rise of potential with a leading, current in the receiver circuit.

Thus the change of potential due to a line of given resistance and inductance depends upon the phase difference in the receiver circuit, and can be varied and controlled by varying this phase difference; that is, by varying the admittance, $Y = g + jb$, of the receiver circuit.

The conductance, g, of the receiver circuit depends upon the consumption of power, — that is, upon the load on the circuit, and thus cannot be varied for the purpose of regulation. Its susceptance, b, however, can be changed by shunting the circuit with a reactance, and will be increased by a shunted inductance, and decreased by a shunted condensance. Hence, for the purpose of investigation, the receiver circuit can be assumed to consist of two branches, a conductance, g, — the non-inductive part of the circuit, — shunted by a susceptance, b, which can be varied without expenditure of energy. The two components of current can thus be considered separately, the energy component as determined by the load on the circuit, and the wattless component, which can be varied for the purpose of regulation.

Obviously, in the same way, the E.M.F. at the receiver circuit may be considered as consisting of two components, the energy component, in phase with the current, and the wattless component, in quadrature with the current. This will correspond to the case of a reactance connected in series to the non-inductive part of the circuit. Since the effect of either resolution into components is the same so far as the line is concerned, we need not make any assumption as to whether the wattless part of the receiver circuit is in shunt, or in series, to the energy part.

Let —

$Z_0 = r_0 - jx_0$ = impedance of the line;

$$z_0 = \sqrt{r_0^2 + x_0^2};$$

$Y = g + jb$ = admittance of receiver circuit;

$$y = \sqrt{g^2 + b^2};$$

$\dot{E}_0 = e_0 + je_0'$ = impressed E.M.F. at generator end of line;

$$E_0 = \sqrt{e_0^2 + e_0'^2};$$

$\dot{E} = e + je'$ = E.M.F. at receiver end of line;

$$E = \sqrt{e^2 + e'^2};$$

$\dot{I}_0 = i_0 + ji_0'$ = current in the line;

$$I_0 = \sqrt{i_0^2 + i_0'^2}.$$

The simplest condition is the non-inductive circuit.

1.) *Non-inductive Receiver Circuit Supplied over an Inductive Line.*

58. In this case, the admittance of the receiver circuit is $Y = g$, since $b = 0$.

We have then —

current, $\dot{I}_o = \dot{E}g$;

impressed E.M.F., $\dot{E}_o = \dot{E} + Z_o\dot{I}_o = \dot{E}(1 + Z_o g)$.

Hence —

E.M.F. at receiver circuit,

$$\dot{E} = \frac{\dot{E}_o}{1 + Z_o g} = \frac{\dot{E}_o}{1 + g r_o - j g x_o};$$

current, $$\dot{I}_o = \frac{\dot{E}_o g}{1 + Z_o g} = \frac{\dot{E}_o g}{1 + g r_o - j g x_o}.$$

Hence, in absolute values —

E.M.F. at receiver circuit, $$E = \frac{E_o}{\sqrt{(1 + g r_o)^2 + g^2 x_o^2}};$$

current, $$I_o = \frac{E_o g}{\sqrt{(1 + g r_o)^2 + g^2 x_o^2}}.$$

The ratio of E.M.Fs. at receiver circuit and at generator, or supply circuit, is —

$$a = \frac{E}{E_o} = \frac{1}{\sqrt{(1 + g r_o)^2 + g^2 x_o^2}};$$

and the power delivered in the non-inductive receiver circuit, or

output, $$P = I_o E = \frac{E_o^2 g}{(1 + g r_o)^2 + g^2 x_o^2}.$$

As a function of g, and with a given E_o, r_o, and x_o, this power is a maximum, if —

$$\frac{dP}{dg} = 0;$$

that is —

$$-1 + g^2 r_o^2 + g^2 x_o^2 = 0;$$

hence —

conductance of receiver circuit for maximum output,

$$g_m = \frac{1}{\sqrt{r_o^2 + x_o^2}} = \frac{1}{z_o}.$$

Resistance of receiver circuit, $r_m = \dfrac{1}{g_m} = z_o$;

and, substituting this in P—

Maximum output, $$P_m = \frac{E_0^2}{2(r_0 + z_0)} = \frac{E_0^2}{2\{r_0 + \sqrt{r_0^2 + x_0^2}\}};$$

and —

ratio of E.M.F. at receiver and at generator end of line,

$$a_m = \frac{E}{E_0} = \frac{1}{\sqrt{2\left(1 + \frac{r_0}{z_0}\right)}};$$

efficiency, $$\frac{r_m}{r_m + r_0} = \frac{z_0}{r_0 + z_0}.$$

That is, the output which can be transmitted over an inductive line of resistance, r_0, and reactance, x_0, — that is, of impedance, z_0, — into a non-inductive receiver circuit, is a maximum, if the resistance of the receiver circuit equals the impedance of the line, $r = z_0$, and is —

$$P_m = \frac{E_0^2}{2(r_0 + z_0)}.$$

The output is transmitted at the efficiency of

$$\frac{z_0}{r_0 + z_0},$$

and with a ratio of E.M.Fs. of

$$a_m = \frac{1}{\sqrt{2\left(1 + \frac{r_0}{z_0}\right)}}.$$

59. We see from this, that the maximum output which can be delivered over an inductive line is less than the output delivered over a non-inductive line of the same resistance — that is, which can be delivered by continuous currents with the same generator potential.

In Fig. 57 are shown, for the constants

$E_0 = 1000$ volts,

$Z_0 = 2.5 - 6j$; that is, $r_0 = 2.5$ ohms, $x_0 = 6$ ohms, $z_0 = 6.5$ ohms,

with the current I_0 as abscissæ, the values —

E.M.F. at Receiver Circuit, E, (Curve I.);
Output of Transmission, P, (Curve II.);
Efficiency of Transmission, (Curve III.).

The same quantities, E and P, for a non-inductive line of resistance, $r_0 = 2.5$ ohms, $x_0 = 0$, are shown in Curves IV., V., and VI.

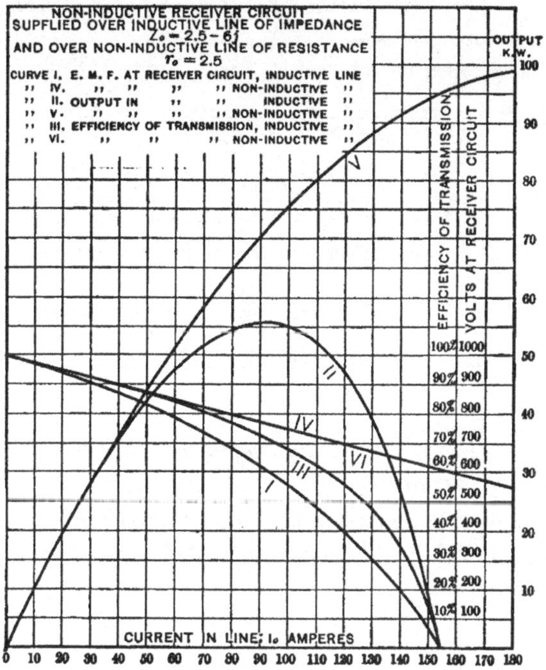

Fig. 57. Non-Inductive Receiver Circuit Supplied Over Inductive Line.

2.) *Maximum Power Supplied over an Inductive Line.*

60. If the receiver circuit contains the susceptance, b, in addition to the conductance, g, its admittance can be written thus:—

$$Y = g + jb, \quad y = \sqrt{g^2 + b^2}.$$

Then —

current, $\dot{I}_0 = \dot{E}\dot{Y}$;

Impressed E.M.F., $\dot{E}_0 = \dot{E} + \dot{I}_0 Z_0 = \dot{E}(1 + \dot{Y}Z_0)$.

Hence —

E.M.F. at receiver terminals,

$$\dot{E} = \frac{\dot{E}_0}{1 + YZ_0} = \frac{\dot{E}_0}{(1 + r_0 g + x_0 b) - j\,(x_0 g - r_0 b)};$$

current,

$$\dot{I}_0 = \frac{\dot{E}_0 Y}{1 + YZ_0} = \frac{\dot{E}_0 (g + jb)}{(1 + r_0 g + x_0 b) - j\,(x_0 g - r_0 b)};$$

or, in absolute values —

E.M.F. at receiver circuit,

$$E = \frac{E_0}{\sqrt{(1 + r_0 g + x_0 b)^2 + (x_0 g - r_0 b)^2}};$$

current,

$$I_0 = E_0 \sqrt{\frac{g^2 + b^2}{(1 + r_0 g + x_0 b)^2 + (x_0 g - r_0 b)^2}};$$

ratio of E.M.Fs. at receiver circuit and at generator circuit,

$$a = \frac{E}{E_0} = \frac{1}{\sqrt{(1 + r_0 g + x_0 b)^2 + (x_0 g - r_0 b)^2}};$$

and the output in the receiver circuit is,

$$P = E^2 g = E_0^2 a^2 g.$$

61. *a.*) *Dependence of the output upon the susceptance of the receiver circuit.*

At a given conductance, g, of the receiver circuit, its output, $P = E_0^2 a^2 g$, is a maximum, if a^2 is a maximum; that is, when —

$$f = \frac{1}{a^2} = (1 + r_0 g + x_0 b)^2 + (x_0 g - r_0 b)^2$$

is a minimum.

The condition necessary is —

$$\frac{df}{db} = 0,$$

or, expanding, $$x_0 (1 + r_0 g + x_0 b) - r_0 (x_0 g - r_0 b) = 0.$$

Hence —

Susceptance of receiver circuit,

$$b = -\frac{x_0}{r_0^2 + x_0^2} = -\frac{x_0}{z_0^2} = -b_0;$$

or $$b + b_0 = 0,$$

that is, if the sum of the susceptances of line and of receiver circuit equals zero.

Substituting this value, we get—

ratio of E.M.Fs. at maximum output,

$$a_1 = \frac{E}{E_o} = \frac{1}{z_o(g + g_o)};$$

maximum output,

$$P_1 = \frac{E_o^2 g}{z_o^2 (g + g_o)^2};$$

current,

$$\dot{I}_o = \frac{\dot{E}_o Y}{1 + Z_o Y} = \frac{\dot{E}_o (g - jb_o)}{1 + (r_o - jx_o)(g - jb_o)}$$

$$= \frac{\dot{E}_o (g - jb_o)}{(1 + r_o g - x_o b_o) - j(r_o b_o + x_o g)};$$

$$I_o = E_o \sqrt{\frac{g^2 + b_o^2}{(1 + r_o g - x_o b_o)^2 + (r_o b_o + x_o g)^2}};$$

and, expanding,

$$I_o = \frac{E_o \sqrt{g^2 + b_o^2}}{z_o (g + g_o)};$$

phase difference in receiver circuit,

$$\tan \hat{\omega} = \frac{b}{g} = -\frac{b_o}{g};$$

phase difference in generator circuit,

$$\tan \hat{\omega}_o = \frac{x + x_o}{r + r_o} = \frac{b_o (y^2 - y_o^2)}{g_o y^2 + g y_o^2}.$$

62. *b.) Dependence of the output upon the conductance of the receiver circuit.*

At a given susceptance, b, of the receiver circuit, its output, $P = E_o^2 a^2 g$, is a maximum, if—

$$\frac{dP}{dg} = 0, \text{ or } \frac{d}{dg}\left(\frac{1}{P}\right) = 0,$$

$$\text{or, } \frac{d}{dg}\left(\frac{1}{a^2 g}\right) = \frac{d}{dg}\left(\frac{(1 + r_o g + x_o b)^2 + (x_o g - r_o b)^2}{g}\right) = 0;$$

that is, expanding, —

$$(1 + r_0 g + x_0 b)^2 + (x_0 g - r_0 b)^2 - 2g(r_0 + r_0^2 g + x_0^2 g) = 0;$$

or, expanding, —

$$(b + b_0)^2 = g^2 - g_0^2; \qquad g = \sqrt{g_0^2 + (b + b_0)^2}.$$

Substituting this value in the equation for a, page 88, we get —

ratio of E.M.Fs.,

$$a_2 = \frac{1}{z_0 \sqrt{2\{g_0^2 + (b + b_0)^2 + g_0\sqrt{g_0^2 + (b + b_0)^2}\}}}$$

$$= \frac{1}{z_0 \sqrt{2g(g + g_0)}} = \frac{y_0}{\sqrt{2g(g + g_0)}};$$

power,

$$P_2 = \frac{E_0^2 y_0^2}{2(g + g_0)} = \frac{E_0^2 y_0^2}{2\{g_0 + \sqrt{g_0^2 + (b + b_0)^2}\}}$$

$$= \frac{E_0^2}{2\left\{r_0 + \sqrt{r_0^2 + \left(x_0 + x\dfrac{z_0^2}{z^2}\right)^2}\right\}}.$$

As a function of the susceptance, b, this power becomes a maximum for $dP_2/db = 0$, that is, according to § 61, if —

$$b = -b_0.$$

Substituting this value, we get —

$$b = -b_0, g = g_0, y = y_0, \text{ hence: } Y = g + jb = g_0 - jb_0;$$

$$x = -x_0, r = r_0, z = z_0, \qquad Z = r - jx = r_0 + jx_0;$$

substituting this value, we get —

ratio of E.M.Fs., $$a_m = \frac{y_0}{2g_0} = \frac{z_0}{2r_0};$$

power, $$P_m = \frac{E_0^2}{4r_0};$$

that is, the same as with a continuous-current circuit; or, in other words, the inductance of the line and of the receiver circuit can be perfectly balanced in its effect upon the output.

63. As a summary, we thus have:

The output delivered over an inductive line of impe

dance, $Z_0 = r_0 - jx_0$, into a non-inductive receiver circuit, is a maximum for the resistance, $r = z_0$, or conductance, $g = y_0$, of the receiver circuit, or —

$$P = \frac{E_0^2}{2(r_0 + z_0)},$$

at the ratio of potentials,

$$a = \frac{1}{\sqrt{2\left(1 + \frac{r_0}{z_0}\right)}}.$$

With a receiver circuit of constant susceptance, b, the output, as a function of the conductance, g, is a maximum for the conductance, —

$$g = \sqrt{g_0^2 + (b + b_0)^2},$$

and is

$$P = \frac{E_0^2 y_0^2}{2(g + g_0)},$$

at the ratio of potentials,

$$a = \frac{y_0}{\sqrt{2g(g + g_0)}}.$$

With a receiver circuit of constant conductance, g, the output, as a function of the susceptance, b, is a maximum for the susceptance, $b = -b_0$, and is

$$P = \frac{E_0^2 g}{z_0^2 (g + g_0)^2},$$

at the ratio of potentials,

$$a = \frac{1}{y_0(g + g_0)}.$$

The maximum output which can be delivered over an inductive line, as a function of the admittance or impedance of the receiver circuit, takes place when $Z = r_0 + jx_0$, or $Y = g_0 - jb_0$; that is, when the resistance or conductance of receiver circuit and line are equal, the reactance or susceptance of the receiver circuit and line are equal but of opposite sign, and is, $P = E_0^2 / 4r_0$, or independent of the reactances, but equal to the output of a continuous-current circuit of equal line resistance. The ratio of potentials is, in this case, $a = z_0 / 2\, r_0$, while in a continuous-current circuit it is equal to ½. The efficiency is equal to 50 per cent.

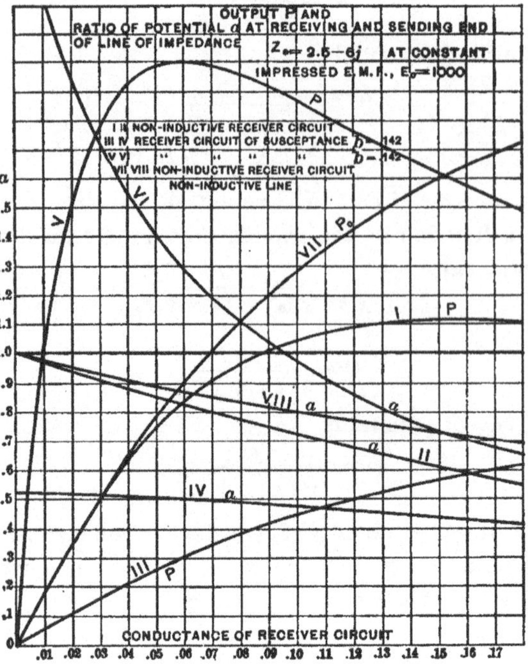

Fig. 58. Variation of the Potential in Line at Different Loads.

64. As an instance, in Fig. 58 are shown, for the constants —

$E_0 = 1000$ volts, and $Z_0 = 2.5 - 6\,j$; that is, for
$r_0 = 2.5$ ohms, $x_0 = 6$ohms, $z_0 = 6.5$ ohms,

and with the variable conductances as abscissæ, the values of the —

output, in Curve I., Curve III., and Curve V.;
ratio of potentials, in Curve II., Curve IV., and Curve VI.;

Curves I. and II. refer to a non-inductive receiver circuit:

Curves III. and IV. refer to a receiver circuit of constant susceptance . . . . $b =$ .142

Curves V. and VI. refer to a receiver circuit of constant susceptance . . . . $b = -.142$;

Curves VII. and VIII. refer to a non-inductive receiver circuit and non-inductive line.

In Fig. 59, the output is shown as Curve I., and the ratio of potentials as Curve II., for the same line constants, for a constant conductance, $g = .0592$ ohms, and for variable susceptances, b, of the receiver circuit.

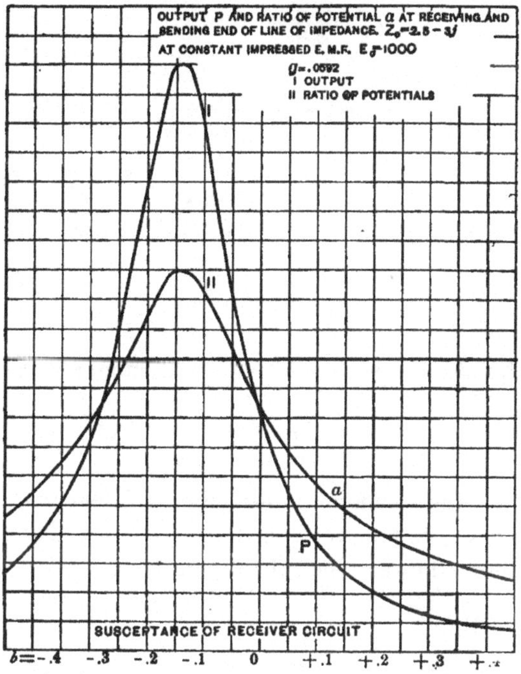

Fig. 59. Variation of Potential in Line at Various Loads.

3.) *Maximum Efficiency.*

65. The output, for a given conductance, g, of a receiver circuit, is a maximum if $b = -b_0$. This, however, is generally not the condition of maximum efficiency.

The loss of energy in the line is constant if the current is constant; the output of the generator for a given current and given generator E.M.F. is a maximum if the current is in phase with the E.M.F. at the generator terminals. Hence the condition of maximum output at given loss, or of maximum efficiency, is —

$$\tan \hat{\omega}_0 = 0.$$

The current is —

$$\dot{I}_0 = \frac{\dot{E}_0}{Z + Z_0}$$

$$= \frac{\dot{E}_0}{(r + r_0) - j(x + x_0)};$$

The current $\dot{I}_0$, is in phase with the E.M.F., $\dot{E}_0$, if its quadrature component — that is, the imaginary term — disappears, or

$$x + x_0 = 0.$$

This, therefore, is the condition of maximum efficiency,

$$x = - x_0.$$

Hence, the condition of maximum efficiency is, that the reactance of the receiver circuit shall be equal, but of opposite sign, to the reactance of the line.

Substituting $x = - x_0$, we have,

ratio of E.M.Fs.,

$$a = \frac{E}{E_0} = \frac{z}{(r + r_0)} = \frac{\sqrt{r^2 + x_0^2}}{(r + r_0)};$$

power, $$P = E_0^2 g a^2 = \frac{E_0^2 r}{(r + r_0)^2},$$

and depending upon the resistance only, and not upon the reactance.

This power is a maximum if $g = g_0$, as shown before; hence, substituting $g = g_0$, $r = r_0$,

maximum power at maximum efficiency, $P_m = \dfrac{E_0^2}{4\,r_0}$,

at a ratio of potentials, $a_m = \dfrac{z_0}{2\,r_0}$,

or the same result as in § 62.

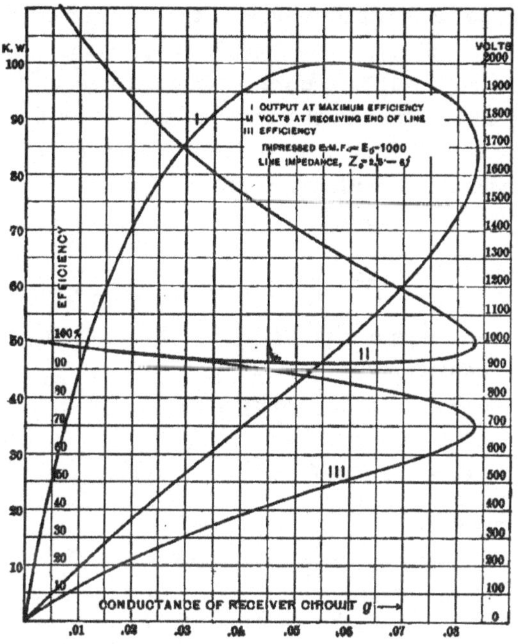

Fig. 60. Load Characteristic of Transmission Line.

In Fig. 60 are shown, for the constants —

$E_0 = 1{,}000$ volts,

$Z_0 = 2.5 - 6j$; $r_0 = 2.5$ ohms, $x_0 = 6$ ohms, $z_0 = 6.5$ ohms, and with the variable conductances, g, of the receiver circuit as abscissæ, the —

Output at maximum efficiency, (Curve I.);
Volts at receiving end of line, (Curve II.);

$$\text{Efficiency} = \frac{r}{r + r_0}, \qquad \text{(Curve III.)}.$$

4.) *Control of Receiver Voltage by Shunted Susceptance.*

66. By varying the susceptance of the receiver circuit, the potential at the receiver terminals is varied greatly. Therefore, since the susceptance of the receiver circuit can be varied at will, it is possible, at a constant generator E.M.F., to adjust the receiver susceptance so as to keep the potential constant at the receiver end of the line, or to vary it in any desired manner, and independently of the generator potential, within certain limits.

The ratio of E.M.Fs. is —

$$a = \frac{E}{E_0} = \frac{1}{\sqrt{(1 + r_0 g + x_0 b)^2 + (x_0 g - r_0 b)^2}}.$$

If at constant generator potential E_0, the receiver potential E shall be constant,

$$a = \text{constant};$$

hence,

$$(1 + r_0 g + x_0 b)^2 + (x_0 g - r_0 b)^2 = \frac{1}{a^2};$$

or, expanding,

$$b = -b_0 + \sqrt{\left(\frac{y_0}{a}\right)^2 - (g + g_0)^2},$$

which is the value of the susceptance, b, as a function of the receiver conductance, — that is, of the load, — which is required to yield constant potential, aE_0, at the receiver circuit.

For increasing g, that is, for increasing load, a point is reached, where, in the expression —

$$b = -b_0 + \sqrt{\left(\frac{y_0}{a}\right)^2 - (g + g_0)^2},$$

the term under the root becomes imaginary, and it thus becomes impossible to maintain a constant potential, aE_0. Therefore, the maximum output which can be transmitted at potential aE_0, is given by the expression —

$$\sqrt{\left(\frac{y_0}{a}\right)^2 - (g + g_0)^2} = 0;$$

hence $b = -b_0$, the susceptance of receiver circuit,

and $g = -g_0 + \frac{y_0}{a}$, the conductance of receiver circuit;

$$P = E_0^2 g a^2$$
$$= a^2 E_0^2 \left(\frac{y_0}{a} - g_0\right), \text{ the output.}$$

67. If $a = 1$, that is, if the voltage at the receiver circuit equals the generator potential —

$$g = y_0 - g_0;$$
$$P = E_0^2 (y_0 - g_0).$$

If $a = 1$ when $g = 0$, $b = 0$

when $g > 0$, $b < 0$;

if $a > 1$ when $g = 0$, or $g > 0$, $b < 0$,

that is, condensance;

if $a < 1$ when $g = 0$, $b > 0$,

when $g = -g_0 + \sqrt{\left(\frac{y_0}{a}\right)^2 - b_0^2}$, $b = 0$;

when $g > -g_0 + \sqrt{\left(\frac{y_0}{a}\right)^2 - b_0^2}$, $b < 0$,

or, in other words, if $a < 1$, the phase difference in the main line must change from lag to lead with increasing load.

68. The value of a giving the maximum possible output in a receiver circuit, is determined by $dP / da = 0$;

expanding: $$2a\left(\frac{y_0}{a} - g_0\right) - \frac{a^2 y_0}{a^2} = 0;$$

hence, $$y_0 = 2 a g_0,$$

and $$a = \frac{y_0}{2 g_0} = \frac{1}{2\sqrt{g_0 r_0}} = \frac{z_0}{2 r_0};$$

the maximum output is determined by —

$$g = -g_0 + \frac{y_0}{a} = g_0;$$

and is,

$$P = \frac{E_0^2}{4r}.$$

From:

$$a = \frac{y_0}{2g_0} = \frac{z_0}{2r_0},$$

the line reactance, x_0, can be found, which delivers a maximum output into the receiver circuit at the ratio of potentials, a,

and

$$z_0 = 2r_0a,$$
$$x_0 = r_0\sqrt{4a^2 - 1};$$

for $a = 1$,

$$z_0 = 2r_0;$$
$$x_0 = r_0\sqrt{3}.$$

If, therefore, the line impedance equals $2a$ times the line resistance, the maximum output, $P = E_0^2 / 4r_0$, is transmitted into the receiver circuit at the ratio of potentials, a.

If $z_0 = 2r_0$, or $x_0 = r_0\sqrt{3}$, the maximum output, $P = E_0^2 / 4r_0$, can be supplied to the receiver circuit, without change of potential at the receiver terminals.

Obviously, in an analogous manner, the law of variation of the susceptance of the receiver circuit can be found which is required to increase the receiver voltage proportionally to the load; or, still more generally, — to cause any desired variation of the potential at the receiver circuit independently of any variation of the generator potential, as, for instance, to keep the potential of a receiver circuit constant, even if the generator potential fluctuates widely.

69. In Figs. 61, 62, and 63, are shown, with the output, $P = E_0^2 g a^2$, as abscissæ, and a constant impressed E.M.F., $E_0 = 1{,}000$ volts, and a constant line impedance, $Z_0 = 2.5 - 6j$, or, $r_0 = 2.5$ ohms, $x_0 = 6$ ohms, $z = 6.5$ ohms, the following values:

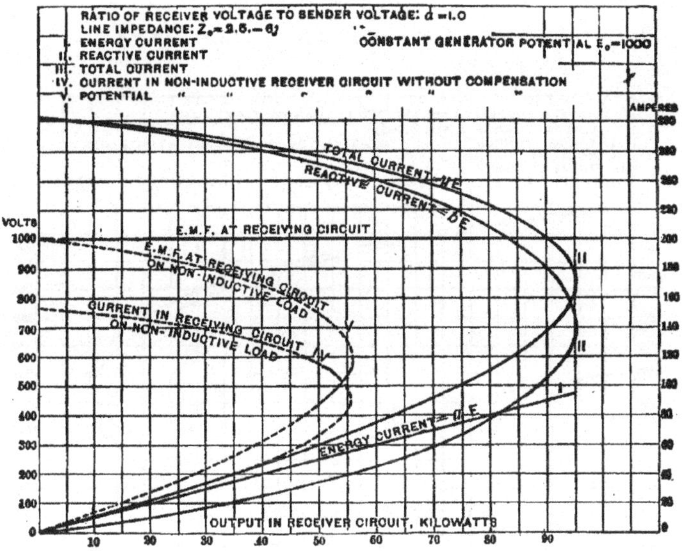

Fig. 61. Variation of Voltage Transmission Lines.

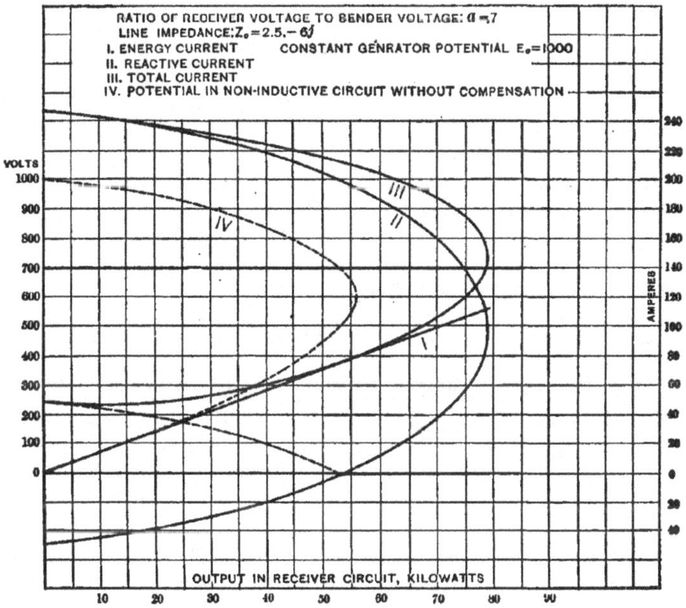

Fig. 62. Variation of Voltage Transmission Lines.

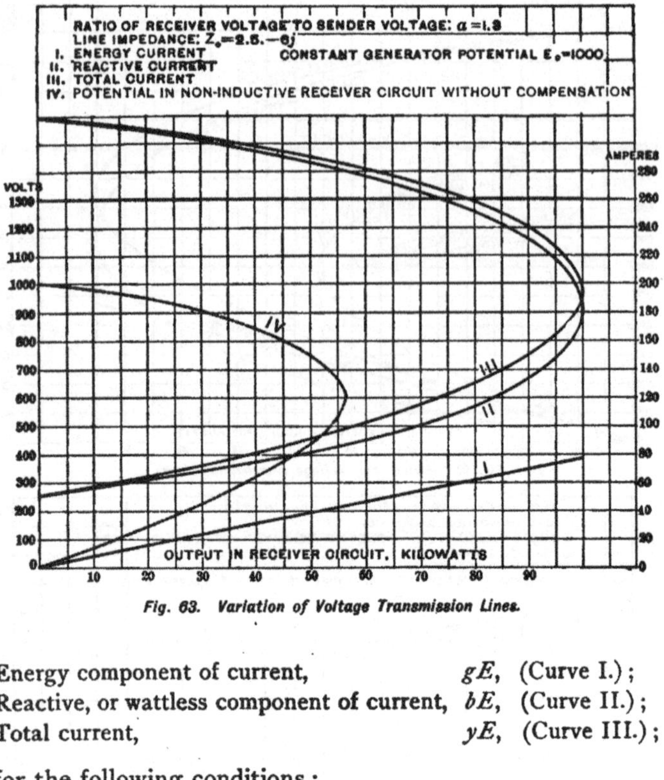

Fig. 63. Variation of Voltage Transmission Lines.

Energy component of current, gE, (Curve I.);
Reactive, or wattless component of current, bE, (Curve II.);
Total current, yE, (Curve III.);

for the following conditions:

$a = 1.0$ (Fig. 61); $a = .7$ (Fig. 62); $a = 1.3$ (Fig. 63).

For the non-inductive receiver circuit (in dotted lines), the curve of E.M.F., E, and of the current, $I = gE$, are added in the three diagrams for comparison, as Curves IV. and V.

As shown, the output can be increased greatly, and the potential at the same time maintained constant, by the judicious use of shunted reactance, so that a much larger output can be transmitted over the line at no drop, or even at a rise, of potential.

5.) *Maximum Rise of Potential at Receiver Circuit.*

70. Since, under certain circumstances, the potential at the receiver circuit may be higher than at the generator, it is of interest to determine what is the maximum value of potential, E, that can be produced at the receiver circuit with a given generator potential, E_0.

The condition is that

$$a = \text{maximum or } \frac{1}{a^2} = \text{minimum};$$

that is,

$$\frac{d(1/a^2)}{dg} = 0, \quad \frac{d(1/a^2)}{db} = 0;$$

substituting,

$$\frac{1}{a^2} = (1 + r_0 g + x_0 b)^2 + (x_0 g - r_0 b)^2$$

and expanding, we get,

$$\frac{d(1/a^2)}{dg} = 0; \quad g = -\frac{r_0}{z_0^2};$$

— a value which is impossible, since neither r_0 nor g can be negative. The next possible value is $g = 0$, — a wattless circuit.

Substituting this value, we get,

$$\frac{1}{a^2} = (1 + x_0 b)^2 + r_0^2 b^2;$$

and by substituting, in

$$\frac{d(1/a^2)}{db} = 0, \ b = -\frac{x_0}{z_0^2} = -b_0,$$

$$b + b_0 = 0;$$

that is, the sum of the susceptances $= 0$, or the condition of resonance is present.

Substituting,

$$b = -b_0,$$

we have

$$a = \frac{1}{\sqrt{r_0 g_0}} = \frac{z_0}{r_0} = \frac{y_0}{g_0}.$$

The current in this case is,

$$I = E_0 g_0 = \frac{E_0}{r_0},$$

or the same as if the line resistance were short-circuited without any inductance.

This is the condition of perfect resonance, with current and E.M.F. in phase.

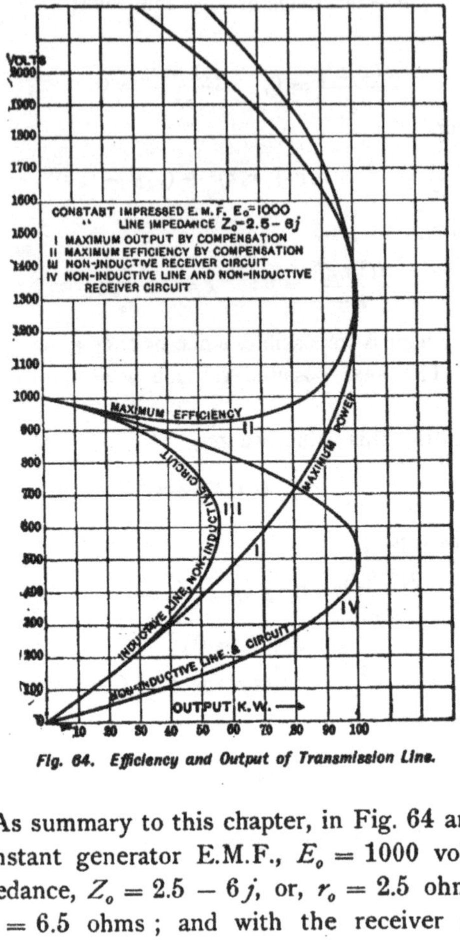

Fig. 64. Efficiency and Output of Transmission Line.

71. As summary to this chapter, in Fig. 64 are plotted, for a constant generator E.M.F., $E_0 = 1000$ volts, and a line impedance, $Z_0 = 2.5 - 6j$, or, $r_0 = 2.5$ ohms, $x_0 = 6$ ohms, $z_0 = 6.5$ ohms; and with the receiver output as abscissæ and the receiver voltages as ordinates, curves representing —

the condition of maximum output, (Curve I.);
the condition of maximum efficiency, (Curve II.);
the condition $b = 0$, or a non-inductive receiver circuit, (Curve III.);
the condition $b = 0$, $b_0 = 0$, or a non-inductive line and non-inductive receiver circuit.

In conclusion, it may be remarked here that of the sources of susceptance, or reactance,

a choking coil or reactive coil corresponds to an inductance;
a condenser corresponds to a condensance;
a polarization cell corresponds to a condensance;
a synchronizing alternator (motor or generator) corresponds to an inductance or a condensance, at will;
an induction motor or generator corresponds to an inductance.

The choking coil and the polarization cell are specially suited for series reactance, and the condenser and synchronizer for shunted susceptance.

CHAPTER X.

EFFECTIVE RESISTANCE AND REACTANCE.

72. The resistance of an electric circuit is determined:—

1.) By direct comparison with a known resistance (Wheatstone bridge method, etc.).

This method gives what may be called the true ohmic resistance of the circuit.

2.) By the ratio:

$$\frac{\text{Volts consumed in circuit}}{\text{Amperes in circuit}}.$$

In an alternating-current circuit, this method gives, not the resistance of the circuit, but the impedance,

$$z = \sqrt{r^2 + x^2}.$$

3.) By the ratio:

$$r = \frac{\text{Power consumed}}{(\text{Current})^2};$$

where, however, the "power" does not include the work done by the circuit, and the counter E.M.Fs. representing it, as, for instance, in the case ot the counter E.M.F. of a motor.

In alternating-current circuits, this value of resistance is the energy coefficient of the E.M.F.,

$$r = \frac{\text{Energy component of E.M.F.}}{\text{Total current}}.$$

It is called the *effective resistance* of the circuit, since it represents the effect, or power, expended by the circuit. The energy coefficient of current,

$$g = \frac{\text{Energy component of current}}{\text{Total E.M.F.}},$$

is called the *effective conductance* of the circuit.

In the same way, the value,

$$x = \frac{\text{Wattless component of E.M.F.}}{\text{Total current}},$$

is the *effective reactance*, and

$$b = \frac{\text{Wattless component of current}}{\text{Total E.M.F.}},$$

is the *effective susceptance* of the circuit.

While the true ohmic resistance represents the expenditure of energy as heat inside of the electric conductor by a current of uniform density, the effective resistance represents the total expenditure of energy.

Since, in an alternating-current circuit in general, energy is expended not only in the conductor, but also outside of it, through hysteresis, secondary currents, etc., the effective resistance frequently differs from the true ohmic resistance in such way as to represent a larger expenditure of energy.

In dealing with alternating-current circuits, it is necessary, therefore, to substitute everywhere the values "effective resistance," "effective reactance," "effective conductance," and "effective susceptance," to make the calculation applicable to general alternating-current circuits, such as inductances, containing iron, etc.

While the true ohmic resistance is a constant of the circuit, depending only upon the temperature, but not upon the E.M.F., etc., the effective resistance and effective reactance are, in general, not constants, but depend upon the E.M.F., current, etc. This dependence is the cause of most of the difficulties met in dealing analytically with alternating-current circuits containing iron.

73. The foremost sources of energy loss in alternating-current circuits, outside of the true ohmic resistance loss, are as follows:

1.) Molecular friction, as,

a.) Magnetic hysteresis;

b.) Dielectric hysteresis.

2.) Primary electric currents, as,

a.) Leakage or escape of current through the insulation, brush discharge; *b.*) Eddy currents in the conductor or unequal current distribution.

3.) Secondary or induced currents, as,

a.) Eddy or Foucault currents in surrounding magnetic materials; *b.*) Eddy or Foucault currents in surrounding conducting materials; *c.*) Secondary currents of mutual inductance in neighboring circuits.

4.) Induced electric charges, electrostatic influence.

While all these losses can be included in the terms effective resistance, etc., only the magnetic hysteresis and the eddy currents in the iron will form the subject of what follows, since they are the most frequent and important sources of energy loss.

Magnetic Hysteresis.

74. In an alternating-current circuit surrounded by iron or other magnetic material, energy is expended outside of the conductor in the iron, by a kind of molecular friction, which, when the energy is supplied electrically, appears as magnetic hysteresis, and is caused by the cyclic reversals of magnetic flux in the iron in the alternating magnetic field.

To examine this phenomenon, first a circuit may be considered, of very high inductance, but negligible true ohmic resistance; that is, a circuit entirely surrounded by iron, as, for instance, the primary circuit of an alternating-current transformer with open secondary circuit.

The wave of current produces in the iron an alternating magnetic flux which induces in the electric circuit an E.M.F., — the counter E.M.F. of self-induction. If the ohmic resistance is negligible, that is, practically no E.M.F. consumed by the resistance, all the impressed E.M.F. must be consumed by the counter E.M.F. of self-induction, that is, the counter E.M.F. equals the impressed E.M.F.; hence, if the impressed E.M.F. is a sine wave, the counter E.M.F., and, therefore, the magnetic flux which induces the counter E.M.F. must follow a sine wave also. The alternating wave of current is not a sine wave in this case, but is distorted by hysteresis. It is possible, however, to plot the current wave in this case from the hysteretic cycle of magnetic flux.

From the number of turns, n, of the electric circuit, the effective counter E.M.F., E, and the frequency, N, of the current, the maximum magnetic flux, Φ, is found by the formula:

$$E = \sqrt{2}\,\pi n N \Phi 10^{-8};$$

hence,

$$\Phi = \frac{E\,10^{8}}{\sqrt{2}\,\pi n N}.$$

A maximum flux, Φ, and magnetic cross-section, S, give the maximum magnetic induction, $\mathfrak{B} = \Phi / S$.

If the magnetic induction varies periodically between $+\mathfrak{B}$ and $-\mathfrak{B}$, the M.M.F. varies between the corresponding values $+\mathfrak{F}$ and $-\mathfrak{F}$, and describes a looped curve, the cycle of hysteresis.

If the ordinates are given in lines of magnetic force, the abscissæ in tens of ampere-turns, then the area of the loop equals the energy consumed by hysteresis in ergs per cycle.

From the hysteretic loop the instantaneous value of M.M.F. is found, corresponding to an instantaneous value of magnetic flux, that is, of induced E.M.F.; and from the M.M.F., $\mathfrak{F}$, in ampere-turns per unit length of magnetic circuit, the length, l, of the magnetic circuit, and the number of turns, n, of the electric circuit, are found the instantaneous values of current, i, corresponding to a M.M.F., $\mathfrak{F}$; that is, magnetic induction $\mathfrak{B}$, and thus induced E.M.F. e, as:

$$i = \frac{\mathfrak{F}\,l}{n}.$$

75. In Fig. 65, four magnetic cycles are plotted, with maximum values of magnetic inductions, $\mathfrak{B} = 2{,}000$, $6{,}000$, $10{,}000$, and $16{,}000$, and corresponding maximum M.M.Fs.,

$\mathfrak{F}$ = 1.8, 2.8, 4.3, 20.0. They show the well-known hysteretic loop, which becomes pointed when magnetic saturation is approached.

These magnetic cycles correspond to average good sheet iron or sheet steel, having a hysteretic coefficient, η = .0033, and are given with ampere-turns per cm as abscissæ, and kilo-lines of magnetic force as ordinates.

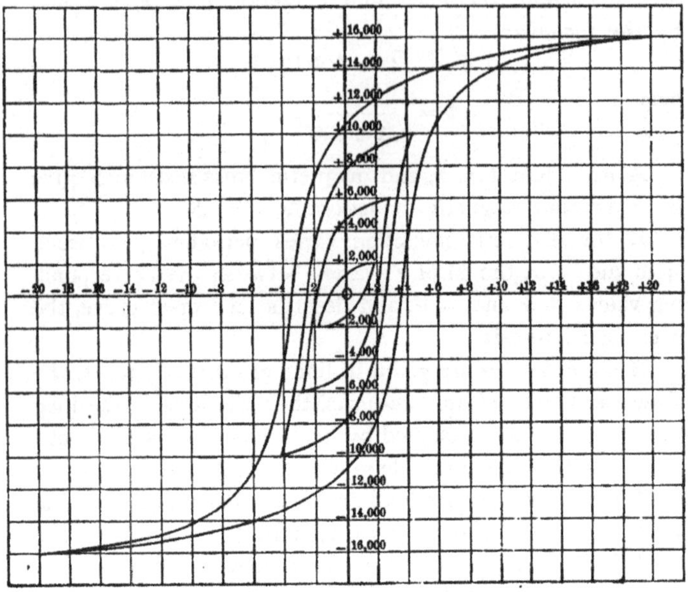

Fig. 65. Hysteretic Cycle of Sheet Iron.

In Figs. 66, 67, 68, and 69, the curve of magnetic induction as derived from the induced E.M.F. is a sine wave. For the different values of magnetic induction of this sine curve, the corresponding values of M.M.F., hence of current, are taken from Fig. 65, and plotted, giving thus the exciting current required to produce the sine wave of magnetism; that is, the wave of current which a sine wave of impressed E.M.F. will send through the circuit.

As shown in Figs. 66, 67, 68, and 69, these waves of alternating current are not sine waves, but are distorted by the superposition of higher harmonics, and are complex harmonic waves. They reach their maximum value at the same time with the maximum of magnetism, that is, 90°

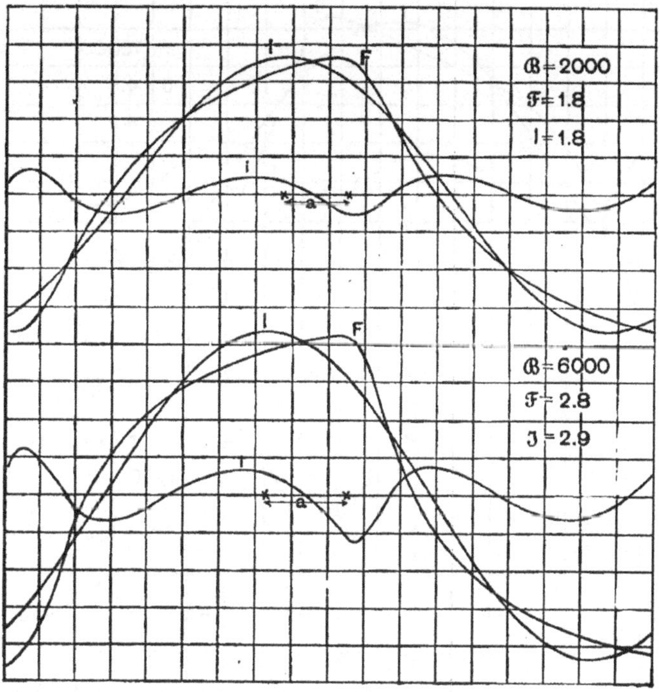

Figs. 66 and 67. Distortion of Current Wave by Hysteresis.

ahead of the maximum induced E.M.F., and hence about 90° behind the maximum impressed E.M.F., but pass the zero line considerably ahead of the zero value of magnetism, or 42°, 52°, 50°, and 41°, respectively.

The general character of these current waves is, that the maximum point of the wave coincides in time with the maximum point of the sine wave of magnetism ; but the current wave is bulged out greatly at the rising, and hollowed in at the decreasing, side. With increasing magnetization, the maximum of the current wave becomes more pointed, as shown by the curve of Fig. 68, for $\mathfrak{B} = 10,000$; and at still

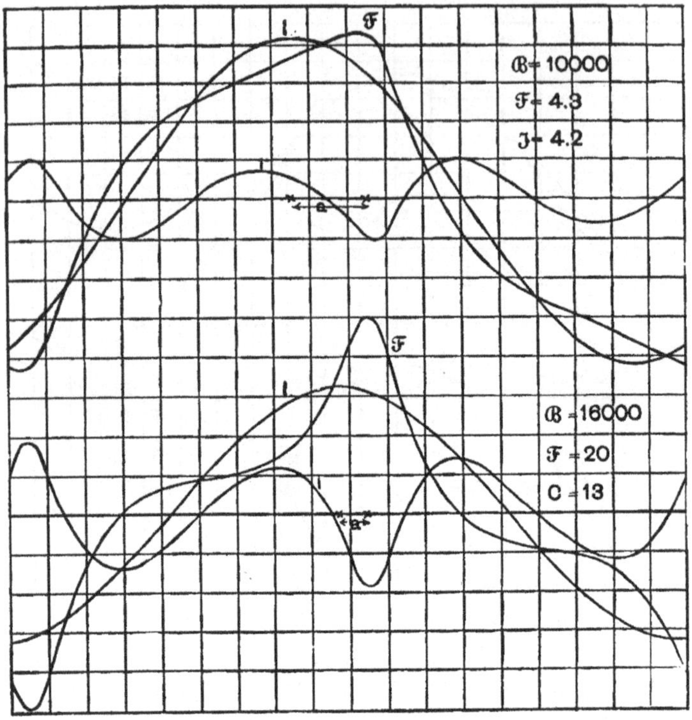

Figs. 68 and 69. Distortion of Current Wave by Hysteresis.

higher saturation a peak is formed at the maximum point, as in the curve of Fig. 69, for $\mathfrak{B} = 16,000$. This is the case when the curve of magnetization reaches within the range of magnetic saturation, since in the proximity of saturation the current near the maximum point of magnetization has to rise abnormally to cause even a small increase of magnetization. The four curves, Figs. 66, 67, 68, and 69, are not drawn to the same scale. The maximum values of M.M.F., corresponding to the maximum values of magnetic induction, $\mathfrak{B}$ = 2,000, 6,000, 10,000, and 16,000 lines of force per cm^2, are $\mathfrak{F}$ = 1.8, 2.8, 4.3, and 20.0 ampere-turns per cm. In the different diagrams these are represented in the ratio of 8 : 6 : 4 : 1, in order to bring the current curves to approximately the same height. The M.M.F., in C.G.S. units, is

$$H = 4\pi / 10\, \mathfrak{F} = 1.257\, \mathfrak{F}.$$

76. The distortion of the wave of magnetizing current is as large as shown here only in an iron-closed magnetic circuit expending energy by hysteresis only, as in an iron-clad transformer on open secondary circuit. As soon as the circuit expends energy in any other way, as in resistance, or by mutual inductance, or if an air-gap is introduced in the magnetic circuit, the distortion of the current wave rapidly decreases and practically disappears, and the current becomes more sinusoidal. That is, while the distorting component remains the same, the sinusoidal component of the current greatly increases, and obscures the distortion. For example, in Figs. 70 and 71, two waves are shown, corresponding in magnetization to the curve of Fig. 67, as the one most distorted. The curve in Fig. 70 is the current wave of a transformer at $\frac{1}{10}$ load. At higher loads the distortion is correspondingly still less, except where the magnetic flux of self-induction, that is, flux passing between primary and secondary, and increasing proportionally to the load, is so large as to reach saturation, in which case a distortion appears again and increases with increasing load. The curve of Fig. 71 is the exciting current of a magnetic circuit containing an air-gap whose length equals $\frac{1}{400}$ the length of the magnetic circuit. These two curves are drawn to $\frac{1}{3}$ the size of the curve in Fig. 67. As shown, both curves are practically sine waves. The sine curves of magnetic flux are shown dotted as ϕ.

77. The distorted wave of current can be resolved into two components: *A true sine wave of equal effective intensity and equal power to the distorted wave,* called the *equivalent* *sine wave*, and a *wattless higher harmonic*, consisting chiefly of a term of triple frequency.

In Figs. 66 to 71 are shown, as *I*, the equivalent sine

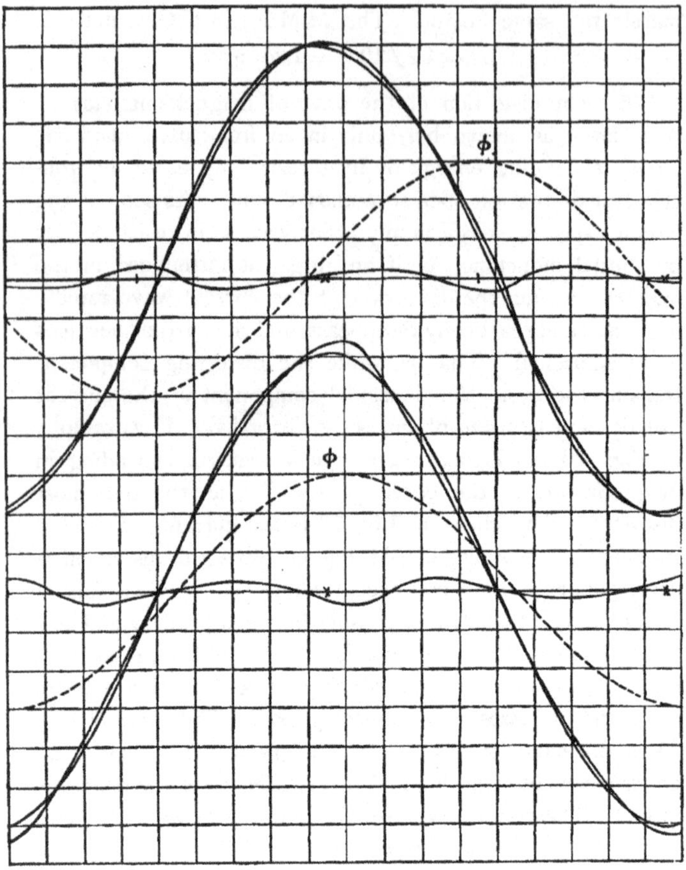

Figs. 70 and 71. Distortion of Current Wave by Hysteresis.

waves and as *i*, the difference between the equivalent sine wave and the real distorted wave, which consists of wattless complex higher harmonics. The equivalent sine wave of M.M.F. or of current, in Figs. 66 to 69, leads the magnetism by 34°, 44°, 38°, and 15°.5, respectively. In Fig. 71 the equivalent sine wave almost coincides with the distorted curve, and leads the magnetism by only 9°.

It is interesting to note, that even in the greatly distorted curves of Figs. 66 to 68, the maximum value of the equivalent sine wave is nearly the same as the maximum value of the original distorted wave of M.M.F., so long as magnetic saturation is not approached, being 1.8, 2.9, and 4.2, respectively, against 1.8, 2.8, and 4.3, the maximum values of the distorted curve. Since, by the definition, the effective value of the equivalent sine wave is the same as that of the distorted wave, it follows, that this distorted wave of exciting current shares with the sine wave the feature, that the maximum value and the effective value have the ratio of $\sqrt{2} \div 1$. Hence, below saturation, the maximum value of the distorted curve can be calculated from the effective value — which is given by the reading of an electro-dynamometer — by using the same ratio that applies to a true sine wave, and the magnetic characteristic can thus be determined by means of alternating currents, with sufficient exactness, by the electro-dynamometer method, in the range below saturation.

78. In Fig. 72 is shown the true magnetic characteristic of a sample of good average sheet iron, as found by the method of slow reversals with the magnetometer; for comparison there is shown in dotted lines the same characteristic, as determined with alternating currents by the electro-dynamometer, with ampere-turns per cm as ordinates, and magnetic inductions as abscissæ. As represented, the two curves practically coincide up to a value of $\mathfrak{B} = 13{,}000$; that is, up to the highest inductions practicable in alternating-current apparatus. For higher saturations, the curves rapidly diverge, and the electro-dynamometer curve shows comparatively small M.M.Fs. producing apparently very high magnetizations.

The same Fig. 72 gives the curve of hysteretic loss, in ergs per cm³ and cycle, as ordinates, and magnetic inductions as abscissæ.

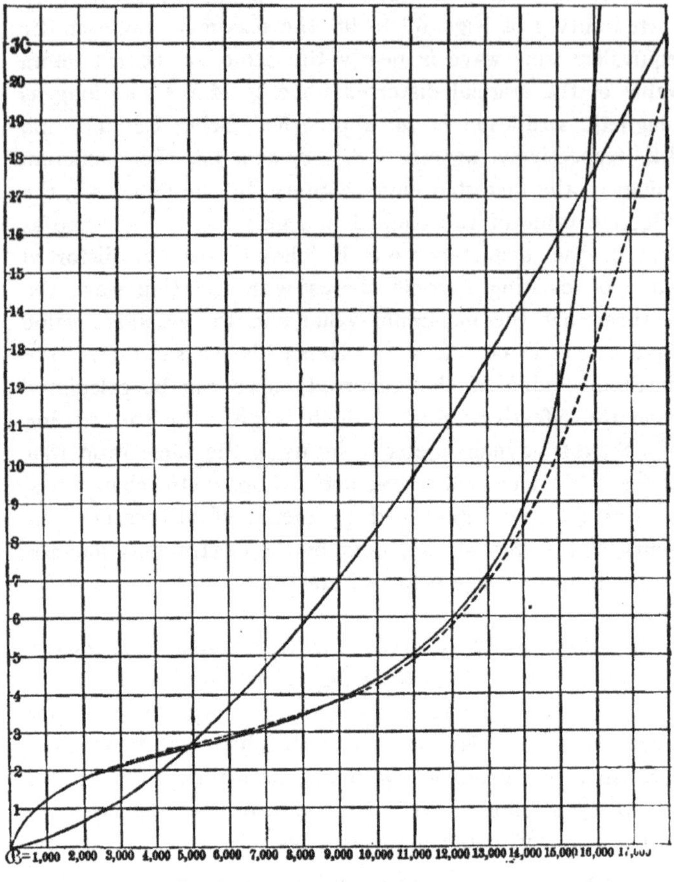

Fig. 72. Magnetization and Hysteresis Curve.

The electro-dynamometer method of determining the magnetic characteristic is preferable for use with alternating-current apparatus, since it is not affected by the phenomenon of magnetic "creeping," which, especially at low densities, may in the magnetometer tests bring the magnetism very much higher, or the M.M.F. lower, than found in practice in alternating-current apparatus.

So far as current strength and energy consumption are concerned, the distorted wave can be replaced by the equivalent sine wave, and the higher harmonics neglected.

All the measurements of alternating currents, with the single exception of instantaneous readings, yield the equivalent sine wave only, and suppress the higher harmonic; since all measuring instruments give either the mean square of the current wave, or the mean product of instantaneous values of current and E.M.F., which, by definition, are the same in the equivalent sine wave as in the distorted wave.

Hence, in all practical applications, it is permissible to neglect the higher harmonic altogether, and replace the distorted wave by its equivalent sine wave, keeping in mind, however, the existence of a higher harmonic as a possible disturbing factor which may become noticeable in those cases where the frequency of the higher harmonic is near the frequency of resonance of the circuit, that is, in circuits containing capacity besides the inductance.

79. The equivalent sine wave of exciting current leads the sine wave of magnetism by an angle α, which is called the *angle of hysteretic advance of phase.* Hence the current lags behind the E.M.F by $\measuredangle\ 90° - \alpha$, and the power is therefore,

$$P = I\dot{E} \cos (90° - \alpha) = IE \sin \alpha.$$

Thus the exciting current, I, consists of an energy component, $I \sin \alpha$, called the *hysteretic or magnetic energy current,* and a wattless component, $I \cos \alpha$, which is called the *magnetizing current.* Or, conversely, the E.M.F. consists of an energy component, $E \sin \alpha$, the *hysteretic energy* E.M.F., and a wattless component, $E \cos \alpha$, the E.M.F. *of self-induction.*

Denoting the absolute value of the impedance of the

circuit, E/I, by z, — where z is determined by the magnetic characteristic of the iron, and the shape of the magnetic and electric circuits, — the impedance is represented, in phase and intensity, by the symbolic expression,

$$Z = r - jx = z \sin \alpha - jz \cos \alpha;$$

and the admittance by,

$$Y = g + jb = \frac{1}{z} \sin \alpha + j\frac{1}{z} \cos \alpha = y \sin \alpha + jy \cos \alpha.$$

The quantities, z, r, x, and y, g, b, are, however, not constants as in the case of the circuit without iron, but depend upon the intensity of magnetization, $\mathfrak{B}$, — that is, upon the E.M.F. This dependence complicates the investigation of circuits containing iron.

In a circuit entirely inclosed by iron, α is quite considerable, ranging from 30° to 50° for values below saturation. Hence, even with negligible true ohmic resistance, no great lag can be produced in ironclad alternating-current circuits.

80. The loss of energy by hysteresis due to molecular friction is, with sufficient exactness, proportional to the 1.6th power of magnetic induction $\mathfrak{B}$. Hence it can be expressed by the formula:

$$W_H = \eta \mathfrak{B}^{1.6}$$

where —

W_H = loss of energy per cycle, in ergs or (C.G.S.) units (= 10^{-7} Joules) per cm³,

$\mathfrak{B}$ = maximum magnetic induction, in lines of force per cm², and

η = the *coefficient of hysteresis.*

This I found to vary in iron from .00124 to .0055. As a fair mean, .0033 * can be accepted for good average annealed sheet iron or sheet steel. In gray cast iron, η averages .013; it varies from .0032 to .028 in cast steel, according to the chemical or physical constitution; and reaches values as high as .08 in hardened steel (tungsten and manganese

* At present, with the improvements in the production and selection of sheet steel for alternating apparatus, .0025 can be considered a fair average in selected material (1899).

steel). Soft nickel and cobalt have about the same coefficient of hysteresis as gray cast iron; in magnetite I found $\eta = .023$.

In the curves of Fig. 62 to 69, $\eta = .0033$.

At the frequency, N, the loss of power in the volume, V, is, by this formula, —

$$P = \eta N V \mathcal{B}^{1.6} 10^{-7} \text{ watts}$$

$$= \eta N V \left(\frac{\Phi}{S} \right)^{1.6} 10^{-7} \text{ watts},$$

where S is the cross-section of the total magnetic flux, Φ.

The maximum magnetic flux, Φ, depends upon the counter E.M.F. of self-induction,

$$E = \sqrt{2}\, \pi N n \Phi 10^{-8},$$

or

$$\Phi = \frac{E\, 10^8}{\sqrt{2}\, \pi N n};$$

where n = number of turns of the electric circuit.

Substituting this in the value of the power, P, and canceling, we get, —

$$P = \eta \frac{E^{1.6}}{N^{.6}} \frac{V 10^{5.8}}{2^{.8} \pi^{1.6} S^{1.6} n^{1.6}} = 58\, \eta \frac{E^{1.6}}{N^{.6}} \frac{V 10^{3}}{S^{1.6} n^{1.6}};$$

or $$P = \frac{A E^{1.6}}{N^{.6}}, \text{ where } A = \eta \frac{V 10^{5.8}}{2^{.8} \pi^{1.6} S^{1.6} n^{1.6}} = 58\, \eta \frac{V 10^{3}}{S^{1.6} n^{1.6}};$$

or, substituting $\eta = .0033$, we have $A = 191.4 \dfrac{V}{S^{1.6} n^{1.6}}$;

or, substituting $V = SL$, where L = length of magnetic circuit,

$$A = \frac{\eta L 10^{5.8}}{2^{.8} \pi^{1.6} S^{.6} n^{1.6}} = \frac{58\, \eta L 10^{3}}{S^{.6} n^{1.6}} = 191.4 \frac{L}{S^{.6} n^{1.6}};$$

and

$$P = \frac{58\, \eta E^{1.6} L 10^{3}}{N^{.6} S^{.6} n^{1.6}} = \frac{191.4\, E^{1.6} L}{N^{.6} S^{.6} n^{1.6}}.$$

In Figs. 73, 74, and 75, is shown a curve of hysteretic loss, with the loss of power as ordinates, and

in curve 73, with the E.M.F., E, as abscissae, for $L = 6$, $S = 20$, $N = 100$, and $n = 100$;

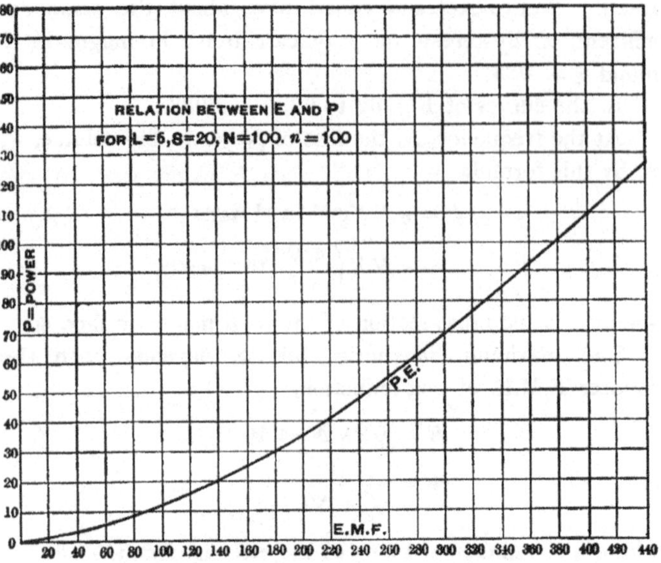

Fig. 73. Hysteresis Loss as Function of E. M. F.

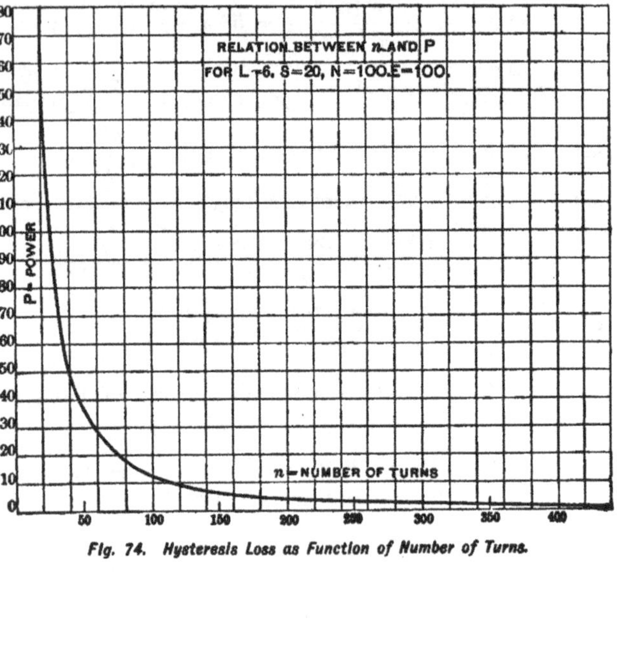

Fig. 74. Hysteresis Loss as Function of Number of Turns.

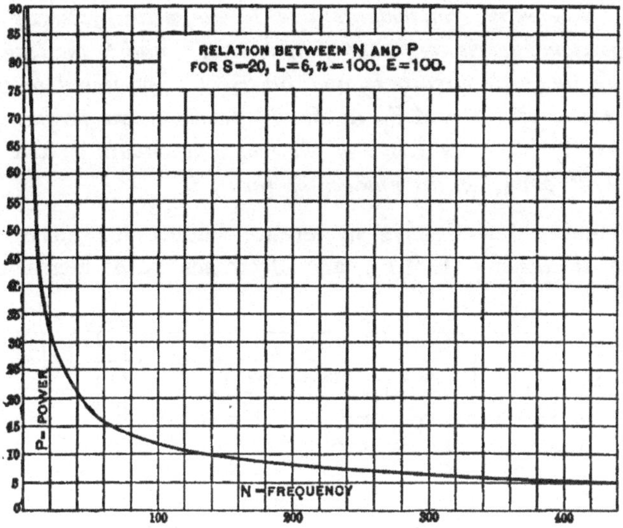

Fig. 75. Hysteresis Loss as Function of Cycles.

in curve 74, with the number of turns as abscissae, for $L = 6$, $S = 20$, $N = 100$, and $E = 100$;

in curve 75, with the frequency, N, or the cross-section, S, as abscissae, for $L = 6$, $n = 100$, and $E = 100$.

As shown, the hysteretic loss is proportional to the 1.6th power of the E.M.F., inversely proportional to the 1.6th power of the number of turns, and inversely proportional to the .6th power of frequency, and of cross-section.

81. If $g =$ effective conductance, the energy component of a current is $I = Eg$, and the energy consumed in a conductance, g, is $P = IE = E^2 g$.

Since, however:

$$P = A \frac{E^{1.6}}{N^{.6}}, \text{ we have } A \frac{E^{1.6}}{N^{.6}} = E^2 g;$$

or

$$g = \frac{A}{N^{.6} E^{.4}} = \frac{58 \, \eta \, L \, 10^8}{E^{.4} N^{.6} S^{.6} n^{1.6}} = 191.4 \frac{L}{E^{.4} N^{.6} S^{.6} n^{1.6}}.$$

From this we have the following deduction:

The effective conductance due to magnetic hysteresis is proportional to the coefficient of hysteresis, η, and to the length of the magnetic circuit, L, and inversely proportional to the .4th power of the E.M.F., *to the .6th power of the frequency, N, and of the cross-section of the magnetic circuit, S, and to the 1.6th power of the number of turns, n.*

Hence, the effective hysteretic conductance increases with decreasing E.M.F., and decreases with increasing

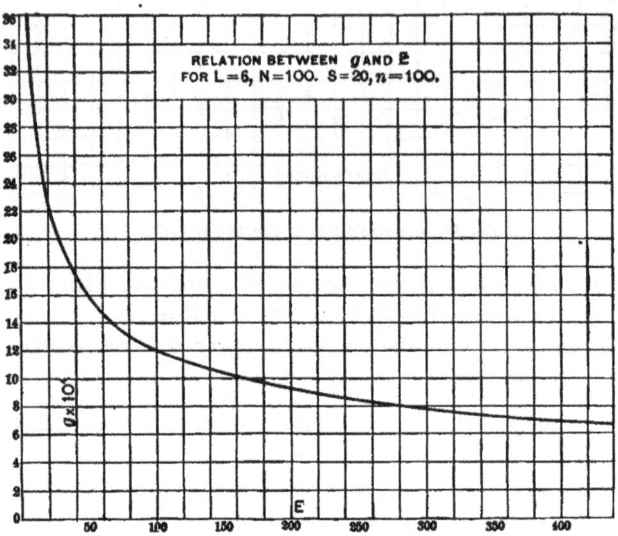

Fig. 76. Hysteresis Conductance as Function of E.M.F.

E.M.F.; it varies, however, much slower than the E.M.F., so that, if the hysteretic conductance represents only a part of the total energy consumption, it can, within a limited range of variation — as, for instance, in constant potential transformers — be assumed as constant without serious error.

In Figs. 76, 77, and 78, the hysteretic conductance, g, is plotted, for $L = 6$, $E = 100$, $N = 100$, $S = 20$ and $n = 100$, respectively, with the conductance, g, as ordinates, and with

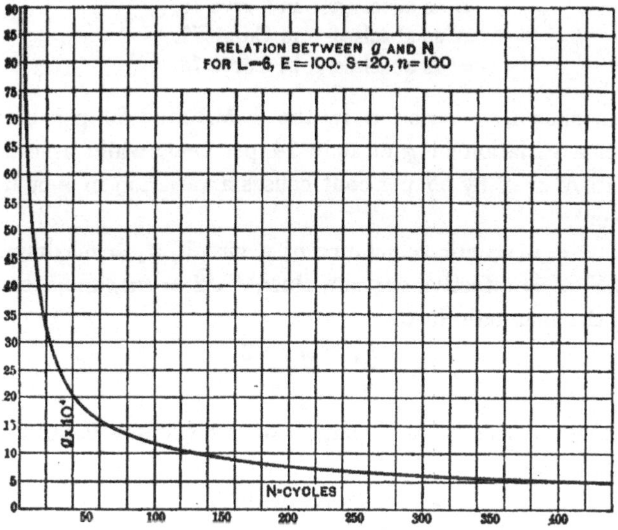

Fig. 77. Hysteresis Conductance as Function of Cycles.

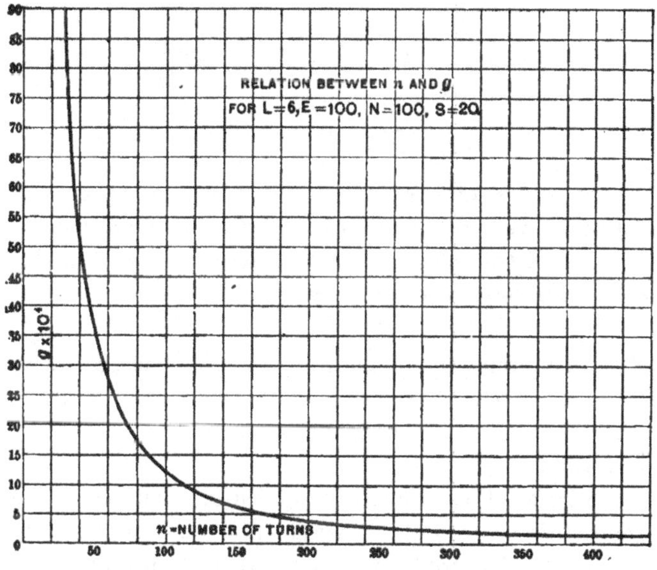

Fig. 78. Hysteresis Conductance as Function of Number of Turns.

E as abscissæ in Curve 76.
N as abscissæ in Curve 77.
n as abscissæ in Curve 78.

As shown, a variation in the E.M.F. of 50 per cent causes a variation in g of only 14 per cent, while a variation in N or S by 50 per cent causes a variation in g of 21 per cent.

If $\mathcal{R}$ = magnetic reluctance of a circuit, $\mathcal{F}_A$ = maximum M.M.F., I = effective current, since $I\sqrt{2}$ = maximum current, the magnetic flux,

$$\Phi = \frac{\mathcal{F}_A}{\mathcal{R}} = \frac{n I \sqrt{2}}{\mathcal{R}}.$$

Substituting this in the equation of the counter E.M.F. of self-induction,

$$E = \sqrt{2}\,\pi N n \Phi 10^{-8},$$

we have

$$E = \frac{2\pi n^2 N I 10^{-8}}{\mathcal{R}};$$

hence, the absolute admittance of the circuit is

$$y = \sqrt{g^2 + b^2} = \frac{I}{E} = \frac{\mathcal{R}\,10^8}{2\pi n^2 N} = \frac{a\mathcal{R}}{N},$$

where

$$a = \frac{10^8}{2\pi n^2}, \text{ a constant.}$$

Therefore, the absolute admittance, y, of a circuit of negligible resistance is proportional to the magnetic reluctance, $\mathcal{R}$, *and inversely proportional to the frequency, N, and to the square of the number of turns, n.*

82. In a circuit containing iron, the reluctance, $\mathcal{R}$, varies with the magnetization; that is, with the E.M.F. Hence the admittance of such a circuit is not a constant, but is also variable.

In an ironclad electric circuit, — that is, a circuit whose magnetic field exists entirely within iron, such as the magnetic circuit of a well-designed alternating-current trans former, — $\mathcal{R}$ is the reluctance of the iron circuit. Hence, if μ = permeability, since —

$$\mathcal{R} = \frac{\mathcal{F}_A}{\Phi},$$

and $$\mathcal{F}_A = LF = \frac{10}{4\pi} L \mathcal{H} = \text{M.M.F.},$$

$$\Phi = S\mathcal{B} = \mu S \mathcal{H} = \text{magnetic flux},$$

and $$\mathcal{R} = \frac{10\,L}{4\pi\mu S};$$

substituting this value in the equation of the admittance,

$$y = \frac{\mathcal{R}\,10^6}{2\pi n^2 N}, \text{ we have } \frac{L\,10^9}{8\pi^2 n^2 \mu SN} = \frac{z}{N\mu},$$

where $$z = \frac{L\,10^9}{8\pi^2 n^2 S} = \frac{127 L\,10^6}{n^2 S}.$$

Therefore, in an ironclad circuit, the absolute admittance, y, is inversely proportional to the frequency, N, to the permeability, μ, to the cross-section, S, and to the square of the number of turns, n; and directly proportional to the length of the magnetic circuit, L.

The conductance is $$g = \frac{A}{N^{.6} E^{.4}};$$

and the admittance, $$y = \frac{z}{N\mu};$$

hence, the angle of hysteretic advance is

$$\sin\alpha = \frac{g}{y} = \frac{A\mu N^{.4}}{zE^{.4}};$$

or, substituting for A and z (p. 117),

$$\sin\alpha = \mu \frac{N^{.4}}{E^{.4}} \quad \frac{\eta L\,10^{5.8}}{2^{.8}\pi^{1.6} S^{.6} n^{1.6}} \quad \frac{8\pi^2 n^2 S}{L\,10^9},$$

$$= \frac{\mu\eta N^{.4} n^{.4} S^{.4} \pi^{.4} 2^{2.2}}{E^{.4}\,10^{3.2}};$$

or, substituting

$$E = 2^{.5}\pi N n S \mathcal{B}\,10^{-8},$$

we have $$\sin\alpha = \frac{4\mu\eta}{\mathcal{B}^{.4}},$$

which is independent of frequency, number of turns, and shape and size of the magnetic and electric circuit.

Therefore, in an ironclad inductance, the angle of hysteretic advance, α, depends upon the magnetic constants, permeability and coefficient of hysteresis, and upon the maximum magnetic induction, but is entirely independent of the frequency, of the shape and other conditions of the magnetic and electric circuit; and, therefore, all ironclad magnetic circuits constructed of the same quality of iron and using the same magnetic density, give the same angle of hysteretic advance.

The angle of hysteretic advance, α, in a closed circuit transformer, depends upon the quality of the iron, and upon the magnetic density only.

The sine of the angle of hysteretic advance equals 4 times the product of the permeability and coefficient of hysteresis, divided by the .4th power of the magnetic density.

83. If the magnetic circuit is not entirely ironclad, and the magnetic structure contains air-gaps, the total reluctance is the sum of the iron reluctance and of the air reluctance, or

$$\mathcal{R} = \mathcal{R}_i + \mathcal{R}_a;$$

hence the admittance is

$$y = \sqrt{g^2 + b^2} = \frac{a}{N}(\mathcal{R}_i + \mathcal{R}_a).$$

Therefore, in a circuit containing iron, the admittance is the sum of the admittance due to the iron part of the circuit, $y_i = a\,\mathcal{R}_i / N$, and of the admittance due to the air part of the circuit, $y_a = a\,\mathcal{R}_a / N$, if the iron and the air are in series in the magnetic circuit.

The conductance, g, represents the loss of energy in the iron, and, since air has no magnetic hysteresis, is not changed by the introduction of an air-gap. Hence the angle of hysteretic advance of phase is

$$\sin\alpha = \frac{g}{y} = \frac{g}{y_i + y_a} = \frac{g}{y_i}\,\frac{\mathcal{R}_i}{\mathcal{R}_i + \mathcal{R}_a},$$

and a maximum, g / y_i, for the ironclad circuit, but decreases with increasing width of the air-gap. The introduction of the air-gap of reluctance, $\mathcal{R}_a$, decreases sin α in the ratio,

$$\frac{\mathcal{R}_i}{\mathcal{R}_i + \mathcal{R}_a}.$$

In the range of practical application, from $\mathcal{B} = 2{,}000$ to $\mathcal{B} = 12{,}000$, the permeability of iron varies between 900 and 2,000 approximately, while sin α in an ironclad circuit varies in this range from .51 to .69. In air, $\mu = 1$.

If, consequently, one per cent of the length of the iron consists of an air-gap, the total reluctance only varies through the above range of densities in the proportion of $1\frac{1}{9}$ to $1\frac{1}{20}$, or about 6 per cent, that is, remains practically constant; while the angle of hysteretic advance varies from sin $\alpha = .035$ to sin $\alpha = .064$. Thus g is negligible compared with b, and b is practically equal to y.

Therefore, in an electric circuit containing iron, but forming an open magnetic circuit whose air gap is not less than $\frac{1}{100}$ the length of the iron, the susceptance is practically constant and equal to the admittance, so long as saturation is not yet approached, or,

$$b = \mathcal{R}_a / N, \text{ or}: x = N / \mathcal{R}_a.$$

The angle of hysteretic advance is small, below 4°, and the hysteretic conductance is,

$$g = \frac{A}{E^{.4} N^{.6}}.$$

The current wave is practically a sine wave.

As an instance, in Fig. 71, Curve II., the current curve of a circuit is shown, containing an air-gap of only $\frac{1}{400}$ of the length of the iron, giving a current wave much resembling the sine shape, with an hysteretic advance of 9°.

84. To determine the electric constants of a circuit containing iron, we shall proceed in the following way:

Let —

E = counter E.M.F. of self-induction;

then from the equation,

$$E = \sqrt{2}\,\pi n N \Phi 10^{-8},$$

where,

$N =$ frequency,
$n =$ number of turns,

we get the magnetism, Φ, and by means of the magnetic cross section, S, the maximum magnetic induction : $\mathcal{B} = \Phi / S$.

From $\mathcal{B}$, we get, by means of the magnetic characteristic of the iron, the M.M.F., $= F$ ampere-turns per cm length, where

$$F = \frac{10}{4\pi}\mathcal{H},$$

if $\mathcal{H} =$ M.M.F. in C.G.S. units.

Hence,

if $L_i =$ length of iron circuit, $\mathfrak{F}_i = L_i F =$ ampere-turns required in the iron ;

if $L_a =$ length of air circuit, $\mathfrak{F}_a = \frac{10\, L_a \mathcal{B}}{4\pi} =$ ampere-turns required in the air ;

hence, $\mathfrak{F} = \mathfrak{F}_i + \mathfrak{F}_a =$ total ampere-turns, maximum value, and $\mathfrak{F} / \sqrt{2} =$ effective value. The exciting current is

$$I = \frac{\mathfrak{F}}{n\sqrt{2}},$$

and the absolute admittance,

$$y = \sqrt{g^2 + b^2} = \frac{I}{E}.$$

If $\mathfrak{F}_i$ is not negligible as compared with $\mathfrak{F}_a$, this admittance, y, is variable with the E.M.F., E.

If —

$V =$ volume of iron,
$\eta =$ coefficient of hysteresis,

the loss of energy by hysteresis due to molecular magnetic friction is,

$$W = \eta N V \mathcal{B}^{1.6};$$

hence the hysteretic conductance is $g = W / E^2$, and variable with the E.M.F., E.

The angle of hysteretic advance is, —

$$\sin \alpha = g / y;$$

the susceptance, $b = \sqrt{y^2 - g^2};$

the effective resistance, $r = g / y^2;$

and the reactance, $x = b / y^2.$

85. As conclusions, we derive from this chapter the following: —

1.) In an alternating-current circuit surrounded by iron, the current produced by a sine wave of E.M.F. is not a true sine wave, but is distorted by hysteresis, and inversely, a sine wave of current requires waves of magnetism and E.M.F. differing from sine shape.

2.) This distortion is excessive only with a closed magnetic circuit transferring no energy into a secondary circuit by mutual inductance.

3.) The distorted wave of current can be replaced by the equivalent sine wave—that is a sine wave of equal effective intensity and equal power—and the superposed higher harmonic, consisting mainly of a term of triple frequency, may be neglected except in resonating circuits.

4.) Below saturation, the distorted curve of current and its equivalent sine wave have approximately the same maximum value.

5.) The angle of hysteretic advance,—that is, the phase difference between the magnetic flux and equivalent sine wave of M.M.F., — is a maximum for the closed magnetic circuit, and depends there only upon the magnetic constants of the iron, upon the permeability, μ, the coefficient of hysteresis, η, and the maximum magnetic induction, as shown in the equation,

$$\sin \alpha = \frac{4 \mu \eta}{\mathcal{B}^{.4}}.$$

6.) The effect of hysteresis can be represented by an admittance, $Y = g + j b$, or an impedance, $Z = r - j x$.

7.) The hysteretic admittance, or impedance, varies with the magnetic induction; that is, with the E.M.F., etc

8.) The hysteretic conductance, g, is proportional to the coefficient of hysteresis, η, and to the length of the magnetic circuit, L, inversely proportional to the .4th power of the E.M.F., E, to the .6th power of frequency, N, and of the cross-section of the magnetic circuit, S, and to the 1.6th power of the number of turns of the electric circuit, n, as expressed in the equation,

$$g = \frac{58\,\eta\, L\, 10^3}{E^{.4} N^{.6} S^{.6} n^{1.6}}.$$

9.) The absolute value of hysteretic admittance,—

$$y = \sqrt{g^2 + b^2},$$

is proportional to the magnetic reluctance: $\mathcal{R} = \mathcal{R}_i + \mathcal{R}_a$, and inversely proportional to the frequency, N, and to the square of the number of turns, n, as expressed in the equation,

$$y = \frac{(\mathcal{R}_i + \mathcal{R}_a)\, 10^8}{2\,\pi\, N n^2}.$$

10.) In an ironclad circuit, the absolute value of admittance is proportional to the length of the magnetic circuit, and inversely proportional to cross-section, S, frequency, N, permeability, μ, and square of the number of turns, n, or

$$y_i = \frac{127\, L\, 10^6}{n^2 S N \mu}.$$

11.) In an open magnetic circuit, the conductance, g, is the same as in a closed magnetic circuit of the same iron part.

12.) In an open magnetic circuit, the admittance, y, is practically constant, if the length of the air-gap is at least $\frac{1}{100}$ of the length of the magnetic circuit, and saturation be not approached.

13.) In a closed magnetic circuit, conductance, susceptance, and admittance can be assumed as constant through a limited range only.

14.) From the shape and the dimensions of the circuits, and the magnetic constants of the iron, all the electric constants, g, b, y; r, x, z, can be calculated.

CHAPTER XI.

FOUCAULT OR EDDY CURRENTS.

86. While magnetic hysteresis or molecular friction is a magnetic phenomenon, eddy currents are rather an electrical phenomenon. When iron passes through a magnetic field, a loss of energy is caused by hysteresis, which loss, however, does not react magnetically upon the field. When cutting an electric conductor, the magnetic field induces a current therein. The M.M.F. of this current reacts upon and affects the magnetic field, more or less; consequently, an alternating magnetic field cannot penetrate deeply into a solid conductor, but a kind of screening effect is produced, which makes solid masses of iron unsuitable for alternating fields, and necessitates the use of laminated iron or iron wire as the carrier of magnetic flux.

Eddy currents are true electric currents, though flowing in minute circuits; and they follow all the laws of electric circuits.

Their E.M.F. is proportional to the intensity of magnetization, $\mathfrak{B}$, and to the frequency, N.

Eddy currents are thus proportional to the magnetization, $\mathfrak{B}$, the frequency, N, and to the electric conductivity, γ, of the iron; hence, can be expressed by

$$i = \beta \gamma \mathfrak{B} N.$$

The power consumed by eddy currents is proportional to their square, and inversely proportional to the electric conductivity, and can be expressed by

$$W = \beta \gamma \mathfrak{B}^2 N^2$$

or, since, $\mathfrak{B}N$ is proportional to the induced E.M.F., E, in the equation

$$E = \sqrt{2}\,\pi\, S\, n\, N \mathfrak{B}\, 10^{-8},$$

it follows that, *The loss of power by eddy currents is proportional to the square of the* E.M.F., *and proportional to the electric conductivity of the iron;* or,

$$W = a E^2 \gamma.$$

Hence, that component of the effective conductance which is due to eddy currents, is

$$g = \frac{W}{E^2} = a \gamma;$$

that is, *The equivalent conductance due to eddy currents in the iron is a constant of the magnetic circuit; it is independent of* E.M.F., *frequency, etc., but proportional to the electric conductivity of the iron,* γ.

87. Eddy currents, like magnetic hysteresis, cause an advance of phase of the current by an *angle of advance,* β; but, unlike hysteresis, eddy currents in general do not distort the current wave.

The angle of advance of phase due to eddy currents is,

$$\sin \beta = \frac{g}{y},$$

where y = absolute admittance of the circuit, g = eddy current conductance.

While the equivalent conductance, g, due to eddy currents, is a constant of the circuit, and independent of E.M.F., frequency, etc., the loss of power by eddy currents is proportional to the square of the E.M.F. of self-induction, and therefore proportional to the square of the frequency and to the square of the magnetization.

Only the energy component, gE, of eddy currents, is of interest, since the wattless component is identical with the wattless component of hysteresis, discussed in a preceding chapter.

88. To calculate the loss of power by eddy currents —

Let V = volume of iron ;
$\mathfrak{B}$ = maximum magnetic induction ;
N = frequency ;
γ = electric conductivity of iron ;
ϵ = coefficient of eddy currents.

The loss of energy per cm³, in ergs per cycle, is

$$w = \epsilon \gamma N \mathfrak{B}^2;$$

hence, the total loss of power by eddy currents is

$$W = \epsilon \gamma V N^2 \mathfrak{B}^2 10^{-7} \text{ watts},$$

and the equivalent conductance due to eddy currents is

$$g = \frac{W}{E^2} = \frac{10 \epsilon \gamma l}{2 \pi^2 S n^2} = \frac{.507 \epsilon \gamma l}{S n^2},$$

where :

l = length of magnetic circuit,
S = section of magnetic circuit,
n = number of turns of electric circuit.

The coefficient of eddy currents, ϵ, depends merely upon the shape of the constituent parts of the magnetic circuit; that is, whether of iron plates or wire, and the thickness of plates or the diameter of wire, etc.

The two most important cases are :

(*a*). Laminated iron.
(*b*). Iron wire.

89. (*a*). *Laminated Iron.*

Let, in Fig. 79,

d = thickness of the iron plates ;
$\mathfrak{B}$ = maximum magnetic induction ;
N = frequency ;
γ = electric conductivity of the iron.

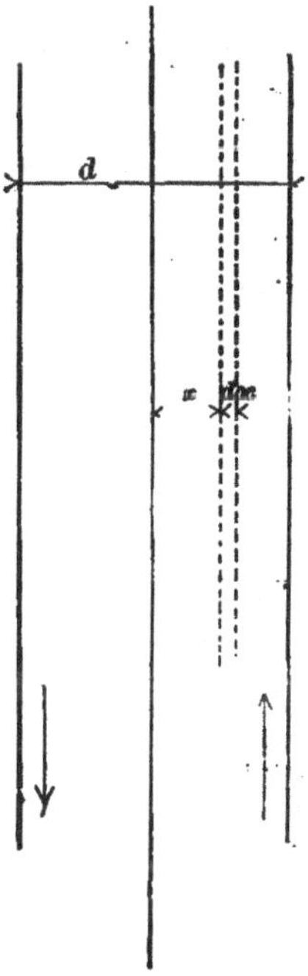

Fig. 79.

Then, if x is the distance of a zone, $d\,x$, from the center of the sheet, the conductance of a zone of thickness, $d\,x$, and of one cm length and width is $\gamma\,d\,x$; and the magnetic flux cut by this zone is $\mathfrak{B}\,x$. Hence, the E.M.F. induced in this zone is

$$\delta E = \sqrt{2}\,\pi N \mathfrak{B}\,x, \text{ in C.G.S. units.}$$

This E.M.F. produces the current:

$$d\,I = \delta E\,\gamma\,d\,x = \sqrt{2}\,\pi N \mathfrak{B}\,\gamma\,x\,d\,x, \text{ in C.G.S. units,}$$

provided the thickness of the plate is negligible as compared with the length, in order that the current may be assumed as flowing parallel to the sheet, and in opposite directions on opposite sides of the sheet.

The power consumed by the induced current in this zone, $d\,x$, is

$$d\,P = \delta E d I = 2\pi^2 N^2 \mathfrak{B}^2\,\gamma\,x^2 d\,x, \text{ in C.G.S. units or ergs per second,}$$

and, consequently, the total power consumed in one cm² of the sheet of thickness, d, is

$$\delta P = \int_{-\frac{d}{2}}^{+\frac{d}{2}} d\,P = 2\,\pi^2 N^2 \mathfrak{B}^2 \gamma \int_{-\frac{d}{2}}^{+\frac{d}{2}} x^2\,d\,x$$

$$= \frac{\pi^2 N^2 \mathfrak{B}^2 \gamma\,d^3}{6}, \text{ in C.G.S. units;}$$

the power consumed per cm³ of iron is, therefore,

$$p = \frac{\delta P}{d} = \frac{\pi^2 N^2 \mathfrak{B}^2 \gamma\,d^2}{6}, \text{ in C.G.S. units or erg-seconds}$$

and the energy consumed per cycle and per cm³ of iron is

$$h = \frac{p}{N} = \frac{\pi^2 \gamma\,d^2 N \mathfrak{B}^2}{6} \text{ ergs.}$$

The coefficient of eddy currents for laminated iron is, therefore,

$$\epsilon = \frac{\pi^2 d^2}{6} = 1.645\,d^2,$$

where γ is expressed in C.G.S. units. Hence, if γ is expressed in practical units or 10^{-9} C.G.S. units,

$$\epsilon = \frac{\pi^2 d^2 10^{-9}}{6} = 1.645\, d^2 10^{-9}.$$

Substituting for the conductivity of sheet iron the approximate value,

$$\gamma = 10^5,$$

we get as the coefficient of eddy currents for laminated iron,

$$\epsilon = \frac{\pi^2}{6} d^2 10^{-9} = 1.645\, d^2 10^{-9}.$$

loss of energy per cm³ and cycle,

$$W = \epsilon \gamma N \mathcal{B}^2 = \frac{\pi^2}{6} d^2 \gamma N \mathcal{B}^2 10^{-9} = 1.645\, d^2 \gamma N \mathcal{B}^2 10^{-9} \text{ ergs}$$
$$= 1.645\, d^2 N \mathcal{B}^2 10^{-4} \text{ ergs};$$

or, $W = \epsilon \gamma N \mathcal{B}^2 10^{-7} = 1.645\, d^2 N \mathcal{B}^2 10^{-11}$ joules;

loss of power per cm³ at frequency, N,

$$p = NW = \epsilon \gamma N^2 \mathcal{B}^2 10^{-7} = 1.645\, d^2 N^2 \mathcal{B}^2 10^{-11} \text{ watts};$$

total loss of power in volume, V,

$$P = Vp = 1.645\, V d^2 N^2 \mathcal{B}^2 10^{-11} \text{ watts}.$$

As an example,

$d = 1$ mm $= .1$ cm; $N = 100$; $\mathcal{B} = 5000$; $V = 1000$ cm³.

$\epsilon = 1{,}645 \times 10^{-11}$;

$W = 4110$ ergs

$= .000411$ joules;

$p = .0411$ watts;

$P = 41.1$ watts.

90. (*b*). *Iron Wire.* Let, in Fig. 80, d = diameter of a piece of iron wire; then if x is the radius of a circular zone of thickness, dx, and one cm in length, the conductance of this

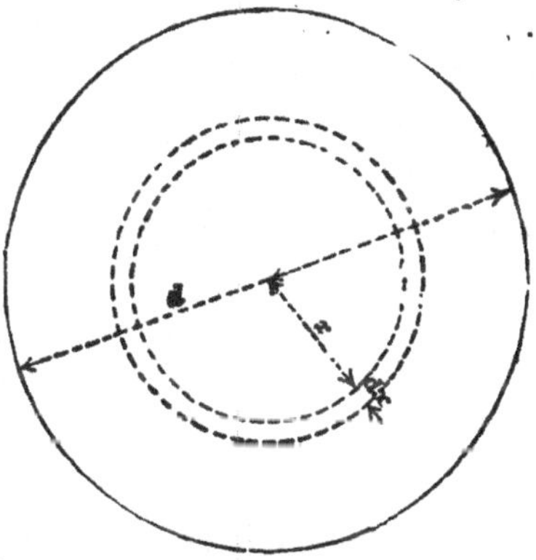

Fig. 80.

zone is, $\gamma\, d\mathrm{x} / 2\,\pi\, \mathrm{x}$, and the magnetic flux inclosed by the zone is $\mathcal{B}\, \mathrm{x}^2 \pi$.

Hence, the E.M.F. induced in this zone is:

$$\delta E = \sqrt{2}\, \pi^2 N \mathcal{B}\, \mathrm{x}^2, \text{ in C.G.S. units,}$$

and the current produced thereby is,

$$d\,I = \frac{\gamma\, d\,x}{2\,\pi\,x} \times \sqrt{2}\, \pi^2 N \mathcal{B}\, x^2$$

$$= \frac{\sqrt{2}\,\pi}{2} \gamma\, N \mathcal{B}\, x\, dx, \text{ in C.G.S. units.}$$

The power consumed in this zone is, therefore,

$$dP = \delta\, E dI = \pi^3\, \gamma\, N^2 \mathcal{B}^2\, x^3 d\,x, \text{ in C.G.S. units}$$

consequently, the total power consumed in one cm length of wire is

$$\delta\, P = \int_0^{\frac{d}{2}} dW = \pi^3\, \gamma\, N^2 \mathcal{B}^2 \int_0^{\frac{d}{2}} x^3\, d\,x$$

$$= \frac{\pi^3}{64} \gamma\, N^2 \mathcal{B}^2 d^4, \text{ in C.G.S. units.}$$

Since the volume of one cm length of wire is

$$v = \frac{d^2 \pi}{4},$$

the power consumed in one cm^3 of iron is

$$p = \frac{\delta\, P}{v} = \frac{\pi^2}{16} \gamma\, N^2 \mathcal{B}^2\, d^2, \text{ in C.G.S. units or erg-seconds,}$$

and the energy consumed per cycle and cm^3 of iron is

$$W = \frac{p}{N} = \frac{\pi^2}{16} \gamma N \mathcal{B}^2\, d^2 \text{ ergs.}$$

Therefore, the coefficient of eddy currents for iron wire is

$$\epsilon = \frac{\pi^2}{16} d^2 = .617\, d^2;$$

or, if γ is expressed in practical units, or 10^{-9} C.G.S. units,

$$\epsilon = \frac{\pi^2}{16} d^2\, 10^{-9} = .617\, d^2\, 10^{-9}$$

Substituting

$$\gamma = 10^5,$$

we get as the coefficient of eddy currents for iron wire,

$$\epsilon = \frac{\pi^2}{16} d^2 10^{-9} = .617\, d^2 10^{-9}.$$

The loss of energy per cm^3 of iron, and per cycle becomes

$$W = \epsilon \gamma N \mathfrak{B}^2 = \frac{\pi^2}{16} d^2 \gamma N \mathfrak{B}^2 10^9 = .617\, d^2 \gamma N \mathfrak{B}^2 10^{-9}$$
$$= .617\, d^2 N \mathfrak{B}^2 10^{-4} \text{ ergs},$$
$$= \epsilon \gamma N \mathfrak{B}^2 10^{-7} = .617\, d^2 N \mathfrak{B}^2 10^{-11} \text{ joules};$$

loss of power per cm^3, at frequency, N,

$$p = Nh = \epsilon \gamma N^2 \mathfrak{B}^2 10^{-7} = .617\, d^2 N^2 \mathfrak{B}^2 10^{-11} \text{ watts};$$

total loss of power in volume, V,

$$P = Vp = .617\, V d^2 N^2 \mathfrak{B}^2 10^{-11} \text{ watts}.$$

As an example,

$d = 1$ mm, $= .1$ cm; $N = 100$; $\mathfrak{B}^2 = 5{,}000$; $V = 1000$ cm³.

Then,

$$\epsilon = .617 \times 10^{-11},$$
$$W = 1540 \text{ ergs} = .000154 \text{ joules},$$
$$p = .0154 \text{ watts},$$
$$P = 15.4 \text{ watts},$$

hence very much less than in sheet iron of equal thickness.

91. *Comparison of sheet iron and iron wire.*

If

d_1 = thickness of lamination of sheet iron, and
d_2 = diameter of iron wire,

the eddy-coefficient of sheet iron being

$$\epsilon_1 = \frac{\pi^2}{6} d_1^2 10^{-9},$$

and the eddy coefficient of iron wire

$$\epsilon_2 = \frac{\pi^2}{16} d_2^2 10^{-9},$$

the loss of power is equal in both — other things being equal — if $\epsilon_1 = \epsilon_2$; that is, if,

$$d_2^2 = \frac{8}{3} d_1^2, \text{ or } d_2 = 1.63\, d_1.$$

It follows that the diameter of iron wire can be 1.63 times, or, roughly, 1⅔ as large as the thickness of laminated iron, to give the same loss of energy through eddy currents, as shown in Fig. 81.

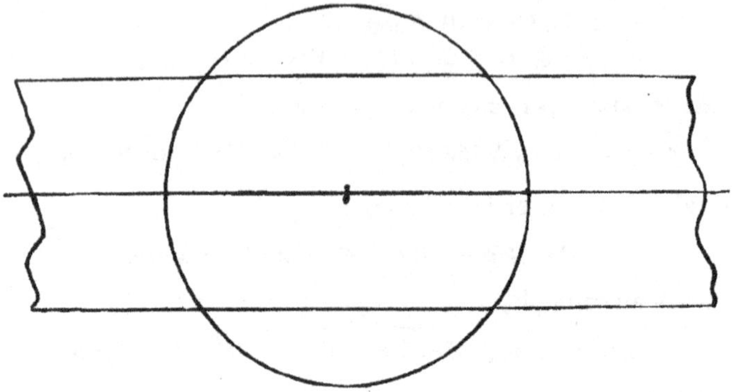

Fig. 81.

92. *Demagnetizing, or screening effect of eddy currents.*

The formulas derived for the coefficient of eddy currents in laminated iron and in iron wire, hold only when the eddy currents are small enough to neglect their magnetizing force. Otherwise the phenomenon becomes more complicated; the magnetic flux in the interior of the lamina, or the wire, is not in phase with the flux at the surface, but lags behind it. The magnetic flux at the surface is due to the impressed M.M.F., while the flux in the interior is due to the resultant of the impressed M.M.F. and to the M.M.F. of eddy currents; since the eddy currents lag 90° behind the flux producing them, their resultant with the impressed M.M.F., and therefore the magnetism in the interior, is made lagging. Thus, progressing from the surface towards the interior, the magnetic flux gradually lags more and more in phase, and at the same time decreases in intensity. While the complete analytical solution of this phenomenon is beyond the scope of this book, a determination of the magnitude of this demagnetization, or screening effect, sufficient to determine whether it is negligible, or whether the subdivision of the iron has to be increased to make it negligible, can be made by calculating the maximum magnetizing effect, which cannot be exceeded by the eddys.

Assuming the magnetic density as uniform over the whole cross-section, and therefore all the eddy currents in phase with each other, their total M.M.F. represents the maximum possible value, since by the phase difference and the lesser magnetic density in the center the resultant M.M.F. is reduced.

In laminated iron of thickness d, the current in a zone of thickness, dx at distance x from center of sheet, is :

$$dI = \sqrt{2}\,\pi N \mathcal{B} j x\,dx \text{ units (C.G.S.)}$$
$$= \sqrt{2}\,\pi N \mathcal{B} j x\,dx\, 10^{-8} \text{ amperes;}$$

hence the total current in sheet is

$$I = \int_0^{\frac{d}{2}} dI = \sqrt{2}\,\pi N \mathcal{B} j\, 10^{-8} \int_0^{\frac{d}{2}} x\,dx$$
$$= \frac{\sqrt{2}\,\pi}{8} N \mathcal{B} j d^2\, 10^{-8} \text{ amperes.}$$

Hence, the maximum possible demagnetizing ampere-turns acting upon the center of the lamina, are

$$I = \frac{\sqrt{2}\,\pi}{8} N \mathcal{B} j d^2\, 10^{-8} = .555\, N \mathcal{B} j d^2\, 10^{-8}$$
$$= .555\, N \mathcal{B} d^2\, 10^{-8} \text{ ampere-turns per cm.}$$

Example: $d = .1$ cm, $N = 100$, $\mathcal{B} = 5{,}000$
or $I = 2.775$ ampere-turns per cm.

93. In iron wire of diameter d, the current in a tubular zone of dx thickness and x radius is

$$dI = \frac{\sqrt{2}}{2} \pi N \mathfrak{B} j x \, dx \, 10^{-8} \text{ amperes;}$$

hence, the total current is

$$I = \int_0^{\frac{d}{2}} dI = \frac{\sqrt{2}}{2} \pi N \mathfrak{B} j \, 10^{-8} \int_0^{\frac{d}{2}} x dx$$

$$= \frac{\sqrt{2}}{16} \pi N \mathfrak{B} j d^2 10^{-8} \text{ amperes.}$$

Hence, the maximum possible demagnetizing ampere-turns, acting upon the center of the wire, are

$$I = \frac{\sqrt{2}\,\pi}{16} N \mathfrak{B} j d^2 10^{-8} = .2775 \, N \mathfrak{B} j d^2 10^{-8}$$

$$= .2775 \, N \mathfrak{B} \, d^2 10^{-8} \text{ ampere-turns per cm.}$$

For example, if $d = .1$ cm, $N = 100$, $\mathfrak{B} = 5{,}000$, then $I = 1{,}338$ ampere-turns per cm; that is, half as much as in a lamina of the thickness d.

94. Besides the eddy, or Foucault, currents proper, which flow as parasitic circuits in the interior of the iron lamina or wire, under certain circumstances eddy currents also flow in larger orbits from lamina to lamina through the whole magnetic structure. Obviously a calculation of these eddy currents is possible only in a particular structure. They are mostly surface currents, due to short circuits existing between the laminæ at the surface of the magnetic structure.

Furthermore, eddy currents are induced outside of the magnetic iron circuit proper, by the magnetic stray field cutting electric conductors in the neighborhood, especially when drawn towards them by iron masses behind, in electric conductors passing through the iron of an alternating field, etc. All these phenomena can be calculated only in particular cases, and are of less interest, since they can and should be avoided.

Eddy Currents in Conductor, and Unequal Current Distribution.

95. If the electric conductor has a considerable size, the alternating magnetic field, in cutting the conductor, may set up differences of potential between the different parts thereof, thus giving rise to local or eddy currents in the copper. This phenomenon can obviously be studied only with reference to a particular case, where the shape of the conductor and the distribution of the magnetic field are known.

Only in the case where the magnetic field is produced by the current flowing in the conductor can a general solution be given. The alternating current in the conductor produces a magnetic field, not only outside of the conductor, but inside of it also; and the lines of magnetic force which close themselves inside of the conductor induce E.M.Fs. in their interior only. Thus the counter E.M.F. of self-inductance is largest at the axis of the conductor, and least at its surface; consequently, the current density at the surface will be larger than at the axis, or, in extreme cases, the current may not penetrate at all to the center, or a reversed current flow there. Hence it follows that only the exterior part of the conductor may be used for the conduction of the current, thereby causing an increase of the ohmic resistance due to unequal current distribution.

The general solution of this problem for round conductors leads to complicated equations, and can be found elsewhere.

In practice, this phenomenon is observed only with very high frequency currents, as lightning discharges; in power distribution circuits it has to be avoided by either keeping the frequency sufficiently low, or having a shape of conductor such that unequal current distribution does not take place, as by using a tubular or a flat conductor, or several conductors in parallel.

96. It will, therefore, be sufficient to determine the largest size of round conductor, or the highest frequency, where this phenomenon is still negligible.

In the interior of the conductor, the current density is not only less than at the surface, but the current lags behind the current at the surface, due to the increased effect of self-inductance. This lag of the current causes the magnetic fluxes in the conductor to be out of phase with each other, making their resultant less than their sum, while the lesser current density in the center reduces the total flux inside of the conductor. Thus, by assuming, as a basis for calculation, a uniform current density and no difference of phase between the currents in the different layers of the conductor, the unequal distribution is found larger than it is in reality. Hence this assumption brings us on the safe side, and at the same time simplifies the calculation greatly.

Let Fig. 82 represent a cross-section of a conductor of radius R, and a uniform current density,

$$i = \frac{I}{R^2 \pi},$$

where I = total current in conductor.

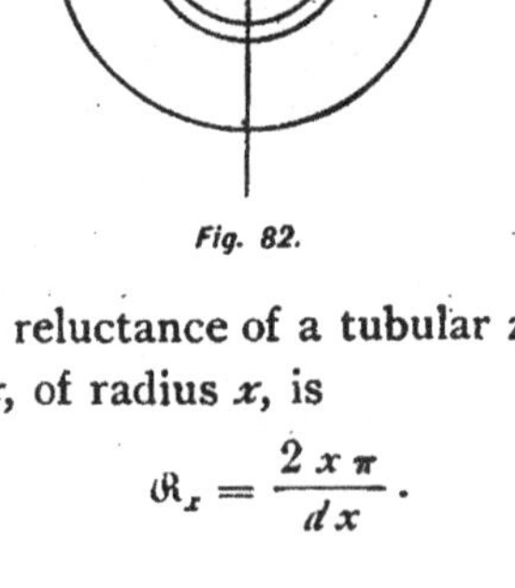

Fig. 82.

The magnetic reluctance of a tubular zone of unit length and thickness dx, of radius x, is

$$\mathcal{R}_x = \frac{2 x \pi}{d x}.$$

The current inclosed by this zone is $I_x = ix^2\pi$, and therefore, the M.M.F. acting upon this zone is

$$\mathfrak{F}_x = 4\pi I_x / 10 = 4\pi^2 ix^2 / 10,$$

and the magnetic flux in this zone is

$$d\Phi = \mathfrak{F}x / \mathfrak{R}x = 2\pi ixdx / 10.$$

Hence, the total magnetic flux inside the conductor is

$$\Phi = \int_0^R d\Phi = \frac{2\pi}{10} i \int_0^R xdx = \frac{\pi i R^2}{10} = \frac{I}{10}.$$

From this we get, as the excess of counter E.M.F. at the axis of the conductor over that at the surface —

$$\Delta E = \sqrt{2}\,\pi N \Phi\, 10^{-8} = \sqrt{2}\,\pi N I\, 10^{-9}, \text{ per unit length,}$$
$$= \sqrt{2}\pi^2 N i R^2 10^{-9};$$

and the reactivity, or specific reactance at the center of the conductor, becomes $k = \Delta E / i = \sqrt{2}\,\pi^2 N R^2\, 10^{-9}$.

Let ρ = resistivity, or specific resistance, of the material of the conductor.

We have then, $k / \rho = \sqrt{2}\,\pi^2 N R^2\, 10^{-9} / \rho$;

and $\rho / \sqrt{k^2 + \rho^2}$,

the ratio of current densities at center and at periphery.

For example, if, in copper, $\rho = 1.7 \times 10^{-6}$, and the percentage decrease of current density at center shall not exceed 5 per cent, that is —

$$\rho \div \sqrt{k^2 + \rho^2} = .95 \div 1,$$

we have, $k = .51 \times 10^{-6}$;

hence $.51 \times 10^{-6} = \sqrt{2}\,\pi^2 N R^2\, 10^{-9}$

or $NR^2 = 36.6$;

hence, when	$N =$	125	100	60	25
	$R =$	.541	.605	.781	1.21 cm.
	$D = 2R =$	1.08	1.21	1.56	2.42 cm.

Hence, even at a frequency of 125 cycles, the effect of unequal current distribution is still negligible at one cm diameter of the conductor. Conductors of this size are, however, excluded from use at this frequency by the external self-induction, which is several times larger than the resistance. We thus see that unequal current distribution is usually negligible in practice. The above calculation was made under the assumption that the conductor consists of unmagnetic material. If this is not the case, but the conductor of iron of permeability μ, then; $d\Phi = \mu \mathfrak{F}_x / \mathfrak{R}_x$ and thus ultimately; $k = \sqrt{2}\, \pi^2 N \mu R^2 10^{-9}$ and; $k / \rho = \sqrt{2}\, \pi^2 N \mu R^2\, 10^{-9} / \rho$. Thus, for instance, for iron wire at $\rho = 10 \times 10^{-6}$, $\mu = 500$ it is, permitting 5% difference between center and outside of wire; $k = 3.2 \times 10^{-6}$ and $NR^2 = .46$,

hence when,	$N =$	125	100	60	25
	$R =$	.061	.068	.088	.136 cm.

thus the effect is noticeable even with relatively small iron wire.

Mutual Inductance.

97. When an alternating magnetic field of force includes a secondary electric conductor, it induces therein an E.M.F. which produces a current, and thereby consumes energy if the circuit of the secondary conductor is closed.

A particular case of such induced secondary currents are the eddy or Foucault currents previously discussed.

Another important case is the induction of secondary E.M.Fs. in neighboring circuits; that is, the interference of circuits running parallel with each other.

In general, it is preferable to consider this phenomenon of mutual inductance as not merely producing an energy component and a wattless component of E.M.F. in the primary conductor, but to consider explicitly both the secondary and the primary circuit, as will be done in the chapter on the alternating-current transformer.

Only in cases where the energy transferred into the secondary circuit constitutes a small part of the total primary energy, as in the discussion of the disturbance caused by one circuit upon a parallel circuit, may the effect on the primary circuit be considered analogously as in the chapter on eddy currents, by the introduction of an energy component, representing the loss of power, and a wattless component, representing the decrease of self-inductance.

Let —

$x = 2\pi NL$ = reactance of main circuit; that is, L = total number of interlinkages with the main conductor, of the lines of magnetic force produced by unit current in that conductor;

$x_1 = 2\pi NL_1$ = reactance of secondary circuit; that is, L_1 = total number of interlinkages with the secondary conductor, of the lines of magnetic force produced by unit current in that conductor;

$x_m = 2\pi NL_m$ = mutual inductance of circuits; that is, L_m = total number of interlinkages with the secondary conductor, of the lines of magnetic force produced by unit current in the main conductor, or total number of interlinkages with the main conductor of the lines of magnetic force produced by unit current in the secondary conductor.

Obviously: $$x_m^2 \leqq x x_1.*$$

---

* As coefficient of self-inductance L, L_1, the total flux surrounding the conductor is here meant. Usually in the discussion of inductive apparatus, especially of transformers, that part of the magnetic flux is denoted self-inductance of the one circuit which surrounds this circuit, but not the other circuit; that is, which passes between both circuits. Hence, the total self-inductance, L, is in this case equal to the sum of the self-inductance, L_1, and the mutual inductance, L_m.

The object of this distinction is to separate the wattless part, L_1, of the total self-inductance, L, from that part, L_m, which represents the transfer of E.M.F. into the secondary circuit, since the action of these two components is essentially different.

Thus, in alternating-current transformers it is customary — and will be done later in this book — to denote as the self-inductance, L, of each circuit only that part of the magnetic flux produced by the circuit which passes between both circuits, and thus acts in "choking" only, but not in transforming; while the flux surrounding both circuits is called mutual inductance, or useful magnetic flux.

With this denotation, in transformers the mutual inductance, L_m, is usually very much greater than the self-inductances, L', and L_1', while, if the self-inductances, L and L_1, represent the total flux, their product is larger than the square of the mutual inductance, L_m; or

$$L L_1 \geqq L_m^2; \qquad (L' + L_m)(L_1' + L_m) > L_m^2.$$

Let r_1 = resistance of secondary circuit. Then the impedance of secondary circuit is

$$Z_1 = r_1 - jx_1, \qquad z_1 = \sqrt{r_1^2 + x_1^2};$$

E.M.F. induced in the secondary circuit, $\dot{E}_1 = jx_m \dot{I}$,

where $\dot{I}$ = primary current. Hence, the secondary current is

$$\dot{I}_1 = \frac{\dot{E}_1}{z_1} = \frac{jx_m}{r_1 - jx_1} \dot{I};$$

and the E.M.F. induced in the primary circuit by the secondary current, $\dot{I}_1$ is

$$\dot{E} = jx_m \dot{I}_1 = \frac{-x_m^2}{r_1 - jx_1} \dot{I};$$

or, expanded,

$$\dot{E} = \left\{ \frac{-x_m^2 r_1}{r_1^2 + x_1^2} - \frac{jx_m^2 x_1}{r_1^2 + x_1^2} \right\} \dot{I}.$$

Hence, the E.M.F. consumed thereby

$$\dot{E}^1 = \left\{ \frac{x_m^2 r_1}{r_1^2 + x_1^2} + \frac{jx_m^2 x_1}{r_1^2 + x_1^2} \right\} \dot{I} = (r - jx)\, \dot{I}.$$

$$r = \frac{x_m^2 r_1}{r_1^2 + x_1^2} = \text{effective resistance of mutual inductance;}$$

$$x = \frac{-x_m^2 x_1}{r_1^2 + x_1^2} = \text{effective reactance of mutual inductance.}$$

The susceptance of mutual inductance is negative, or of opposite sign from the reactance of self-inductance. Or,

Mutual inductance consumes energy and decreases the self-inductance.

Dielectric and Electrostatic Phenomena.

98. While magnetic hysteresis and eddy currents can be considered as the energy component of inductance, condensance has an energy component also, namely, dielectric hysteresis. In an alternating magnetic field, energy is consumed in hysteresis due to molecular friction, and similarly, energy is also consumed in an alternating electrostatic field in the dielectric medium, in what is called electrostatic or dielectric hysteresis.

While the laws of the loss of energy by magnetic hysteresis are fairly well understood, and the magnitude of the effect known, the phenomenon of dielectric hysteresis is still almost entirely unknown as concerns its laws and the magnitude of the effect.

It is quite probable that the loss of power in the dielectric in an alternating electrostatic field consists of two distinctly different components, of which the one is directly proportional to the frequency, — analogous to magnetic hysteresis, and thus a constant loss of energy per cycle, independent of the frequency; while the other component is proportional to the square of the frequency, — analogous to the loss of power by eddy currents in the iron, and thus a loss of energy per cycle proportional to the frequency.

The existence of a loss of power in the dielectric, proportional to the square of the frequency, I observed some time ago in paraffined paper in a high electrostatic field and at high frequency, by the electro-dynamometer method, and other observers under similar conditions have found the same result.

Arno of Turin found at low frequencies and low field strength in a larger number of dielectrics, a loss of energy per cycle independent of the frequency, but proportional to the 1.6th power of the field strength, — that is, following the same law as the magnetic hysteresis,

$$W_{\mathfrak{H}} = \eta \mathfrak{B}^{1.6}.$$

This loss, probably true dielectric static hysteresis, was observed under conditions such that a loss proportional to the square of density and frequency must be small, while at high densities and frequencies, as in condensers, the true dielectric hysteresis may be entirely obscured by a viscous loss, represented by $W_{\mathfrak{H}} = \epsilon N \mathfrak{B}^2$.

99. If the loss of power by electrostatic hysteresis is proportional to the square of the frequency and of the field intensity, — as it probably nearly is under the working conditions of alternating-current condensers, — then it is proportional to the square of the E.M.F., that is, the effective conductance, g, due to dielectric hysteresis is a constant; and, since the condenser susceptance, $-b = b'$, is a constant also, — unlike the magnetic inductance, — the ratio of conductance and susceptance, that is, the angle of difference of phase due to dielectric hysteresis, is a constant. This I found proved by experiment. This would mean that the dielectric hysteretic admittance of a condenser,

$$Y = g + jb = g - jb',$$

where: g = hysteretic conductance, b' = hysteretic susceptance; and the dielectric hysteretic impedance of a condenser,

$$Z = r - jx = r + jx_c,$$

where: r = hysteretic resistance, x_c = hysteretic condensance; and the angle of dielectric hysteretic lag, $\tan \alpha = b' / g = x_c / r$, are constants of the circuit, independent of E.M.F. and frequency. The E.M.F. is obviously inversely proportional to the frequency.

The true static dielectric hysteresis, observed by Arno as proportional to the 1.6th power of the density, will enter the admittance and the impedance as a term variable and dependent upon E.M.F. and frequency, in the same manner as discussed in the chapter on magnetic hysteresis.

To the magnetic hysteresis corresponds, in the electrostatic field, the static component of dielectric hysteresis, following, probably, the same law of 1.6th power.

To the eddy currents in the iron corresponds, in the electrostatic field, the viscous component of dielectric hysteresis, following the square law.

As a rule however, these hysteresis losses in the alternating electrostatic field of a condenser are very much smaller than the losses in an alternating magnetic field, so that while the latter exert a very marked effect on the design of apparatus, representing frequently the largest of all the losses of energy, the dielectric losses are so small as to be very difficult to observe.

To the phenomenon of mutual inductance corresponds, in the electrostatic field, the electrostatic induction, or influence.

100. The alternating electrostatic field of force of an electric circuit induces, in conductors within the field of force, electrostatic charges by what is called electrostatic influence. These charges are proportional to the field strength; that is, to the E.M.F. in the main circuit.

If a flow of current is produced by the induced charges, energy is consumed proportional to the square of the charge; that is, to the square of the E.M.F.

These induced charges, reacting upon the main conductor, influence therein charges of equal but opposite phase, and hence lagging behind the main E.M.F. by the angle of lag between induced charge and inducing field. They require the expenditure of a charging current in the main conductor in quadrature with the induced charge thereon; that is, nearly in quadrature with the E.M.F., and hence consisting of an energy component in phase with the E.M.F. — representing the power consumed by electrostatic influence — and a wattless component, which increases the capacity of the conductor, or, in other words, reduces its capacity reactance, or condensance.

Thus, the electrostatic influence introduces an effective conductance, g, and an effective susceptance, b, — of the same sign with condenser susceptance, — into the equations of the electric circuit.

While theoretically g and b should be constants of the circuit, frequently they are very far from such, due to disruptive phenomena beginning to appear at high electrostatic stresses.

Even the capacity condensance changes at very high potentials; escape of electricity into the air and over the surfaces of the supporting insulators by brush discharge or electrostatic glow takes place. As far as this electrostatic corona reaches, the space is in electric connection with the conductor, and thus the capacity of the circuit is determined, not by the surface of the metallic conductor, but by the exterior surface of the electrostatic glow surrounding the conductor. This means that with increasing potential, the capacity increases as soon as the electrostatic corona appears; hence, the condensance decreases, and at the same time an energy component appears, representing the loss of power in the corona.

This phenomenon thus shows some analogy with the decrease of magnetic inductance due to saturation.

At moderate potentials, the condensance due to capacity can be considered as a constant, consisting of a wattless component, the condensance proper, and an energy component, the dielectric hysteresis.

The condensance of a polarization cell, however, begins to decrease at very low potentials, as soon as the counter E.M.F. of chemical dissociation is approached.

The condensance of a synchronizing alternator is of the nature of a variable quantity; that is, the effective reactance changes gradually, according to the relation of impressed and of counter E.M.F., from inductance over zero to condensance.

Besides the phenomena discussed in the foregoing as terms of the energy components and the wattless components of current and of E.M.F., the electric leakage is to be considered as a further energy component; that is, the direct escape of current from conductor to return conductor through the surrounding medium, due to imperfect insulating qualities. This leakage current represents an effective conductance, g, theoretically independent of the E.M.F., but in reality frequently increasing greatly with the E.M.F., owing to the decrease of the insulating strength of the medium upon approaching the limits of its disruptive strength.

101. In the foregoing, the phenomena causing loss of energy in an alternating-current circuit have been discussed; and it has been shown that the mutual relation between current and E.M.F. can be expressed by two of the four constants:

Energy component of E.M.F., in phase with current, and = current × effective resistance, or r;
wattless component of E.M.F., in quadrature with current, and = current × effective reactance, or x;
energy component of current, in phase with E.M.F., and = E.M.F. × effective conductance, or g;
wattless component of current, in quadrature with E.M.F., and = E.M.F. × effective susceptance, or b.

In many cases the exact calculation of the quantities, r, x, g, b, is not possible in the present state of the art.

In general, r, x, g, b, are not constants of the circuit, but depend — besides upon the frequency — more or less upon E.M.F., current, etc. Thus, in each particular case it becomes necessary to discuss the variation of r, x, g, b, or to determine whether, and through what range, they can be assumed as constant.

In what follows, the quantities r, x, g, b, will always be considered as the coefficients of the energy and wattless components of current and E.M.F., — that is, as the *effective* quantities, — so that the results are directly applicable to the general electric circuit containing iron and dielectric losses.

Introducing now, in Chapters VII. to IX., instead of "ohmic resistance," the term "effective resistance," etc., as discussed in the preceding chapter, the results apply also — within the range discussed in the preceding chapter — to circuits containing iron and other materials producing energy losses outside of the electric conductor.

CHAPTER XII.

POWER, AND DOUBLE FREQUENCY QUANTITIES IN GENERAL.

102. Graphically alternating currents and E.M.F's are represented by vectors, of which the length represents the intensity, the direction the phase of the alternating wave. The vectors generally issue from the center of co-ordinates.

In the topographical method, however, which is more convenient for complex networks, as interlinked polyphase circuits, the alternating wave is represented by the straight line between two points, these points representing the absolute values of potential (with regard to any reference point chosen as co-ordinate center) and their connection the difference of potential in phase and intensity.

Algebraically these vectors are represented by complex quantities. The impedance, admittance, etc., of the circuit is a complex quantity also, in symbolic denotation.

Thus current, E.M.F., impedance, and admittance are related by multiplication and division of complex quantities similar as current, E.M.F., resistance, and conductance are related by Ohms law in direct current circuits.

In direct current circuits, power is the product of current into E.M.F. In alternating current circuits, if

$$\underset{\cdot}{E} = e^1 + je^{11}$$
$$\underset{\cdot}{I} = i^1 + ji^{11}$$

The product,

$$P_0 = \underset{\cdot}{E}\underset{\cdot}{I} = (e^1 i^1 - e^{11} i^{11}) + j\,(e^{11} i^1 + e^1 i^{11})$$

is not the power; that is, multiplication and division, which are correct in the inter-relation of current, E.M.F., impedance, do not give a correct result in the inter-relation of E.M.F., current, power. The reason is, that E I are vectors of the same frequency, and Z a constant numerical factor which thus does not change the frequency.

The power P, however, is of double frequency compared with $\dot{E}$ and $\dot{I}$, that is, makes a complete wave for every half wave of $\dot{E}$ or $\dot{I}$, and thus cannot be represented by a vector in the same diagram with $\dot{E}$ and $\dot{I}$.

$P_0 = \dot{E}\dot{I}$ is a quantity of the same frequency with $\dot{E}$ and $\dot{I}$, and thus cannot represent the power.

103. Since the power is a quantity of double frequency of $\dot{E}$ and $\dot{I}$, and thus a phase angle ω in $\dot{E}$ and $\dot{I}$ corresponds to a phase angle $2\,\omega$ in the power, it is of interest to investigate the product $\dot{E}\dot{I}$ formed by doubling the phase angle.

Algebraically it is,

$$\dot{P} = \dot{E}\dot{I} = (e^1 + je^{11})(i^1 + ji^{11}) =$$
$$(e^1 i^1 + j^2 e^{11} i^{11}) + (je^{11} i^{11} + e^1 j i^{11})$$

Since $j^2 = -1$, that is $180°$ rotation for $\dot{E}$ and $\dot{I}$, for the double frequency vector, P, $j^2 = +1$, or $360°$ rotation, and

$$j \times 1 = j$$
$$1 \times j = -j$$

That is, multiplication with j reverses the sign, since it denotes a rotation by $180°$ for the power, corresponding to a rotation of $90°$ for $\dot{E}$ and $\dot{I}$.

Hence, substituting these values, we have,

$$\dot{P} = [\dot{E}\dot{I}] = (e^1 i^1 + e^{11} i^{11}) + j\,(e^{11} i^1 \quad e^1 i^{11})$$

The symbol $[\dot{E}\dot{I}]$ here denotes the transfer from the frequency of $\dot{E}$ and $\dot{I}$ to the double frequency of $\dot{P}$.

The product, $P = [E\ I]$ consists of two components; the real component,

$$P^1 = [EI]^1 = (e^1 i^1 + e^{11} i^{11})$$

and the imaginary component,

$$jP^j = j\,[EI]^j = j\,(e^{11} i^1 - e^1 i^{11})$$

The component,

$$P^1 = [EI]^1 = (e^1 i^1 + e^{11} i^{11})$$

is the power of the circuit, $= E\ I \cos\ (E\ I)$

The component,

$$P^j = [EI]^j = (e^{11} i^1 - e^1 i^{11})$$

is what may be called the "wattless power," or the powerless or quadrature volt-amperes of the circuit, $= E\ I$ sin $(E\ I)$.

The real component will be distinguished by the index 1, the imaginary or wattless component by the index j.

By introducing this symbolism, the power of an alternating circuit can be represented in the same way as in the direct current circuit, as the symbolic product of current and E.M.F.

Just as the symbolic expression of current and E.M.F. as complex quantity does not only give the mere intensity, but also the phase,

$$E = e^1 + je^{11}$$

$$E = \sqrt{e^{1^2} + e^{11^2}}$$

$$\tan \phi = \frac{e^{11}}{e^1}$$

so the double frequency vector product $P = [E\ I]$ denotes more than the mere power, by giving with its two components $P^1 = [E\ I]^1$ and $P^j = [E\ I]^j$, the true energy volt-amperes, and the wattless volt-amperes.

If

$$E = e^1 + je^{11}$$

$$I = i^1 + ji^{11}$$

then

$$E = \sqrt{{e^1}^2 + {e^{11}}^2}$$

$$I = \sqrt{{i^1}^2 + {i^{11}}^2}$$

and

$$P^1 = [\underset{..}{EI}]^1 = e^1 i^1 + e^{11} i^{11})$$

$$P^j = [\underset{..}{EI}]^j = (e^{11} i^1 - e^1 i^{11})$$

or

$$\begin{aligned} {P^1}^2 + {P^j}^2 &= {e^1}^2 {i^1}^2 + {e^{11}}^2 {i^{11}}^2 + {e^{11}}^2 {i^1}^2 + {e^1}^2 {i^{11}}^2 \\ &= ({e^1}^2 + {e^{11}}^2)({i^1}^2 + {i^{11}}^2) \\ &= (EI)^2 \\ &= Q^2 \end{aligned}$$

where Q = total volt amperes of circuit. That is,

The true power P^1 and the wattless power P^j are the two rectangular components of the total apparent power Q of the circuit.

Consequently,

In symbolic representation as double frequency vector products, powers can be combined and resolved by the parallelogram of vectors just as currents and E.M.F's in graphical or symbolic representation.

The graphical methods of treatment of alternating current phenomena are here extended to include double frequency quantities as power, torque, etc.

$$\frac{P^1}{Q} = p = \cos\omega = \text{power factor.}$$

$$\frac{P^j}{Q} = q = \sin\omega = \text{inductance factor}$$

of the circuit, and the general expression of power is,

$$\begin{aligned} \dot{P} &= Q\,(p + jq) \\ &= Q\,(\cos\omega + j\sin\omega) \end{aligned}$$

104. The introduction of the double frequency vector product $\dot{P} = [\underset{..}{E\,I}]$ brings us outside of the limits of algebra, however, and the commutative principle of algebra, $a \times b = b \times a$, does not apply any more, but we have,

$$[\underset{.}{E}\underset{.}{I}] \text{ unlike } [\underset{.}{I}\underset{.}{E}]$$

since

$$[\underset{.}{E}\underset{.}{I}] = [\underset{.}{E}\underset{.}{I}]^1 + j\,[\underset{.}{E}\underset{.}{I}]^j$$
$$[\underset{.}{I}\underset{.}{E}] = [\underset{.}{I}\underset{.}{E}]^1 + j\,[\underset{.}{I}\underset{.}{E}]^j$$
$$= [\underset{.}{E}\underset{.}{I}]^1 - j\,[\underset{.}{E}\underset{.}{I}]^j$$

we have

$$[\underset{.}{E}\underset{.}{I}]^1 = [\underset{.}{I}\underset{.}{E}]^1$$
$$[\underset{.}{E}\underset{.}{I}]^j = -\,[\underset{.}{I}\underset{.}{E}]^j$$

that is, the imaginary component reverses its sign by the interchange of factors.

The physical meaning is, that if the wattless power $[\underset{.}{E}\,\underset{.}{I}]^j$ is lagging with regard to $\underset{.}{E}$, it is leading with regard to $\underset{.}{I}$.

The wattless component of power is absent, or the total apparent power is true power, if

$$[\underset{.}{E}\underset{.}{I}]^j = (e^{11}i^1 - e^1 i^{11}) = 0.$$

that is,

$$\frac{e^{11}}{e^1} = \frac{i^{11}}{i^1}$$

or,

$$\tan(\underset{.}{E}) = \tan(\underset{.}{I}),$$

that is, $\underset{.}{E}$ and $\underset{.}{I}$ are in phase or in opposition.

The true power is absent, or the total apparent power wattless, if

$$[\underset{.}{E}\underset{.}{I}]^1 = (e^1 i^1 + e^{11}i^{11} = 0$$

that is,

$$\frac{e^{11}}{e^1} = -\frac{i^1}{i^{11}}$$

or,

$$\tan \underset{.}{E} = -\cot \underset{.}{I}$$

that is, $\underset{.}{E}$ and $\underset{.}{I}$ are in quadrature,

The wattless power is lagging (with regard to E or leading with regard to I) if,

$$[\dot{E}\dot{I}]^j > 0$$

and leading if,

$$[\dot{E}\dot{I}]^j < 0$$

The true power is negative, that is, power returns, if,

$$[\dot{E}\dot{I}]^1 < 0$$

We have,

$$[\dot{E}, -\dot{I}] = [-\dot{E}, \dot{I}] = -[\dot{E}\dot{I}]$$
$$[-\dot{E}, -\dot{I}] = +[\dot{E}\dot{I}]$$

that is, when representing the power of a circuit or a part of a circuit, current and E.M.F. must be considered in their proper *relative* phases, but their phase relation with the remaining part of the circuit is immaterial.

We have further

$$[\dot{E}, j\dot{I}] = -j[\dot{E}, \dot{I}] = [\dot{E}, \dot{I}]^j - j[\dot{E}, \dot{I}]^1$$
$$[j\dot{E}, \dot{I}] = j[\dot{E}, \dot{I}] = -[\dot{E}, \dot{I}]^j + j[\dot{E}, \dot{I}]^1$$
$$[j\dot{E}, j\dot{I}] = [\dot{E}, \dot{I}] = [\dot{E}\dot{I}]^1 + j[\dot{E}, \dot{I}]^j$$

105. If $\dot{P}_1 = [\dot{E}_1\dot{I}_1]$, $\dot{P}_2 = [\dot{E}_2\dot{I}_2] \ldots \dot{P}_n = [\dot{E}_n\dot{I}_n]$

are the symbolic expressions of the power of the different parts of a circuit or network of circuits, the total power of the whole circuit or network of circuits is

$$\dot{P} = \dot{P}_1 + \dot{P}_2 + \cdots + \dot{P}_n$$
$$P^1 = P_1^1 + P_2^1 + \cdots + P_n^1$$
$$P^j = P_1^j + P_2^j \cdots + P_n^j$$

In other words, the total power in symbolic expression (true as well as wattless) of a circuit or system is the sum of the powers of its individual components in symbolic expression.

The first equation is obviously directly a result from the law of conservation of energy.

One result derived herefrom is for instance:

If in a generator supplying power to a system the current is out of phase with the E.M.F. so as to give the wattless power P^j, the current can be brought into phase with the generator E.M.F., or the load on the generator made non-inductive by inserting anywhere in the circuit an apparatus producing the wattless power $-P^j$; that is, compensation for wattless currents in a system takes place regardless of the location of the compensating device.

Obviously between the compensating device and the source of wattless currents to be compensated for, wattless currents will flow, and for this reason it may be advisable to bring the compensator as near as possible to the circuit to be compensated.

106. Like power, torque in alternating apparatus is a double frequency vector product also, of magnetism and M.M.F. or current, and thus can be treated in the same way.

In an induction motor, for instance, the torque is the product of the magnetic flux in one direction into the component of secondary induced current in phase with the magnetic flux in time, but in quadrature position therewith in space, times the number of turns of this current, or since the induced E.M.F. is in quadrature and proportional to the magnetic flux and the number of turns, the torque of the induction motor is the product of the induced E.M.F. into the component of secondary current in quadrature therewith in time and space, or the product of the induced current into the component of induced E.M.F. in quadrature therewith in time and space.

Thus if

$E^1 = e^1 + je^{11}$ = induced E.M.F. in one direction in space.

$I_2 = i^1 + j\, i^{11}$ = secondary current in the quadrature direction in space, the torque is

$$T = [EI]^j = e^{11}i^1 - e^1 i^{11}.$$

By this equation the torque is given in watts, the meaning being that $T = [E\, I]^j$ is the power which would be exerted by the torque at synchronous speed, or the torque in synchronous watts.

The torque proper is then

$$T_0 = \frac{T}{2\pi Np}$$

where

p = number of pairs of poles of the motor.

In the polyphase induction motor, if $I_2 = i^1 + ji^{11}$ is the secondary current in quadrature position, in space, to E.M.F. E_1.

The current in the same direction in space as E_1 is $I_1 = jI_2 = -i^{11} + ji^1$; thus the torque can also be expressed as

$$T = [E_1 I_1]^1 = e^{11}i^1 - e^1 i^{11}$$

CHAPTER XIII.

DISTRIBUTED CAPACITY, INDUCTANCE, RESISTANCE, AND LEAKAGE.

107. As far as capacity has been considered in the foregoing chapters, the assumption has been made that the condenser or other source of negative reactance is shunted across the circuit at a definite point. In many cases, however, the capacity is distributed over the whole length of the conductor, so that the circuit can be considered as shunted by an infinite number of infinitely small condensers infinitely near together, as diagrammatically shown in Fig. 83.

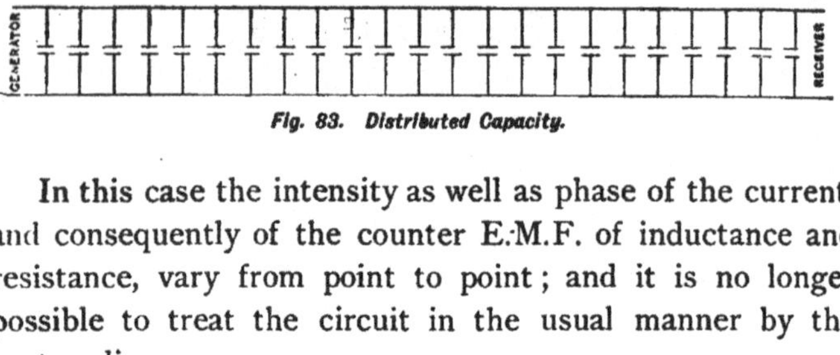

Fig. 83. Distributed Capacity.

In this case the intensity as well as phase of the current, and consequently of the counter E.M.F. of inductance and resistance, vary from point to point; and it is no longer possible to treat the circuit in the usual manner by the vector diagram.

This phenomenon is especially noticeable in long-distance lines, in underground cables, and to a certain degree in the high-potential coils of alternating-current transformers for very high voltage. It has the effect that not only the E.M.Fs., but also the currents, at the beginning, end, and different points of the conductor, are different in intensity and in phase.

Where the capacity effect of the line is small, it may with sufficient approximation be represented by one condenser of the same capacity as the line, shunted across the line. Frequently it makes no difference either, whether this condenser is considered as connected across the line at the generator end, or at the receiver end, or at the middle.

The best approximation is to consider the line as shunted at the generator and at the motor end, by two condensers of $\frac{1}{6}$ the line capacity each, and in the middle by a condenser of $\frac{2}{3}$ the line capacity. This approximation, based on Simpson's rule, assumes the variation of the electric quantities in the line as parabolic. If, however, the capacity of the line is considerable, and the condenser current is of the same magnitude as the main current, such an approximation is not permissible, but each line element has to be considered as an infinitely small condenser, and the differential equations based thereon integrated. Or the phenomena occurring in the circuit can be investigated graphically by the method given in Chapter VI. § 37, by dividing the circuit into a sufficiently large number of sections or line elements, and then passing from line element to line element, to construct the topographic circuit characteristics.

108. It is thus desirable to first investigate the limits of applicability of the approximate representation of the line by one or by three condensers.

Assuming, for instance, that the line conductors are of 1 cm. diameter, and at a distance from each other of 50 cm., and that the length of transmission is 50 km., we get the capacity of the transmission line from the formula —

$$C = 1.11 \times 10^{-6} \kappa l \div 4 \log_\epsilon 2d / \delta \text{ microfarads,}$$

where

κ = dielectric constant of the surrounding medium = 1 in air;
l = length of conductor = 5×10^6 cm.;
d = distance of conductors from each other = 50 cm.;
δ = diameter of conductor = 1 cm.

Since $C = .3$ microfarads,
the capacity reactance is $x = 10^6 / 2\pi NC$ ohms,

where $N =$ frequency; hence, at $N = 60$ cycles,

$$x = 8{,}900 \text{ ohms};$$

and the charging current of the line, at $E = 20{,}000$ volts, becomes,

$$i_o = E / x = 2.25 \text{ amperes.}$$

The resistance of 100 km of line of 1 cm diameter is 22 ohms; therefore, at 10 per cent = 2,000 volts loss in the line, the main current transmitted over the line is

$$I = \frac{2{,}000}{22} = 91 \text{ amperes,}$$

representing about 1,800 kw.

In this case, the condenser current thus amounts to less than 2½ per cent., and hence can still be represented by the approximation of one condenser shunted across the line.

If the length of transmission is 150 km., and the voltage, 30,000,

capacity reactance at 60 cycles,	$x = 2{,}970$ ohms;
charging current,	$i_o = 10.1$ amperes;
line resistance,	$r = 66$ ohms;
main current at 10 per cent loss,	$I = 45.5$ amperes.

The condenser current is thus about 22 per cent of the main current, and the approximate calculation of the effect of line capacity still fairly accurate.

At 300 km length of transmission it will, at 10 per cent. loss and with the same size of conductor, rise to nearly 90 per cent. of the main current, thus making a more explicit investigation of the phenomena in the line necessary.

In most cases of practical engineering, however, the capacity effect is small enough to be represented by the approximation of one; viz., three condensers shunted across the line.

109. *A.) Line capacity represented by one condenser shunted across middle of line.*

Let —

$Y = g + jb =$ admittance of receiving circuit;

$z = r - jx =$ impedance of line;

$b_c =$ condenser susceptance of line.

Denoting, in Fig. 84,

the E.M.F., viz., current in receiving circuit by E, I,
the E.M.F. at middle of line by E',
the E.M.F., viz., current at generator by E_0, I_0;

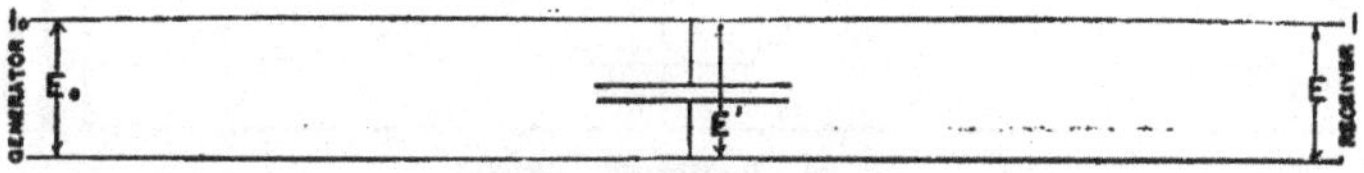

Fig. 84. Capacity Shunted across Middle of Line.

We have,

$$I = E(g + jb);$$

$$E' = E + \frac{r - jx}{2} I$$

$$= E\left\{1 + \frac{(r - jx)(g + jb)}{2}\right\};$$

$$I_0 = I - jb_c E'$$

$$= E\left\{g + jb - jb_c\left[1 + \frac{(r - jx)(g + jb)}{2}\right]\right\}$$

$$E_0 = E' + \frac{r - jx}{2} I_0$$

$$= E\left\{1 + \frac{(r - jx)(g + jb)}{2} + \frac{(r - jx)(g + jb)}{2} - \frac{jb_c(r - jx)}{2} - jb_c\frac{(r - jx)^2(g + jb)}{4}\right\};$$

or, expanding,

$$I_0 = E\{[g + b_c(rb - xg)] + j[(b - b_c) - (rg + xb)]\};$$

$$E_0 = E\left\{1 + (r - jx)(g + jb) - \frac{jb_c}{2}(r - jx)\left[1 + \frac{(r - jx)(g + jb)}{2}\right]\right\}$$

$$= E\left\{1 + (r - jx)\left(g + jb - \frac{jb_c}{2}\right) - \frac{jb_c}{4}(r - jx)^2 (g + jb)\right\}.$$

110. *B.) Line capacity represented by three condensers, in the middle and at the ends of the line.*

Denoting, in Fig. 85,

the E.M.F. and current in receiving circuit by E, I,
the E.M.F. at middle of line by E',

the current on receiving side of line by I',
the current on generator side of line by I'',
the E.M.F., viz., current at generator by E_0, I_0,

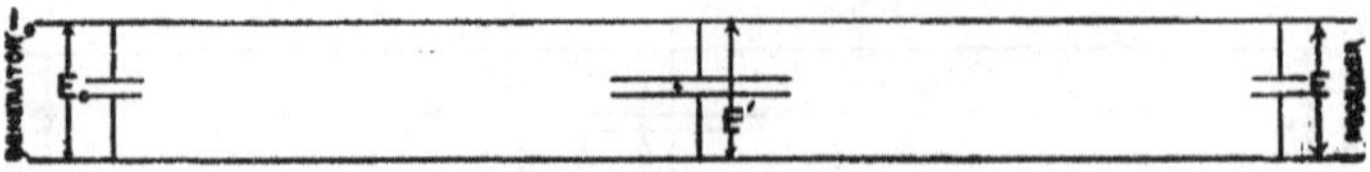

Fig. 85. Distributed Capacity.

otherwise retaining the same denotations as in *A.*),

We have,

$$I = E\,(g + jb);$$

$$I' = I - \frac{jb_c}{6}E$$

$$= E\left\{g + jb - \frac{jb_c}{6}\right\};$$

$$E' = E + \frac{r - jx}{2}I'$$

$$= E\left\{1 + \frac{r - jx}{2}\left(g + jb - \frac{jb_c}{6}\right)\right\};$$

$$I'' = I' - \frac{4\,jb_c}{6}E'$$

$$= E\left\{g + jb - \frac{5\,jb_c}{6} - \frac{jb_c}{3}(r - jx)\left(g + jb - \frac{jb_c}{6}\right)\right\};$$

$$E_0 = E' + \frac{r - jx}{2}I'';$$

$$E_0 = E\left\{1 + (r - jx)\left(g + jb - \frac{jb_c}{2}\right) - \frac{jb_c}{6}(r - jx)^2 \left(g + jb - \frac{jb_c}{6}\right)\right\};$$

$$I_0 = I'' - \frac{jb_c}{6}E_0;$$

$$I_0 = E\left\{(g + jb - jb_c) - \frac{jb_c}{6}(r - jx)\left(3g + 3jb - \frac{5\,jb_c}{6}\right) - \frac{b_c^2}{36}(r - jx)^2\left(g + jb - \frac{jb_c}{6}\right)\right\}.$$

As will be seen, the first terms in the expression of E_0 and of I_0 are the same in *A.*) and in *B.*).

111. *C.) Complete investigation of distributed capacity, inductance, leakage, and resistance.*

In some cases, especially in very long circuits, as in lines conveying alternating power currents at high potential over extremely long distances by overhead conductors or underground cables, or with very feeble currents at extremely high frequency, such as telephone currents, the consideration of the *line resistance* — which consumes E.M.Fs. in phase with the current — and of the *line reactance* — which consumes E.M.Fs. in quadrature with the current — is not sufficient for the explanation of the phenomena taking place in the line, but several other factors have to be taken into account.

In long lines, especially at high potentials, the *electrostatic capacity* of the line is sufficient to consume noticeable currents. The charging current of the line condenser is proportional to the difference of potential, and is one-fourth period ahead of the E.M.F. Hence, it will either increase or decrease the main current, according to the relative phase of the main current and the E.M.F.

As a consequence, the current will change in intensity as well as in phase, in the line from point to point; and the E.M.Fs. consumed by the resistance and inductance will therefore also change in phase and intensity from point to point, being dependent upon the current.

Since no insulator has an infinite resistance, and as at high potentials not only leakage, but even direct *escape of electricity* into the air, takes place by "silent discharge," we have to recognize the existence of a current approximately proportional and in phase with the E.M.F. of the line. This current represents consumption of energy, and is therefore analogous to the E.M.F. consumed by resistance, while the condenser current and the E.M.F. of inductance are wattless.

Furthermore, the alternate current passing over the line induces in all neighboring conductors secondary currents, which react upon the primary current, and thereby introduce E.M.Fs. of *mutual inductance* into the primary circuit. Mutual inductance is neither in phase nor in quadrature with the current, and can therefore be resolved into an *energy component* of mutual inductance in phase with the current, which acts as an increase of resistance, and into a *wattless component* in quadrature with the current, which decreases the self-inductance.

This mutual inductance is not always negligible, as, for instance, its disturbing influence in telephone circuits shows.

The alternating potential of the line induces, by *electrostatic influence,* electric charges in neighboring conductors outside of the circuit, which retain corresponding opposite charges on the line wires. This electrostatic influence requires the expenditure of a current proportional to the E.M.F., and consisting of an *energy component,* in phase with the E.M.F., and a *wattless component,* in quadrature thereto.

The alternating electromagnetic field of force set up by the line current produces in some materials a loss of energy by magnetic hysteresis, or an expenditure of E.M.F. in phase with the current, which acts as an increase of resistance. This electromagnetic hysteretic loss may take place in the conductor proper if iron wires are used, and will then be very serious at high frequencies, such as those of telephone currents.

The effect of *eddy currents* has already been referred to under "mutual inductance," of which it is an energy component.

The alternating electrostatic field of force expends energy in dielectrics by what is called *dielectric hysteresis.* In concentric cables, where the electrostatic gradient in the dielectric is comparatively large, the dielectric hysteresis may at high potentials consume considerable amounts of energy. The dielectric hysteresis appears in the circuit as consumption of a current, whose component in phase with the E.M.F. is the *dielectric energy current,* which may be considered as the power component of the capacity current.

Besides this, there is the increase of ohmic resistance due to *unequal distribution of current,* which, however, is usually not large enough to be noticeable.

112. This gives, as the most general case, and per unit length of line:

E.M.Fs. *consumed in phase with the current I,* and $= rI$, representing consumption of energy, and due to:

Resistance, and its increase by unequal current distribution; to the energy component of *mutual inductance;* to *induced currents;* to the energy component of *self-inductance;* or to *electromagnetic hysteresis.*

E.M.Fs. *consumed in quadrature with the current I,* and $= xI$, wattless, and due to:

Self-inductance, and *Mutual inductance.*

Currents consumed in phase with the E.M.F., E, and $= gE$, representing consumption of energy, and due to:

Leakage through the insulating material, including silent discharge; energy component of *electrostatic influence;* energy component of *capacity,* or of *dielectric hysteresis.*

Currents consumed in quadrature to the E.M.F., E, and $= bE$, being wattless, and due to:

Capacity and *Electrostatic influence.*

Hence we get four constants: —

Effective resistance, r,
Effective reactance, x,
Effective conductance, g,
Effective susceptance, $b = -b_c$,

per unit length of line, which represent the coefficients, per unit length of line, of

E.M.F. consumed in phase with current;
E.M.F. consumed in quadrature with current;
Current consumed in phase with E.M.F.;
Current consumed in quadrature with E.M.F.

113. This line we may assume now as feeding into a *receiver circuit of any description*, and determine the current and E.M.F. at any point of the circuit.

That is, an E.M.F. and current (differing in phase by any desired angle) may be given at the terminals of receiving circuit. To be determined are the E.M.F. and current at any point of the line; for instance, at the generator terminals.

Or,
$$Z_1 = r_1 - j x_1;$$
the impedance of receiver circuit, or admittance,
$$Y_1 = g_1 + j b_1,$$
and E.M.F., $\dot{E}_0$, at generator terminals are given. Current and E.M.F. at any point of circuit to be determined, etc.

114. Counting now the distance, x, from a point, 0, of the line which has the E.M.F.,

$$\dot{E}_1 = e_1 + j e_1', \text{ and the current: } \dot{I}_1 = i_1 + j i_1',$$

and counting x positive in the direction of rising energy, and negative in the direction of decreasing energy, we have at any point, x, in the line differential, dx:

Leakage current: $\dot{E} g \, dx$;
Capacity current: $- j \dot{E} b_c \, dx$;

hence, the total current consumed by the line element, dx, is

$$d\dot{I} = \dot{E}(g - j b_c)\, dx, \text{ or,}$$

$$\frac{d\dot{I}}{dx} = \dot{E}(g - j b_c). \qquad (1)$$

E.M.F. consumed by resistance, $\dot{I} r \, dx$;
E.M.F. consumed by reactance, $- j \dot{I} x \, dx$;

hence, the total E.M.F. consumed in the line element, $d\text{x}$, is

$$dE = I(r - jx)\,d\text{x}, \text{ or,}$$

$$\frac{dE}{d\text{x}} = I(r - jx). \qquad (2)$$

These *fundamental differential equations:*

$$\left.\begin{aligned} \frac{dI}{d\text{x}} &= E(g - jb_c), \qquad (1) \\ \frac{dE}{d\text{x}} &= I(r - jx), \qquad (2) \end{aligned}\right\}$$

are symmetrical with respect to I and E.

Differentiating these equations:

$$\left.\begin{aligned} \frac{d^2I}{d\text{x}^2} &= \frac{dE}{d\text{x}}(g - jb_c), \\ \frac{d^2E}{d\text{x}^2} &= \frac{dI}{d\text{x}}(r - jx); \end{aligned}\right\} \qquad (3)$$

and substituting (1) and (2) in (3), we get:

$$\left.\begin{aligned} \frac{d^2E}{d\text{x}^2} &= E(g - jb_c)(r - jx), \qquad (4) \\ \frac{d^2I}{d\text{x}^2} &= I(g - jb_c)(r - jx), \qquad (5) \end{aligned}\right\}$$

the *differential equations of E and I.*

115. *These differential equations are identical, and consequently I and E are functions differing by their limiting conditions only.*

These equations, (4) and (5), are of the form:

$$\frac{d^2w}{d\text{x}^2} = w(g - jb_c)(r - jx), \qquad (6)$$

and are integrated by

$$w = a\epsilon^{v\text{x}},$$

where ϵ is the basis of natural logarithms; for, differen tiating this, we get,

$$\frac{d^2w}{d\text{x}^2} = v^2a\epsilon^{v\text{x}} = v^2w;$$

hence, $$v^2 = (g - j b_c)(r - j x);$$ (7)

or, $$v = \pm \sqrt{(g - j b_c)(r - j x)};$$

hence, *the general integral* is :

$$w = a \epsilon^{+vx} + b \epsilon^{-vx}, \tag{8}$$

where a and b are the two constants of integration; Substituting

$$v = \alpha - j\beta \tag{9}$$

into (7), we have,

$$(\alpha - j\beta)^2 = (g - j b_c)(r - j x);$$

or,

$$\left.\begin{aligned} \alpha^2 - \beta^2 &= g r - x b_c; \\ 2\alpha\beta &= g x + b_c r; \end{aligned}\right\} \quad (b)$$

therefore,

$$\alpha^2 + \beta^2 = \sqrt{(g^2 + b_c^2)(r^2 + x^2)}; \tag{10}$$

$$\left.\begin{aligned} \alpha &= \sqrt{1/2\,\{\sqrt{(g^2 + b_c^2)(r^2 + x^2)} + (g r - b_c x)\}}; \\ \beta &= \sqrt{1/2\,\{\sqrt{(g^2 + b_c^2)(r^2 + x^2)} - (g r - b_c x)\}}; \end{aligned}\right\} \quad (11)$$

substituting (9) into (8) :

$$\begin{aligned} w &= a \epsilon^{(\alpha - j\beta)x} + b \epsilon^{(-\alpha + j\beta)x} \\ &= a\epsilon^{\alpha x}(\cos\beta x - j\sin\beta x) + b\epsilon^{-\alpha x}(\cos\beta x + j\sin\beta x); \end{aligned}$$

$$w = (a\epsilon^{\alpha x} + b\epsilon^{-\alpha x})\cos\beta x - j(a\epsilon^{\alpha x} - b\epsilon^{-\alpha x})\sin\beta x \tag{12}$$

which is the general solution of differential equations (4) and (5)

Differentiating (8) gives :

$$dw/dx = v(a\epsilon^{vx} - b\epsilon^{-vx});$$

hence, substituting, (9) :

$$dw/dx = (\alpha - j\beta)\{(a\epsilon^{\alpha x} - b\epsilon^{-\alpha x})\cos\beta x - j(a\epsilon^{\alpha x} + b\epsilon^{-\alpha x})\sin\beta x\}. \tag{13}$$

Substituting now $\dot{I}$ for w, and substituting (13) in (1), and writing,

$$(\alpha - j\beta)\, a = \dot{A},$$
$$(\alpha - j\beta)\, b = \dot{B},$$

we get,

$$\left.\begin{aligned} I &= \frac{1}{a - j\beta}\{(A\epsilon^{ax} + B\epsilon^{-ax})\cos\beta x - j(A\epsilon^{ax} - B\epsilon^{-ax}) \\ &\qquad \sin\beta x\}; \\ E &= \frac{1}{g - jb_c}\{(A\epsilon^{ax} - B\epsilon^{-ax})\cos\beta x - j(A\epsilon^{ax} + B\epsilon^{-ax}) \\ &\qquad \sin\beta x\}; \end{aligned}\right\} \quad (14)$$

where A and B are the constants of integration.

Transformed, we get,

$$\left.\begin{aligned} I &= \frac{1}{a - j\beta}\left\{A\epsilon^{ax}(\cos\beta x - j\sin\beta x) + B\epsilon^{-ax}\right. \\ &\qquad \left.(\cos\beta x + j\sin\beta x)\right\} \\ E &= \frac{1}{g - jb_c}\left\{A\epsilon^{ax}(\cos\beta x - j\sin\beta x) - B\epsilon^{-ax}\right. \\ &\qquad \left.(\cos\beta x + j\sin\beta x)\right\} \end{aligned}\right\} \quad (14b.)$$

Thus the waves consist of two components, one, with factor $A\epsilon^{ax}$, increasing in amplitude toward the generator, the other, with factor $B\epsilon^{-ax}$, decreasing toward the generator. The latter may be considered as a reflected wave.

At the point $x = 0$.

$$I^1 = \frac{\dot{A} + \dot{B}}{a - j\beta}$$

$$E^1 = \frac{\dot{A} - \dot{B}}{g - jb_c}$$

Thus $m\,(\cos\hat{\omega} - j\sin\hat{\omega}) = \dfrac{B}{A}$

and,

m = amplitude.

$\hat{\omega}$ = angle of reflection.

These are the general integral equations of the problem.

116. If —

$$\left.\begin{aligned} I_1 &= i_1 + j i_1' \text{ is the current} \\ E_1 &= e_1 + j e_1' \text{ is the E.M.F.} \end{aligned}\right\} \text{ at point, } x = 0. \qquad (15)$$

by substituting (15) in (14), we get:

$$\left.\begin{aligned} 2\dot{A} &= \{(\alpha i_1 + \beta i_1') + (g e_1 + b_c e_1')\} \\ &\quad + j\{(\alpha i_1' - \beta i_1) + (g e_1' - b_c e_1)\}, \\ 2\dot{B} &= \{(\alpha i_1 + \beta i_1') - (g e_1 + b_c e_1')\} \\ &\quad + j\{(\alpha i_1' - \beta i_1) - (g e_1' - b_c e_1)\}, \end{aligned}\right\} \qquad (16)$$

α and β being determined by equations (11).

117. If $Z = R - jX$ is the impedance of the receiver circuit, $\dot{E}_0 = e_0 + j e_0'$ is the E.M.F. at dynamo terminals (17), and l = length of line, we get

at $\qquad x = 0,$

$$\dot{I} = \frac{\dot{A} + \dot{B}}{\alpha - j\beta},$$

$$\dot{E} = \frac{\dot{A} - \dot{B}}{g - j b_c};$$

hence $\qquad Z = \frac{\dot{E}}{\dot{I}} = \frac{\dot{A} - \dot{B}}{\dot{A} + \dot{B}} \cdot \frac{\alpha - j\beta}{g - j b_c};$

or $\qquad \frac{\dot{A} - \dot{B}}{\dot{A} + \dot{B}} = Z \frac{g - j b_c}{\alpha - j\beta}. \qquad (18)$

At $x = l$,

$$\dot{E}_0 = \frac{1}{g - j b_c} \{(\dot{A}\epsilon^{\alpha l} - \dot{B}\epsilon^{-\alpha l}) \cos \beta l - j(\dot{A}\epsilon^{\alpha l} + \dot{B}\epsilon^{-\alpha l}) \sin \beta l\}. \qquad (19)$$

Equations (18) and (19) determine the constants A and B, which, substituted in (14), give the final integral equations.

The length, $x_0 = 2\pi / \beta$ is a complete wave length (20), which means, that in the distance $2\pi / \beta$ the phases of the components of current and E.M.F. repeat, and that in half this distance, they are just opposite.

Hence the remarkable condition exists that, in a very long line, at different points the currents at the same time flow in opposite directions, and the E.M.Fs. are opposite.

118. The difference of phase, $\hat{\omega}$, between current, $\dot{I}$, and E.M.F., $\dot{E}$, at any point, x, of the line, is determined by

the equation,

$$D(\cos\hat{\omega} + j\sin\hat{\omega}) = \frac{\dot{E}}{\dot{I}},$$

where D is a constant.

Hence, $\hat{\omega}$ varies from point to point, oscillating around a medium position, $\hat{\omega}_\infty$, which it approaches at infinity.

This difference of phase, $\hat{\omega}_\infty$, towards which current and E.M.F. tend at infinity, is determined by the expression,

$$D(\cos\hat{\omega}_\infty + j\sin\hat{\omega}_\infty) = \left|\frac{\dot{E}}{\dot{I}}\right|_{x=\infty}$$

or, substituting for $\dot{E}$ and $\dot{I}$ their values, and since $\epsilon^{-ax} = 0$ and $A\,\epsilon^{ax}(\cos\beta x - j\sin\beta x)$, cancels, and

$$D(\cos\hat{\omega}_\infty + j\sin\hat{\omega}_\infty) = \frac{a - j\beta}{g - jb_c}$$

$$= \frac{(ag + \beta b_c) - j(ab_c - \beta g)}{b_c^2 + g^2};$$

hence, $$\tan\hat{\omega}_\infty = \frac{-ab_c + \beta g}{ag + \beta b_c}. \qquad (21)$$

This angle, $\hat{\omega}_\infty$, $= 0$; that is, current and E.M.F. come more and more in phase with each other, when

$$ab_c - \beta g = 0; \text{ that is,}$$

$$a \div \beta = g \div b_c, \text{ or,}$$

$$\frac{a^2 - \beta^2}{2a\beta} = \frac{g^2 - b_c^2}{2gb_c};$$

substituting (10), gives,

$$\frac{gr - b_c x}{gx + b_c r} = \frac{g^2 - b_c^2}{2gb_c};$$

hence, expanding, $$r \div x = g \div b_c; \qquad (22)$$

that is, *the ratio of resistance to inductance equals the ratio of leakage to capacity.*

This angle, $\hat{\omega}_\infty$, $= 45°$; that is, current and E.M.F. differ by $\frac{1}{8}$th period, if $-ab_c + \beta g = ag + \beta b_c$; or,

$$\frac{a}{\beta} = \frac{b_c + g}{b_c + g};$$

which gives: $$rg + xb_c = 0. \qquad (23)$$

That is, *two of the four line constants must be zero; either g and x, or g and b_c.*

The case where $g = 0 = x$, that is a line having only resistance and distributed capacity, but no self-induction is approximately realized in concentric or multiple conductor cables, and in these the phase angle tends towards 45° lead for infinite length.

119. As an instance, in Fig. 86 a line diagram is shown, with the distances from the receiver end as abscissæ. The diagram represents one and one-half complete waves, and gives total effective current, total E.M.F., and differ-

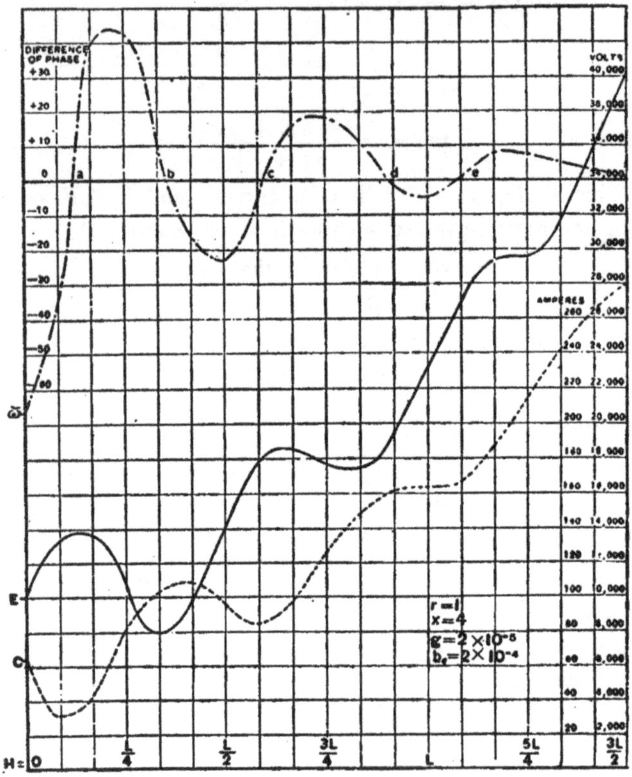

Fig. 86.

ence of phase between both as function of the distance from receiver circuit; under the conditions,

E.M.F. at receiving end, 10,000 volts; hence, $E_1 = e_1 = 10{,}000$; current at receiving end, 65 amperes, with a power factor of .385.

that is, $I = i_1 + j i_1' = 25 + 60 j$;

line constants per unit length,

$$\left.\begin{array}{ll} r = 1, & g = 2 \times 10^{-5}, \\ x = 4, & b_c = 20 \times 10^{-5}; \end{array}\right\}$$

hence,

$$\left.\begin{array}{r} \alpha = 4.95 \times 10^{-3}, \\ \beta = 28.36 \times 10^{-3}, \\ \alpha^2 + \beta^2 = .829 \times 10^{-3}; \end{array}\right\}$$

$$x_0 = L = \frac{2\pi}{\beta} = 221.5 = \left\{ \text{length of line corresponding to one complete period of the wave of propagation.} \right.$$

$$\left.\begin{array}{l} A = 1.012 - 1.206\, j, \\ B = \ .812 + \ .794\, j. \end{array}\right\}$$

These values, substituted, give,

$$\begin{aligned} I = & \{\epsilon^{\alpha x}(47.3 \cos \beta x + 27.4 \sin \beta x) - \epsilon^{-\alpha x} \\ & (22.3 \cos \beta x + 32.6 \sin \beta x)\} \\ & + j\{\epsilon^{\alpha x}(27.4 \cos \beta x - 47.3 \sin \beta x) + \epsilon^{-\alpha x} \\ & (32.6 \cos \beta x - 22.3 \sin \beta x)\}; \end{aligned}$$

$$\begin{aligned} E = & \{\epsilon^{\alpha x}(6450 \cos \beta x + 4410 \sin \beta x) + \epsilon^{-\alpha x} \\ & (3530 \cos \beta x + 4410 \sin \beta x)\} \\ & + j\{\epsilon^{\alpha x}(4410 \cos \beta x - 6450 \sin \beta x) + \epsilon^{-\alpha x} \\ & (4410 \cos \beta x - 3530 \sin \beta x)\}; \end{aligned}$$

$$\tan \hat{\omega}_\infty = \frac{-\alpha b_c + \beta g}{\alpha g + \beta b_c} = -.073, \qquad \hat{\omega}_\infty = -4.2^\circ.$$

120. As a further instance are shown the characteristic curves of a transmission line of the relative constants,

$r : x : g : b = 8 : 32 : 1.25 \times 10^{-4} : 25 \times 10^{-4}$; and $e = 25{,}000$, $i = 200$ at the receiving circuit, for the conditions,

a, non-inductive load in the receiving circuit, Fig. 87.

b, wattless receiving circuit of 90° lag, Fig. 88.

c, wattless receiving circuit of 90° lead, Fig. 89.

These curves are determined graphically by constructing the topographic circuit characteristics in polar coördinates as explained in Chapter VI., paragraphs 36 and 37, and deriving corresponding values of current, potential difference and phase angle therefrom.

As seen from these diagrams, for wattless receiving circuit, current and E.M.F. oscillate in intensity inversely to

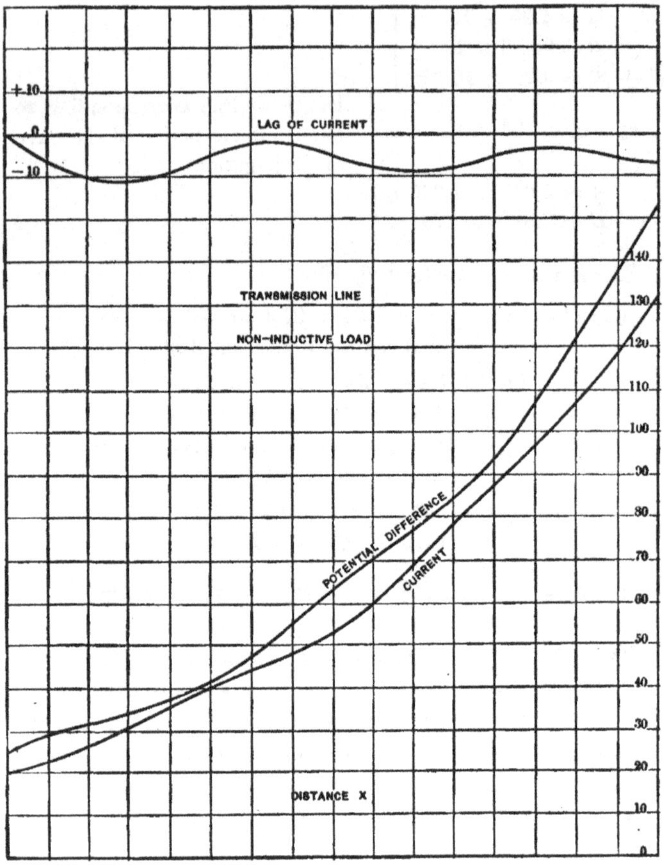

Fig. 87.

each other, with an amplitude of oscillation gradually decreasing when passing from the receiving circuit towards the generator, while the phase angle between current and E.M.F. oscillates between lag and lead with decreasing amplitude. Approximately maxima and minima of current coincide with minima and maxima of E.M.F. and zero phase angles.

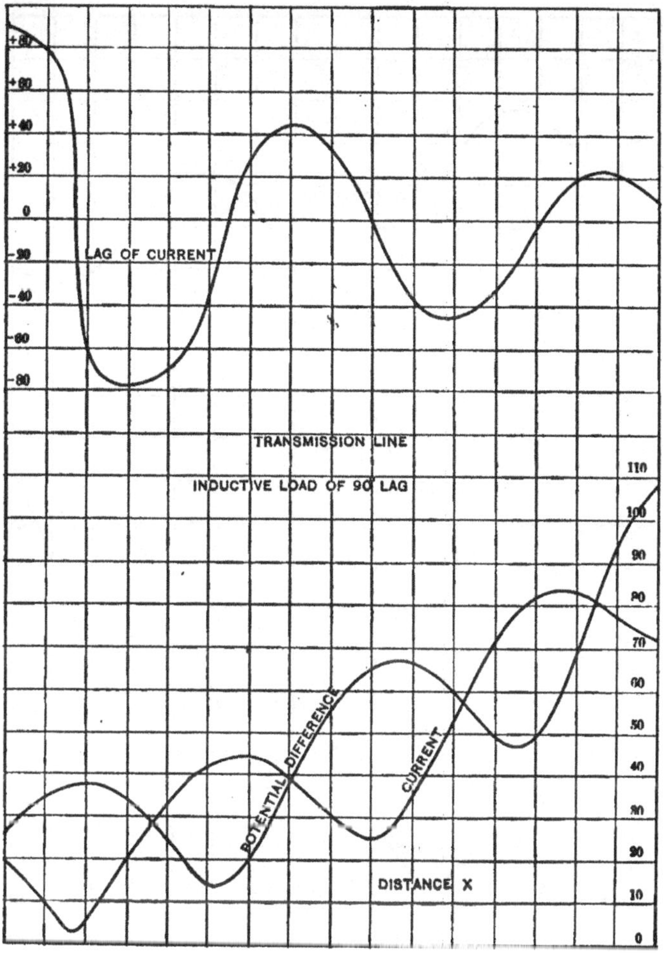

Fig. 88.

For such graphical constructions, polar coördinate paper and two angles α and δ are desirable, the angle α being the angle between current and change of E.M.F., $\tan \alpha = \frac{x}{r} = 4$, and the angle δ the angle between E.M.F. and change of current, $\tan \delta = \frac{b}{g} = 20$ in above instance.

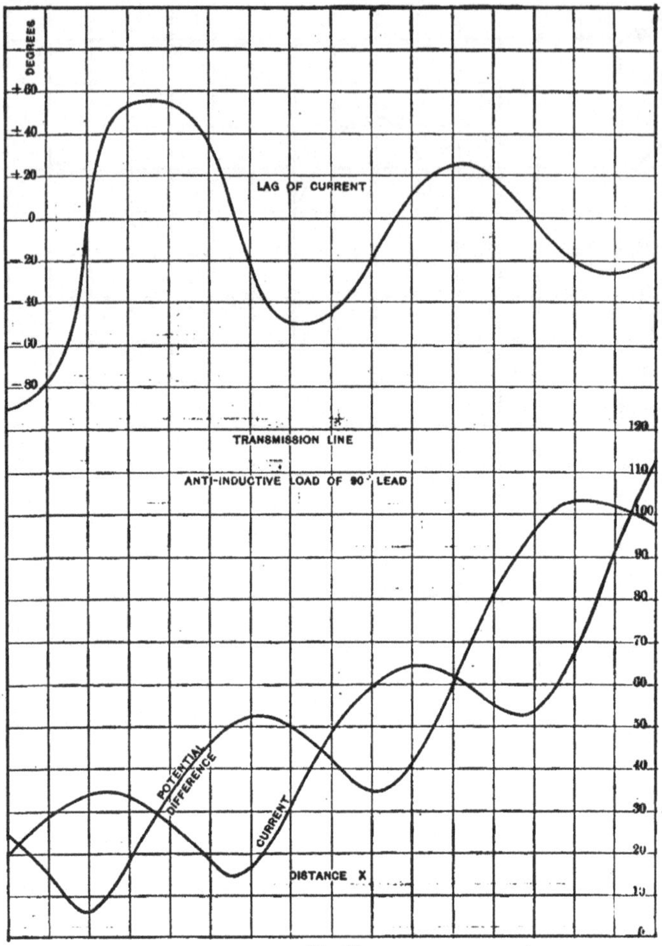

Fig. 89.

With non-inductive load, Fig. 87, these oscillations of intensity have almost disappeared, and only traces of them are noticeable in the fluctuations of the phase angle and the relative values of current and E.M.F. along the line.

Towards the generator end of the line, that is towards rising power, the curves can be extended indefinitely, approaching more and more the conditions of non-inductive circuit, towards decreasing power, however, all curves ultimately reach the conditions of a wattless receiving circuit, as Figs. 88 and 89, at the point where the total energy in-

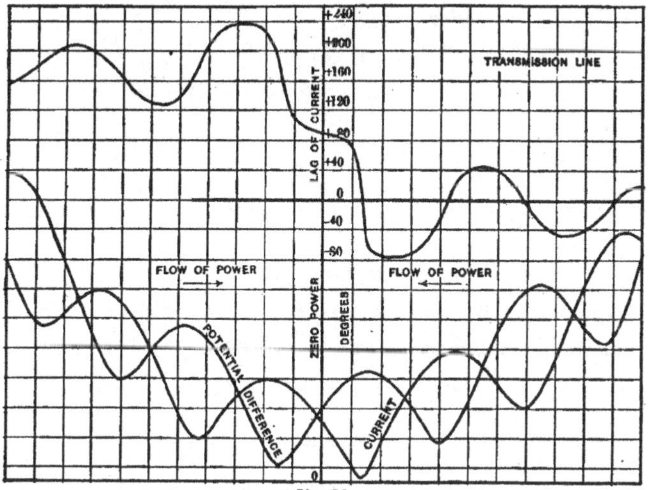

Fig. 90.

put into the line has been consumed therein, and at this point the two curves for lead and for lag join each other as shown in Fig. 90, the one being a prolongation of the other, and the flow of power in the line reverses. Thus in Fig. 90 power flows from both sides of the line towards the point of zero power marked by 0, where current and E.M.F. are in quadrature with each other, the current being leading with regard to the flow of power from the left, and lagging with regard to the flow of power from the right side of the diagram.

121. The following are some particular cases:

A.) Open circuit at end of lines:

$$x = 0: I_1 = 0.$$

$$A = (ge_1 + b_c e_1') + j(ge_1' - b_c e_1) = -B;$$

hence,

$$\left.\begin{aligned} E &= \frac{1}{g - j b_c} A\{(\epsilon^{ax} + \epsilon^{-ax}) \cos \beta x - j(\epsilon^{ax} - \epsilon^{-ax}) \sin \beta x\}; \\ I &= \frac{1}{a - j\beta} A\{(\epsilon^{ax} - \epsilon^{-ax}) \cos \beta x - j(\epsilon^{ax} + \epsilon^{-ax}) \sin \beta x\}. \end{aligned}\right\}$$

B.) Line grounded at end:

$$x = 0: E_1 = 0.$$

$$A = (a i_1 + \beta i_1') + j(a i_1' - \beta i_1) = B$$

$$\left.\begin{aligned} E &= \frac{1}{g - jb_c} A\{(\epsilon^{ax} - \epsilon^{-ax}) \cos \beta x - j(\epsilon^{ax} + \epsilon^{-ax}) \sin \beta x\}; \\ I &= \frac{1}{a - j\beta} A\{(\epsilon^{ax} + \epsilon^{-ax}) \cos \beta x - j(\epsilon^{ax} - \epsilon^{-ax}) \sin \beta x\}. \end{aligned}\right\}$$

C.) Infinitely long conductor:

Replacing x by – x, that is, counting the distance positive in the direction of decreasing energy, we have,

$$x = \infty : I = 0, E = 0;$$

hence

$$B = 0,$$

and

$$\left.\begin{aligned} E &= \frac{1}{g - j b_c} A \epsilon^{-ax} (\cos \beta x + j \sin \beta x); \\ I &= \frac{1}{a - j\beta} A \epsilon^{-ax} (\cos \beta x + j \sin \beta x), \end{aligned}\right\}$$

revolving decay of the electric wave, that is the reflected wave does not exist.

The total impedance of the infinitely long conductor is

$$Z = \frac{\dot{E}}{I} = \frac{a - j\beta}{g - jb_c}$$

$$= \frac{(a - j\beta)(g + jb_c)}{g^2 + b_c^2} = \frac{(ag + \beta b_c) - j(\beta g - ab_c)}{g^2 + b_c^2}.$$

The infinitely long conductor acts like an impedance

$$Z = \frac{\alpha g + \beta b_c}{g^2 + b_c^2} - j \frac{\beta g - \alpha b_c}{g^2 + b_c^2};$$

that is, like a resistance

$$R = \frac{\alpha g + \beta b_c}{g^2 + b_c^2},$$

combined with a reactance

$$X = \frac{\beta g - \alpha b_c}{g^2 + b_c^2}.$$

We thus get the difference of phase between E.M.F. and current,

$$\tan \hat{\omega} = \frac{X}{R} = \frac{\beta g - \alpha b_c}{\alpha g + \beta b_c},$$

which is constant at all points of the line.

If $g = 0$, $x = 0$, we have,

$$\alpha = \beta = \sqrt{\frac{b_c r}{2}};$$

hence,

$$\tan \hat{\omega} = 1, \text{ or,}$$
$$\hat{\omega} = 45°;$$

that is, current and E.M.F. differ by ⅛th period.

D.) Generator feeding into closed circuit:

Let $x = 0$ be the center of cable; then,

$$\dot{E}_x = - \dot{E}_{-x}; \qquad \text{hence: } \dot{E} = 0 \text{ at } x = 0;$$
$$\dot{I}_x = \dot{I}_{-x};$$

which equations are the same as in *B*, where the line is grounded at $x = 0$.

E.) Let the length of a line be one-quarter wave length,

$$\beta l = \frac{\pi}{2}$$

and assume the resistance r and conductance g as negligible compared with x and b_c.

$$r = 0 = g$$

These values substituted in (11) give

$$\alpha = 0.$$
$$\beta = \sqrt{b_c x}$$

Let the E.M.F. at the receiving end of the line be assumed zero vector

$$\dot{E}_1 = e_1 = \text{E.M.F. and}$$
$$\dot{I}_1 = i_1 + j i_1' = \text{current at end of line} \quad \text{x} = 0$$
$$\dot{E}_0 = \text{E.M.F. and}$$
$$\dot{I}_0 = \text{current at beginning of line}$$
$$\text{x} = l = \frac{\pi}{2\beta} = \frac{\pi}{2\sqrt{b_c x}}$$

Substituting in (16) these values of $\dot{E}_1$ and $\dot{I}_1$ and also $r = 0 = g$, we have

$$2\dot{A} = -j\beta\,(i_1 + j i_1') - j b_c e_1$$
$$2\dot{B} = -j\beta\,(i_1 + j i_1') + j b_c e_1$$

From these equations it follows that

$$\dot{A} + \dot{B} = -j\beta\,(i_1 + j i_1')$$
$$\dot{A} - \dot{B} = -j\, b_c e_1$$

which values, together with the foregoing values of $\dot{E}_1$, $\dot{I}_1$, r, g, α, and β, substituted in (14) reduce these equations to

$$\dot{I} = (i_1 + j i_1') \cos \sqrt{b_c x}\,\text{x} - j e_1 \sqrt{\frac{b_c}{x}} \sin \sqrt{b_c x}\,\text{x}$$

$$\dot{E} = e_1 \cos \sqrt{b_c x}\,\text{x} - j\,(i_1 + j i_1') \sqrt{\frac{x}{b_c}} \sin \sqrt{b_c x}\,\text{x}$$

Then at $x = \frac{\pi}{2\sqrt{b_c x}}$

$$I_0 = -je_1\sqrt{\frac{b_c}{x}} \qquad I_0 = e_1\sqrt{\frac{b_c}{x}}$$

$$E_0 = -j(i_1 + ji_1')\sqrt{\frac{x}{b_c}} \qquad E_0 = I_1\sqrt{\frac{x}{b_c}}$$

Hence also

$$E_0 I_0 = e_1 I_1$$

E_0 and I_0 are both in quadrature ahead of e_1 and I_1 respectively.

$I_1 = E_0\sqrt{\frac{b_c}{x}} =$ constant, if $E_0 =$ constant. That is, at constant impressed E.M.F. E_0, the current I_1 in the receiving circuit of a line of one-quarter wave length is constant, and inversely (constant potential — constant current transformation by inductive line). In this case, the current I_0 at the beginning of the line is proportional to the load e_1 at the end of the line.

If $\qquad x_0 = lx =$ total reactance,

$\qquad b_0 = lb_c =$ total susceptance of line, then

$$x_0 b_0 = \frac{\pi^2}{4}$$

Instance $x = 4$, $b_c = 20 \times 10^{-6}$, $E_0 = 10{,}000$ *V.* Hence $l = 55.5$, $x_0 = 222$, $b_0 = .0111$, $I_1 = 70.7$, $I_0 = .00707\,e$.

122. An interesting application of this method is the determination of the ***natural period of a transmission line;*** that is the frequency at which such a line discharges an accumulated charge of atmospheric electricity (lightning), or oscillates at a sudden change of load, as a break of circuit.

The discharge of a condenser through a circuit containing self-induction and resistance is oscillating (provided that the resistance does not exceed a certain critical value depending upon the capacity and the self-induction). That is, the discharge current alternates with constantly decreasing intensity. The frequency of this oscillating discharge depends upon the capacity, C, and the self-induction, L, of the circuit, and to a much lesser extent upon the resistance, so that if the resistance of the circuit is not excessive the frequency of oscillation can, by neglecting the resistance, be expressed with fair, or even close, approximation by the formula

$$N = \frac{1}{2\pi\sqrt{CL}}.$$

An electric transmission line represents a capacity as well as a self-induction; and thus when charged to a certain potential, for instance, by atmospheric electricity, as by induction from a thunder-cloud passing over or near the line, the transmission line discharges by an oscillating current.

Such a transmission line differs, however, from an ordinary condenser, in that with the former the capacity and the self-induction are distributed along the circuit.

In determining the frequency of the oscillating discharge of such a transmission line, a sufficiently close approximation is obtained by neglecting the resistance of the line, which, at the relatively high frequency of oscillating discharges, is small compared with the reactance. This assumption means that the dying out of the discharge current through the influence of the resistance of the circuit is neglected, and the current assumed as an alternating current of approximately the same frequency and the same intensity as the initial waves of the oscillating discharge current. By this means the problem is essentially simplified.

Let l = total length of a transmission line,

r = resistance per unit length,

x = reactance per unit length $= 2\pi NL$.

where L = coefficient of self-induction or inductance per unit length;

g = conductance from line to return (leakage and discharge into the air) per unit length;

b = capacity susceptance per unit length $= 2\pi NC$

where C = capacity per unit length.

x = the distance from the beginning of the line,

We have then the equations:

The E.M.F.,

$$\dot{E} = \frac{1}{g - jb}\left\{ (\dot{A}\epsilon^{a\mathrm{x}} - \dot{B}\epsilon^{-a\mathrm{x}}) \cos \beta\mathrm{x} - j\,(\dot{A}\epsilon^{a\mathrm{x}} + \dot{B}\epsilon^{-a\mathrm{x}}) \sin \beta\mathrm{x} \right\}$$

the current, (14.)

$$\dot{I} = \frac{1}{a - j\beta}\left\{ (\dot{A}\epsilon^{a\mathrm{x}} + \dot{B}\epsilon^{-a\mathrm{x}}) \cos \beta\mathrm{x} - j\,(\dot{A}\epsilon^{a\mathrm{x}} - \dot{B}\epsilon^{-a\mathrm{x}}) \sin \beta\mathrm{x} \right\}$$

where,

$$\left.\begin{aligned} a &= \sqrt{\frac{1}{2}\left\{ \sqrt{(g^2 + b^2)(r^2 + x^2)} + (gr - bx) \right\}} \\ \beta &= \sqrt{\frac{1}{2}\left\{ \sqrt{(g^2 + b^2)(r^2 + x^2)} - (gr - bx) \right\}} \end{aligned}\right\} \quad (11.)$$

ϵ = base of the natural logarithms, and $\dot{A}$ and $\dot{B}$ integration constants.

Neglecting the line resistance, $r = 0$, and the conductance (leakage, etc.), $g = 0$, gives,

$$\left.\begin{aligned} a &= 0 \\ \beta &= \sqrt{bx} \end{aligned}\right\} \quad (2[illegible]$$

These values substituted in (14) give,

$$\left.\begin{aligned} \dot{E} &= \frac{j}{b}\left\{ (\dot{A} - \dot{B}) \cos \sqrt{bx}\,\mathrm{x} - j\,(\dot{A} + \dot{B}) \sin \sqrt{bx}\,\mathrm{x} \right\} \\ \dot{I} &= \frac{j}{\sqrt{bx}}\left\{ (\dot{A} + \dot{B}) \cos \sqrt{bx}\,\mathrm{x} - j\,(\dot{A} - \dot{B}) \sin \sqrt{bx}\,\mathrm{x} \right\} \end{aligned}\right\} \quad (25.)$$

If the discharge takes place at the point: $x = 0$, that is, if the distance is counted from the discharge point to the end of the line; $x = l$, hence:

$$\text{At } x = 0,\ E = 0,$$
$$\text{At } x = l,\ I = 0.$$

Substituting these values in (25) gives,
For $x = 0$,

$$\dot{A} - \dot{B} = 0 \qquad \dot{A} = \dot{B}$$

which reduces these equations to,

$$\left.\begin{aligned} \dot{E} &= \frac{2\dot{A}}{b} \sin \sqrt{bx}\, x \\ \dot{I} &= \frac{2j\dot{A}}{\sqrt{bx}} \cos \sqrt{bx}\, x \end{aligned}\right\} \quad (26.)$$

and at $x = 0$,

$$\dot{I}_o = \frac{2j\dot{A}}{\sqrt{bx}} \quad (27.)$$

At $x = l$, $I = 0$, thus, substituted in (26),

$$\cos \sqrt{bx}l = 0 \quad (28.)$$

hence:

$$\sqrt{bx}\, l = \frac{(2k+1)\pi}{2}, k = 0, 1, 2, \ldots \quad (29.)$$

that is, $\sqrt{bx}\, l$ is an odd multiple of $\frac{\pi}{2}$. And at $x = l$,

$$\dot{E}_c = \pm \frac{2\dot{A}}{b} \quad (30.)$$

Substituting in (29) the values,

$$b = 2\pi NC$$
$$x = 2\pi NL$$

we have,

$$2\pi N \sqrt{CL}\, l = \frac{(2k+1)\pi}{2}$$

hence,

$$N = \frac{2k+1}{4l\sqrt{CL}} \quad (31.)$$

the frequency of the oscillating discharge,

where $k = 0, 1, 2. \ldots$

That is, the oscillating discharge of a transmission line of distributed capacity does not occur at one definite frequency (as that of a condenser), but the line can discharge at any one of an infinite number of frequencies, which are the odd multiples of the fundamental discharge frequency,

$$N_1 = \frac{1}{4l\sqrt{CL}} \qquad (32.)$$

Since

$$\left. \begin{array}{l} C_o = lC = \text{total capacity of transmission line,} \\ L_o = lL = \text{total self-inductance of transmission line,} \end{array} \right\} \qquad (33.)$$

we have,

$$N = \frac{2k+1}{4\sqrt{C_o L_o}} \text{ the frequency of oscillation,} \qquad (34.)$$

or natural period of the line, and

$$N_1 = \frac{1}{4\sqrt{C_o L_o}} \text{ the fundamental,} \qquad (35.)$$

or lowest natural period of the line.

From (30), (33), and (34),

$$b = 2\pi NC = \frac{(2k+1)\pi}{2l}\sqrt{\frac{C_o}{L_o}} \qquad (36.)$$

and from (29),

$$\sqrt{bx} = \frac{(2k+1)\pi}{2l}. \qquad (37.)$$

These substituted in (26) give,

$$\left. \begin{array}{l} \dot{E} = \dfrac{4l}{(2k+1)\pi}\sqrt{\dfrac{L_o}{C_o}}\,\dot{A} \sin \dfrac{(2k+1)\pi}{2}\dfrac{x}{l} \\ \dot{I} = \dfrac{4jl}{(2k+1)\pi}\,\dot{A} \cos \dfrac{(2k+1)\pi}{2}\dfrac{x}{l} \end{array} \right\} \qquad (38.)$$

The oscillating discharge of a line can thus follow any of the forms given by making $k = 0, 1, 2, 3 \ldots$ in equation (38).

Reduced from symbolic representation to absolute values

by multiplying E with $\cos 2\pi Nt$ and I with $\sin 2\pi Nt$ and omitting j, and substituting N from equation (34), we have,

$$E = \frac{4l}{(2k+1)\pi}\sqrt{\frac{L_0}{C_0}}\,A \sin\frac{(2k+1)\pi}{2}\frac{x}{l}\cos\frac{(2k+1)\pi}{2}\frac{t}{\sqrt{C_0L_0}}$$
$$I = \frac{4l}{(2k+1)\pi}\,A \cos\frac{(2k+1)\pi}{2}\frac{x}{l}\sin\frac{(2k+1)\pi}{2}\frac{t}{\sqrt{C_0L_0}} \qquad (39.)$$

where A is an integration constant, depending upon the initial distribution of voltage, before the discharge, and t = time after discharge.

123. The fundamental discharge wave is thus, for $k = 0$,

$$\left.\begin{aligned} E_1 &= \frac{4l}{\pi}\sqrt{\frac{L_0}{C_0}}\,A \sin\frac{\pi x}{2l}\cos\frac{\pi t}{2\sqrt{C_0L_0}} \\ I_1 &= \frac{4l}{\pi}\,A \cos\frac{\pi x}{2l}\sin\frac{\pi t}{2\sqrt{C_0L_0}} \end{aligned}\right\}(40).$$

With this wave the current is a maximum at the beginning of the line: $x = 0$, and gradually decreases to zero at the end of the line: $x = l$.

The voltage is zero at the beginning of the line, and rises to a maximum at the end of the line.

Thus the relative intensities of current and potential along the line are as represented by Fig. 91, where the cur- is shown as I, the potential as E.

The next higher discharge frequency, for: $k = 1$, gives:

$$\left.\begin{aligned} E_3 &= \frac{4l}{3\pi}\sqrt{\frac{L_0}{C_0}}\,A \sin\frac{3\pi x}{2l}\cos\frac{3\pi t}{\sqrt{C_0L_0}} \\ I_3 &= \frac{4l}{3\pi}\,A \cos\frac{3\pi x}{2l}\sin\frac{3\pi t}{2\sqrt{C_0L_0}} \end{aligned}\right\}(41.)$$

Here the current is again a maximum at the beginning of the line: $x = 0$, and gradually decreases, but reaches zero at one-third of the line: $x = \frac{l}{3}$, then increases again, in

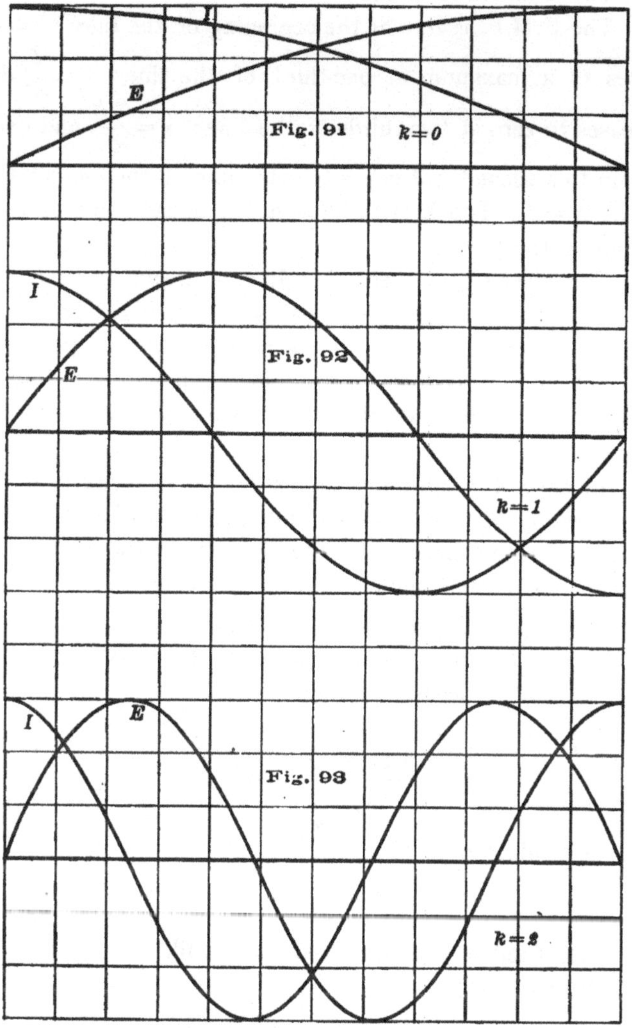

Figs. 91–93.

the opposite direction, reaches a second but opposite maximum at two-thirds of the line: $x = \frac{2l}{3}$, and decreases to zero at the end of the line. There is thus a nodal point of current at one-third of the line.

The E.M.F. is zero at the beginning of the line: $x = 0$, rises to a maximum at one-third of the line: $x = \frac{l}{3}$, decreases to zero at two-thirds of the line: $x = \frac{2l}{3}$, and rises again to a second but opposite maximum at the end of the line: $x = l$. The E.M.F. thus has a nodal point at two-thirds of the line.

The discharge waves: $k = 1$, are shown in Fig. 92, those with $k = 2$, with two nodal points, in Fig. 93.

Thus k is the number of nodal points or zero points of current and of E.M.F. existing in the line (not counting zero points at the ends of the line, which of course are not nodes).

In case of a lightning discharge the capacity C_o is the capacity of the line against ground, and thus has no direct relation to the capacity of the line conductor against its return. The same applies to the inductance L_o.

If d = diameter of line conductor,
D = distance of conductor above ground,
and l = length of conductor,

the capacity is,

$$C_o = \frac{1.11 \times 10^{-6} l}{2\, lg \frac{4D}{d}} \, mf$$

the self-inductance,

$$L_o = 2 \times 10^{-6} l g \frac{4D}{d} \, mh \qquad (42.)$$

The fundamental frequency of oscillation is thus, by substituting (42) in (35),

$$N_1 = \frac{1}{4\sqrt{C_o L_o}} = \frac{7.5 \times 10^9}{l} \qquad (43.)$$

That is, the frequency of oscillation of a line discharging to ground is independent of the size of line wire and its distance from the ground, and merely depends upon the length l of the line, being inversely proportional thereto.

We thus get the numerical values,

Length of line

	10	20	30	40	50	60	80	100 miles.
=	1.6	3.2	4.8	6.4	8	9.6	12.8	16×10^6 cm.

hence frequency,

$N_1 =$ 4680 2340 1560 1170 937.5 780 585 475 cycles.

As seen, these frequencies are comparatively low, and especially with very long lines almost approach alternator frequencies.

The higher harmonics of the oscillation are the odd multiples of these frequencies.

Obviously all these waves of different frequencies represented in equation (39) can occur simultaneously in the oscillating discharge of a transmission line, and in general the oscillating discharge of a transmission line is thus of the form,

$$\left(\text{by substituting: } a_k = \frac{A_k}{2k+1}\right)$$

$$\left.\begin{aligned} E &= \frac{4l}{\pi}\sqrt{\frac{L_0}{C_0}}\left\{a_1 \sin\frac{\pi x}{2l}\cos\frac{\pi t}{2\sqrt{C_0 L_0}} + a_3 \sin\frac{3\pi x}{2l}\right. \\ &\qquad \left.\cos\frac{3\pi t}{2\sqrt{C_0 L_0}} + \cdots\right\} \\ I &= \frac{4l}{\pi}\left\{a_1 \cos\frac{\pi x}{2l}\sin\frac{\pi t}{2\sqrt{C_0 L_0}} + a_3 \cos\frac{3\pi x}{2l}\sin\frac{3\pi t}{2\sqrt{C_0 L_0}}\right. \\ &\qquad \left. + \cdots\right\} \end{aligned}\right\} \quad (44.)$$

where $a_1\ a_3\ a_5 \ldots$ are constants depending upon the initial distribution of potential in the transmission line, at the moment of discharge, or at $t = 0$, and calculated therefrom.

124. As an instance the following discharge equation of a line charged to a uniform potential e over its entire length, and then discharging at $x = 0$, has been calculated.

The harmonics are determined up to the 11—that is, a_1, a_3, a_5, a_7, a_9, a_{11}.

These six unknown quantities require six equations, which are given by assuming $E = e$ for $x = \frac{l}{6}, \frac{2l}{6}, \frac{3l}{6}, \frac{4l}{6}, \frac{5l}{6}, \frac{6l}{6}$.

At $t = 0$, $E = e$, equation (44) assumes the form

$$e = \frac{4l}{\pi}\sqrt{\frac{L_0}{C_0}}\left\{a_1 \sin\frac{\pi x}{2l} + a_3 \sin\frac{3\pi x}{2l} + \ldots + a_{11} \sin\frac{11\pi x}{2l}\right\} \tag{45.}$$

Substituting herein for x the values: $\frac{l}{6}, \frac{2l}{6}, \ldots \frac{6l}{6}$

gives six equations for the determination of $a_1, a_3 \ldots a_{11}$. These equations solved give,

$$\left.\begin{aligned} E &= e\,(1.26 \sin\omega \cos\phi + .40 \sin 3\omega \cos 3\phi + .22 \sin 5\omega \cos 5\phi + .12 \sin 7\omega \cos 7\phi + .07 \sin 9\omega \cos 9\phi + .02 \sin 11\omega \cos 11\phi) \\ I &= e\sqrt{\frac{C_0}{L_0}}\,(1.26 \cos\omega \sin\phi + .40 \cos 3\omega \sin 3\phi + .22 \cos 5\omega \sin 5\phi + .12 \cos 7\omega \sin 7\phi + .07 \cos 9\omega \sin 9\phi + .02 \cos 11\omega \sin 11\phi) \end{aligned}\right\} \tag{46.}$$

where,

$$\left.\begin{aligned} \omega &= \frac{\pi x}{2l} \\ \phi &= \frac{\pi t}{2\sqrt{C_0 L_0}} \end{aligned}\right\} \tag{47.}$$

Instance,

Length of line, $l = 25$ miles $= 4 \times 10^6$ cm.

Size of wire: No. 000 B. & S. G., thus: $d = 1$ cm.

Height above ground: $D = 18$ feet $= 550$ cm.

Let $e = 25{,}000$ volts $=$ potential of line in the moment of discharge.

We then have,

$E = 31{,}500 \sin \omega \cos \phi + 10{,}000 \sin 3\omega \cos 3\phi + 5500 \sin 5\omega \cos 5\phi + 3000 \sin 7\omega \cos 7\phi + 1750 \sin 9\omega \cos 9\phi + 500 \sin 11\omega \cos 11\phi.$

$I = 61.7 \cos \omega \sin \phi + 19.6 \cos 3\omega \sin 3\phi + 10.8 \cos 5\omega \sin 5\phi + 5.9 \cos 7\omega \sin 7\phi + 3.4 \cos 9\omega \sin 9\phi + 1.0 \cos 11\omega \sin 11\phi.$

$$\omega = .39 x \, 10^{-6}$$
$$\phi = 1.18 t \, 10^{+4}$$

A simple harmonic oscillation as a line discharge would require a sinoidal distribution of potential on the transmission line at the instant of discharge, which is not probable, so that probably all lightning discharges of transmission lines or oscillations produced by sudden changes of circuit conditions are complex waves of many harmonics, which in their relative magnitude depend upon the initial charge and its distribution—that is, in the case of the lightning discharge, upon the atmospheric electrostatic field of force.

The fundamental frequency of the oscillating discharge of a transmission line is relatively low, and of not much higher magnitude than frequencies in commercial use in alternating current circuits. Obviously, the more nearly sinusoidal the distribution of potential before the discharge, the more the low harmonics predominate, while a very unequal distribution of potential, that is a very rapid change along the line, as caused for instance by a sudden short circuit rupturing itself instantly, causes the higher harmonics to predominate, which as a rule are more liable to cause excessive rises of voltage by resonance.

125. As has been shown, the electric distribution in a transmission line containing distributed capacity, self-induction, etc., can be represented either by a polar diagram with the phase as amplitude, and the intensity as radius vector, as in Fig. 34, or by a rectangular diagram with the distance as abscissae, and the intensity as ordinate, as in Fig. 35 and in the preceding paragraphs.

In the former case, the consecutive points of the circuit characteristic refer to consecutive points along the transmission line, and thus to give a complete representation of the phenomenon, should not be plotted in one plane but in front of each other by their distance along the transmission line. That is, if 0, 1, 2, etc., are the polar vectors in Fig. 34, corresponding to equi-distant points of the transmission line, 1 should be in a plane vertically in front of the plane of 0, 2 by the same distance in front of 1, etc.

In Fig. 35 the consecutive points of the circuit characteristic represent vectors of different phase, and thus should be rotated out of the plane around the zero axis by the angles of phase difference, and then give a length view of the same space diagram, of which Fig. 34 gives a view along the axis.

Thus, the electric distribution in a transmission line can be represented completely only by a space diagram, and as complete circuit characteristic we get for each of the lines a screw shaped space curve, of which the distance along the axis of the screw represents the distance along the transmission line, and the distance of each point from the axis represents by its direction the phase, and by its length the intensity.

Hence the electric distribution in a transmission line leads to a space problem of which Figs. 34 and 35 are partial views. The single-phase line is represented by a double screw, the three-phase line by a triple screw, and the quarter-phase four-wire line by a quadruple screw. In the symbolic expression of the electric distribution in the transmission line, the real part of the symbolic equation represents a projection on a plane passing through the axis of the screw, and the imaginary part a projection on a plane perpendicular to the first, and also passing through the axis of the screw.

CHAPTER XIV.

THE ALTERNATING-CURRENT TRANSFORMER.

126. The simplest alternating-current apparatus is the transformer. It consists of a magnetic circuit interlinked with two electric circuits, a primary and a secondary. The primary circuit is excited by an impressed E.M.F., while in the secondary circuit an E.M.F. is induced. Thus, in the primary circuit power is consumed, and in the secondary a corresponding amount of power is produced.

Since the same magnetic circuit is interlinked with both electric circuits, the E.M.F. induced per turn must be the same in the secondary as in the primary circuit; hence, the primary induced E.M.F. being approximately equal to the impressed E.M.F., the E.M.Fs. at primary and at secondary terminals have approximately the ratio of their respective turns. Since the power produced in the secondary is approximately the same as that consumed in the primary, the primary and secondary currents are approximately in inverse ratio to the turns.

127. Besides the magnetic flux interlinked with both electric circuits — which flux, in a closed magnetic circuit transformer, has a circuit of low reluctance — a magnetic cross-flux passes between the primary and secondary coils, surrounding one coil only, without being interlinked with the other. This magnetic cross-flux is proportional to the current flowing in the electric circuit, or rather, the ampere-turns or M.M.F. increase with the increasing load on the transformer, and constitute what is called the self-inductance of the transformer; while the flux surrounding both coils may be considered as mutual inductance. This cross-flux of self-induction does not induce E.M.F. in the secondary circuit, and is thus, in general, objectionable, by causing a drop of voltage and a decrease of output. It is this cross-flux, however, or flux of self-inductance, which is utilized in special transformers, to secure automatic regulation, for constant power, or for constant current, and in this case is exaggerated by separating primary and secondary coils. In the constant potential transformer however, the primary and secondary coils are brought as near together as possible, or even interspersed, to reduce the cross-flux.

As will be seen by the self-inductance of a circuit, not the total flux produced by, and interlinked with, the circuit is understood, but only that (usually small) part of the flux which surrounds one circuit without interlinking with the other circuit.

128. The alternating magnetic flux of the magnetic circuit surrounding both electric circuits is produced by the combined magnetizing action of the primary and of the secondary current.

This magnetic flux is determined by the E.M.F. of the transformer, by the number of turns, and by the frequency.

If

Φ = maximum magnetic flux,

N = frequency,

n = number of turns of the coil;

the E.M.F. induced in this coil is

$$E = \sqrt{2}\,\pi N n \Phi\, 10^{-8} = 4.44\, N n \Phi\, 10^{-8} \text{ volts};$$

hence, if the E.M.F., frequency, and number of turns are determined, the maximum magnetic flux is

$$\Phi = E\, 10^{8} / \sqrt{2}\,\pi N n.$$

To produce the magnetism, Φ, of the transformer, a M.M.F. of $\mathfrak{F}$ ampere-turns is required, which is determined by the shape and the magnetic characteristic of the iron, in the manner discussed in Chapter X.

For instance, in the closed magnet circuit transformer, the maximum magnetic induction is $\mathfrak{B} = \Phi / S$, where S = the cross-section of magnetic circuit.

129. To induce a magnetic density, $\mathfrak{B}$, a M.M.F. of $\mathfrak{H}_m$ ampere-turns maximum is required, or, $\mathfrak{H}_m / \sqrt{2}$ ampere-turns effective, per unit length of the magnetic circuit; hence, for the total magnetic circuit, of length, l,

$$\mathfrak{F} = \frac{l\mathfrak{H}_m}{\sqrt{2}} \text{ ampere-turns;}$$

or

$$I = \frac{\mathfrak{F}}{n} = \frac{l\mathfrak{H}_m}{n\sqrt{2}} \text{ amps. eff.}$$

where n = number of turns.

At no load, or open secondary circuit, this M.M.F., $\mathfrak{F}$, is furnished by the *exciting current*, I_{00}, improperly called the *leakage current*, of the transformer; that is, that small amount of primary current which passes through the transformer at open secondary circuit.

In a transformer with open magnetic circuit, such as the "hedgehog" transformer, the M.M.F., $\mathfrak{F}$, is the sum of the M.M.F. consumed in the iron and in the air part of the magnetic circuit (see Chapter X.).

The energy of the exciting current is the energy consumed by hysteresis and eddy currents and the small ohmic loss.

The exciting current is not a sine wave, but is, at least in the closed magnetic circuit transformer, greatly distorted by hysteresis, though less so in the open magnetic circuit transformer. It can, however, be represented by an equivalent sine wave, I_{00}, of equal intensity and equal power with the distorted wave, and a wattless higher harmonic, mainly of triple frequency.

Since the higher harmonic is small compared with the

total exciting current, and the exciting current is only a small part of the total primary current, the higher harmonic can, for most practical cases, be neglected, and the exciting current represented by the equivalent sine wave.

This equivalent sine wave, I_{00}, leads the wave of magnetism, Φ, by an angle, α, the angle of hysteretic advance of phase, and consists of two components, — the hysteretic energy current, in quadrature with the magnetic flux, and therefore in phase with the induced E.M.F. $= I_{00} \sin \alpha$; and the magnetizing current, in phase with the magnetic flux, and therefore in quadrature with the induced E.M.F., and so wattless, $= I_{00} \cos \alpha$.

The exciting current, I_{00}, is determined from the shape and magnetic characteristic of the iron, and number of turns; the hysteretic energy current is —

$$I_{00} \sin \alpha = \frac{\text{Power consumed in the iron}}{\text{Induced E.M.F.}}.$$

130. Graphically, the polar diagram of M.M.Fs. of a transformer is constructed thus:

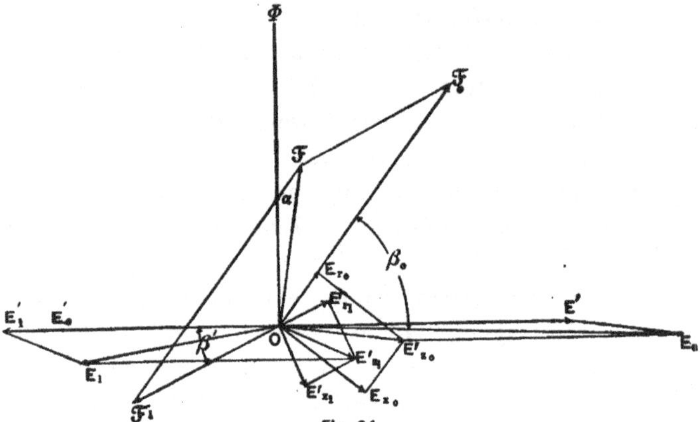

Fig. 94.

Let, in Fig. 94, $\overline{O\Phi}$ = the magnetic flux in intensity and phase (for convenience, as intensities, the effective values are used throughout), assuming its phase as the vertical;

that is, counting the time from the moment where the rising magnetism passes its zero value.

Then the resultant M.M.F. is represented by the vector $\overline{O\mathfrak{F}}$, leading $\overline{O\Phi}$ by the angle $\mathfrak{F}O\Phi = \alpha$.

The induced E.M.Fs. have the phase 180°, that is, are plotted towards the left, and represented by the vectors $\overline{OE_0'}$ and $\overline{OE_1'}$.

If, now, β' = angle of lag in the secondary circuit, due to the total (internal and external) secondary reactance, the secondary current I_1, and hence the secondary M.M.F., $\mathfrak{F}_1 = n_1 I_1$, will lag behind E_1' by an angle β', and have the phase, $180° + \beta'$, represented by the vector $\overline{O\mathfrak{F}_1}$. Constructing a parallelogram of M.M.Fs., with $\overline{O\mathfrak{F}}$ as a diagonal and $\overline{O\mathfrak{F}_1}$ as one side, the other side or $\overline{O\mathfrak{F}_0}$ is the primary M.M.F., in intensity and phase, and hence, dividing by the number of primary turns, n_0, the primary current is $I_0 = \mathfrak{F}_0 / n_0$.

To complete the diagram of E.M.Fs., we have now, —

In the primary circuit:

E.M.F. consumed by resistance is $I_0 r_0$, in phase with I_0, and represented by the vector $\overline{OEr_0}$;

E.M.F. consumed by reactance is $I_0 x_0$, 90° ahead of I_0, and represented by the vector $\overline{OEx_0}$;

E.M.F. consumed by induced E.M.F. is E', equal and opposite to E_0, and represented by the vector $\overline{OE'}$.

Hence, the total primary impressed E.M.F. by combination of $\overline{OEr_0}$, $\overline{OEx_0}$, and $\overline{OE'}$ by means of the parallelogram of E.M.Fs. is,

$$E_0 = \overline{OE_0},$$

and the difference of phase between the primary impressed E.M.F. and the primary current is

$$\beta_0 = E_0 O \mathfrak{F}_0.$$

In the secondary circuit:

Counter E.M.F. of resistance is $I_1 r_1$ in opposition with I_1, and represented by the vector $\overline{OE'r_1}$;

Counter E.M.F. of reactance is $I_1 x_1$, 90° behind I_1, and represented by the vector $\overline{OE_1 x_1'}$.

Induced E.M.Fs., E_1' represented by the vector $\overline{OE_1'}$.

Hence, the secondary terminal voltage, by combination of $\overline{OEr_1'}$, $\overline{OEx_1'}$ and $\overline{OE_1'}$ by means of the parallelogram of E.M.Fs. is

$$E_1 = \overline{OE_1},$$

and the difference of phase between the secondary terminal voltage and the secondary current is

$$\beta_1 = E_1 O \mathfrak{F}_1.$$

As will be seen in the primary circuit the "components of impressed E.M.F. required to overcome the counter E.M.Fs." were used for convenience, and in the secondary circuit the "counter E.M.Fs."

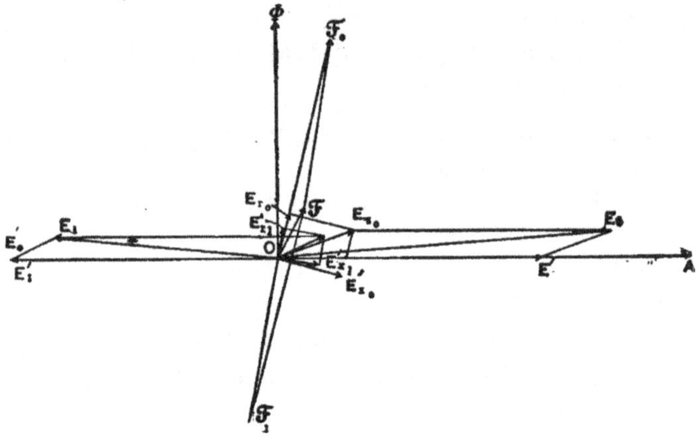

Fig. 95. Transformer Diagram with 90° Lag in Secondary Circuit.

131. In the construction of the transformer diagram, it is usually preferable not to plot the secondary quantities, current and E.M.F., direct, but to reduce them to correspondence with the primary circuit by multiplying by the ratio of turns, $a = n_0 / n_1$, for the reason that frequently primary and secondary E.M.Fs., etc., are of such different magnitude as not to be easily represented on the same scale; or the primary circuit may be reduced to the secondary in the same way. In either case, the vectors representing the two induced E.M.Fs. coincide, or $\overline{OE_1}' = \overline{OE_0}'$.

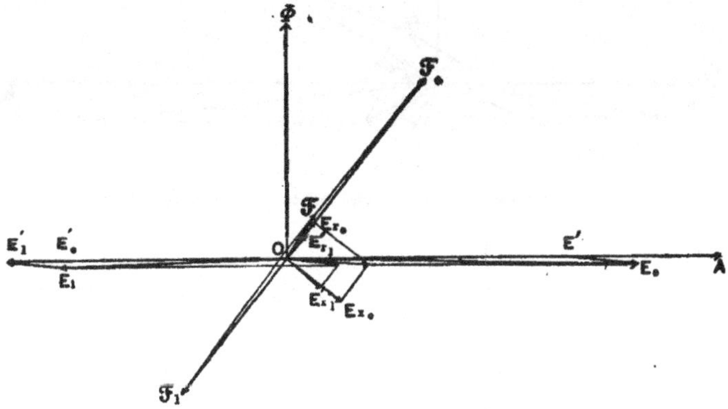

Fig. 96. Transformer Diagram with 50° Lag in Secondary Circuit.

Figs. 96 to 107 give the polar diagram of a transformer having the constants —

$r_0 = .2$ ohms,	$b_0 = .0173$ mhos,
$x_0 = .33$ ohms,	$E_1' = 100$ volts,
$r_1 = .00167$ ohms,	$I_1 = 60$ amperes,
$x_1 = .0025$ ohms,	$a = 10$ degrees.
$g_0 = .0100$ mhos,	

for the conditions of secondary circuit,

$\beta_1' =$ 80° lag	in Fig. 95.	$\beta_1' =$ 20° lead	in Fig.	99.
50° lag	" 96.	50° lead	"	100.
20° lag	" 97.	80° lead	"	101.
O, or in phase;	" 98.			

As shown with a change of β_1' the other quantities E_0, I_1 I_0, etc., change in intensity and direction. The loci described by them are circles, and are shown in Fig. 102, with the point corresponding to non-inductive load marked. The part of the locus corresponding to a lagging secondary

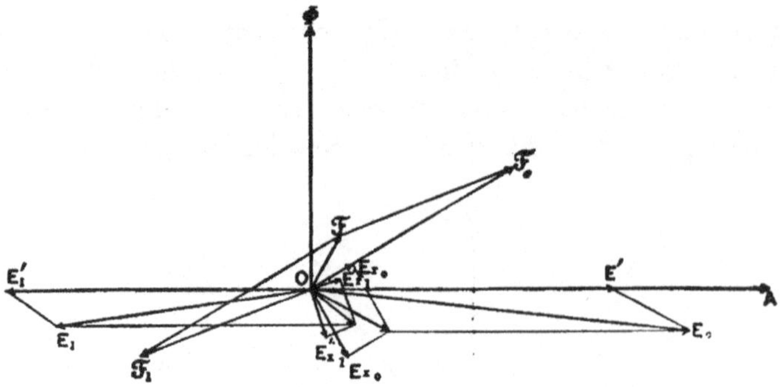

Fig. 97. Transformer Diagram with 20° Lag in Secondary Circuit.

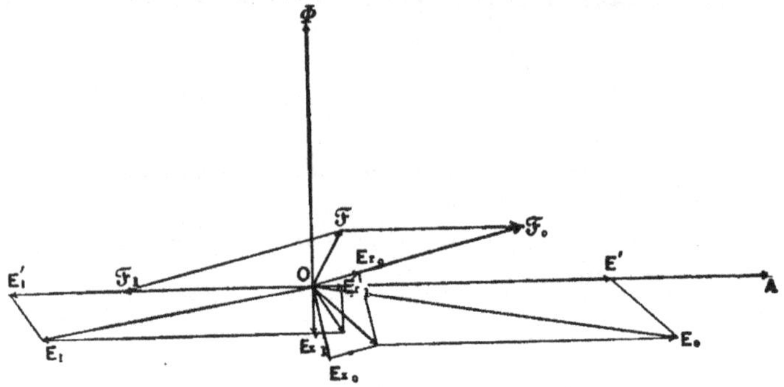

Fig. 98. Transformer Diagram with Secondary Current in Phase with E.M.F.

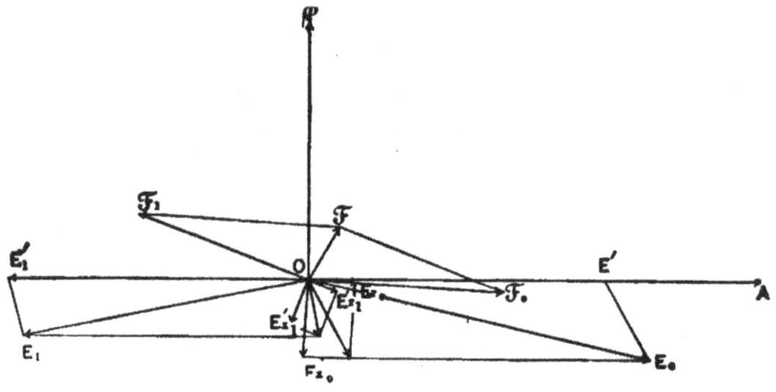

Fig. 99. Transformer Diagram with 20° Lead in Secondary Current.

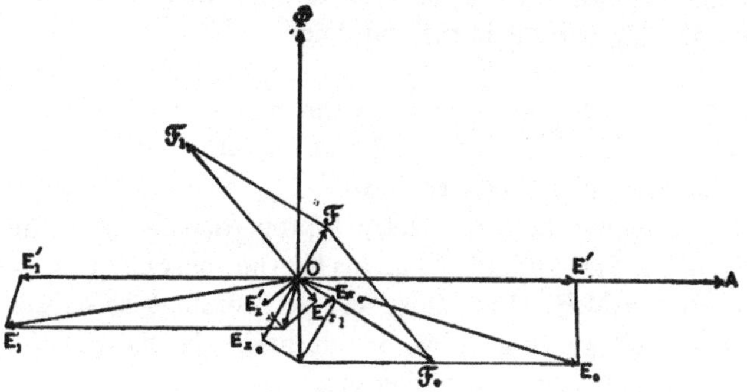

Fig. 100. Transformer Diagram with 50° Lead in Secondary Circuit.

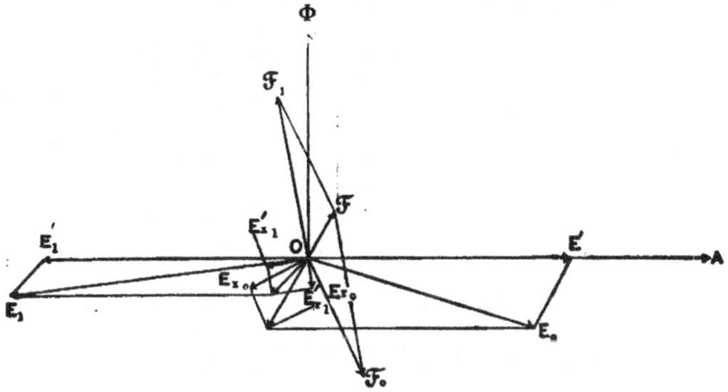

Fig. 101. Transformer Diagram with 80° Lead in Secondary Circuit.

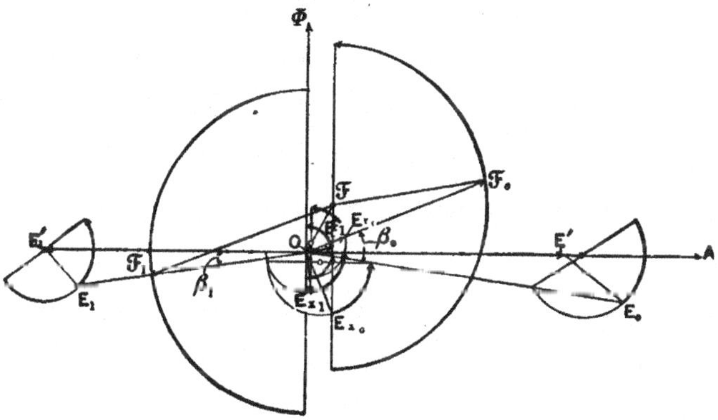

Fig. 102.

current is shown in thick full lines, and the part corresponding to leading current in thin full lines.

132. This diagram represents the condition of constant secondary induced E.M.F., E_1', that is, corresponding to a constant maximum magnetic flux.

By changing all the quantities proportionally from the diagram of Fig. 102, the diagrams for the constant primary impressed E.M.F. (Fig. 103), and for constant secondary terminal voltage (Fig. 104), are derived. In these cases, the locus gives curves of higher order.

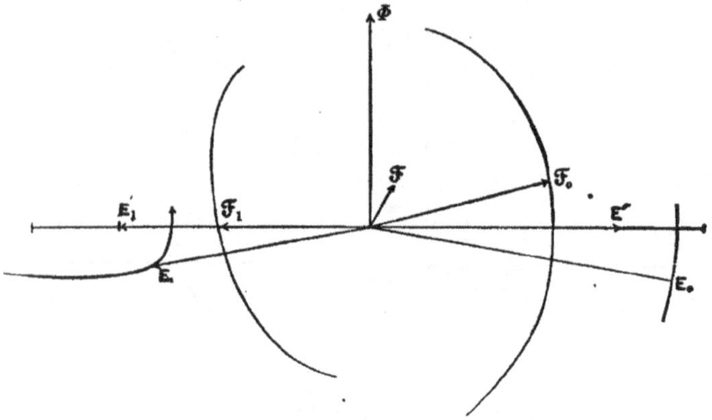

Fig. 103.

Fig. 105 gives the locus of the various quantities when the load is changed from full load, $I_1 = 60$ amperes in a non-inductive secondary external circuit to no load or open circuit.

a.) By increase of secondary resistance; *b.*) by increase of secondary inductive reactance; *c.*) by increase of secondary capacity reactance.

As shown in *a.*), the locus of the secondary terminal voltage, E_1, and thus of E_o, etc., are straight lines; and in *b.*) and *c.*), parts of one and the same circle *a.*) is shown in full lines, *b.*) in heavy full lines, and *c.*) in light full lines. This diagram corresponds to constant maximum magnetic flux; that is, to constant secondary induced E.M.F. The diagrams representing constant primary impressed E.M.F. and constant secondary terminal voltage can be derived from the above by proportionality.

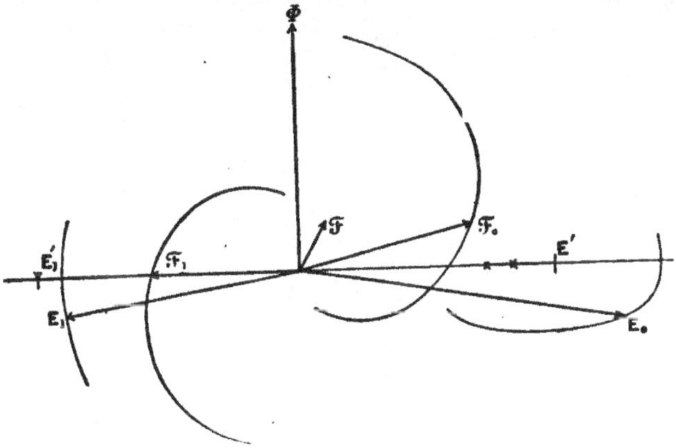

Fig. 10:.

133. It must be understood, however, that for the purpose of making the diagrams plainer, by bringing the different values to somewhat nearer the same magnitude, the constants chosen for these diagrams represent, not the magnitudes found in actual transformers, but refer to greatly exaggerated internal losses.

In practice, about the following magnitudes would be found:

$r_0 = .01$ ohms;	$x_1 = .00025$ ohms;
$x_0 = .033$ ohms;	$g_0 = .001$ ohms;
$r_1 = .00008$ ohms;	$b_0 = .00173$ ohms;

that is, about one-tenth as large as assumed. Thus the changes of the values of E_0, E_1, etc., under the different conditions will be very much smaller.

Symbolic Method.

134. In symbolic representation by complex quantities the transformer problem appears as follows:

The exciting current, I_{00}, of the transformer depends upon the primary E.M.F., which dependance can be represented by an admittance, the "primary admittance," $Y_0 = g_0 + jb_0$, of the transformer.

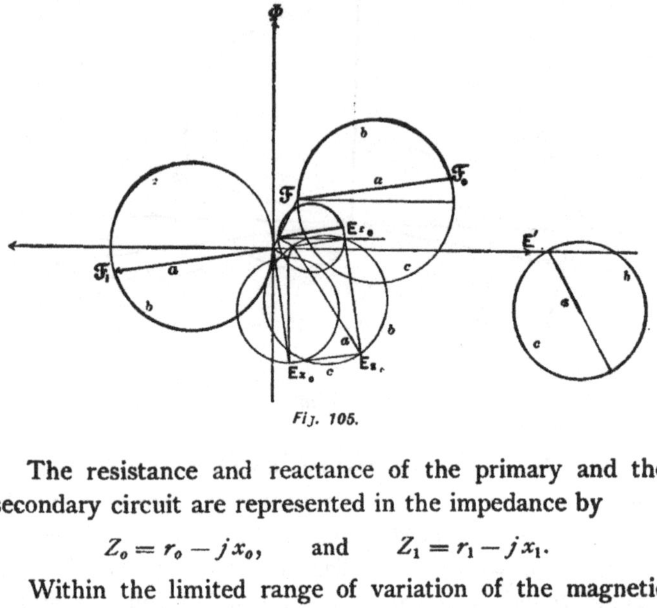

Fig. 105.

The resistance and reactance of the primary and the secondary circuit are represented in the impedance by

$$Z_0 = r_0 - jx_0, \qquad \text{and} \qquad Z_1 = r_1 - jx_1.$$

Within the limited range of variation of the magnetic density in a constant potential transformer, admittance and impedance can usually, and with sufficient exactness, be considered as constant.

Let

n_0 = number of primary turns in series;

n_1 = number of secondary turns in series;

$a = \frac{n_0}{n_1}$ = ratio of turns;

$Y_0 = g_0 + jb_0$ = primary admittance

$= \frac{\text{Exciting current}}{\text{Primary counter E.M.F.}};$

$Z_0 = r_0 - jx_0 =$ primary impedance

$= \frac{\text{E.M.F. consumed in primary coil by resistance and reactance}}{\text{Primary current}};$

$Z_1 = r_1 - jx_1 =$ secondary impedance

$= \frac{\text{E.M.F. consumed in secondary coil by resistance and reactance}}{\text{Secondary current}};$

where the reactances, x_0 and x_1, refer to the true self-inductance only, or to the cross-flux passing between primary and secondary coils; that is, interlinked with one coil only.

Let also

$Y = g + jb =$ total admittance of secondary circuit, including the internal impedance;

$\dot{E}_0 =$ primary impressed E.M.F.;

$\dot{E}' =$ E.M.F. consumed by primary counter E.M.F.;

$\dot{E}_1 =$ secondary terminal voltage;

$\dot{E}_1' =$ secondary induced E.M.F.;

$\dot{I}_0 =$ primary current, total;

$\dot{I}_{00} =$ primary exciting current;

$\dot{I}_1 =$ secondary current.

Since the primary counter E.M.F., $\dot{E}'$, and the secondary induced E.M.F., $\dot{E}_1'$, are proportional by the ratio of turns, a,

$$\dot{E}' = - a\dot{E}_1'. \tag{1}$$

The secondary current is:

$$\dot{I}_1 = Y\dot{E}_1', \tag{2}$$

consisting of an energy component, $g\dot{E}_1'$, and a reactive component, $b\dot{E}_1'$.

To this secondary current corresponds the component of primary current,

$$\dot{I}_0' = \frac{-Y\dot{E}_1'}{a} = \frac{Y\dot{E}'}{a^2}. \tag{3}$$

The primary exciting current is—

$$\dot{I}_{00} = Y_0\dot{E}'. \tag{4}$$

Hence, the total primary current is:

$$\dot{I}_0 = \dot{I}_0' + \dot{I}_{00} \tag{5}$$

$$= \frac{Y\dot{E}'}{a^2} + Y_0\dot{E}',$$

or, $$\dot{I}_0 = \frac{\dot{E}'}{a^2}\{Y + a^2 Y_0\} \tag{6}$$
$$= -\frac{\dot{E}_1'}{a}\{Y + a^2 Y_0\}$$

The E.M.F. consumed in the secondary coil by the internal impedance is $Z_1 \dot{I}_1$.

The E.M.F. induced in the secondary coil by the magnetic flux is $\dot{E}_1'$.

Therefore, the secondary terminal voltage is

$$\dot{E}_1 = \dot{E}_1' - Z_1 \dot{I}_1,$$

or, substituting (2), we have

$$\dot{E}_1 = \dot{E}_1'\{1 - Z_1 Y\} \tag{7}$$

The E.M.F. consumed in the primary coil by the internal impedance is $Z_0 \dot{I}_0$.

The E.M.F. consumed in the primary coil by the counter E.M.F. is $\dot{E}'$.

Therefore, the primary impressed E.M.F. is

$$\dot{E}_0 = \dot{E}' + Z_0 \dot{I}_0,$$

or, substituting (6),

$$\left.\begin{aligned} \dot{E}_0 &= \dot{E}'\left\{1 + Z_0 Y_0 + \frac{Z_0 Y}{a^2}\right\} \\ &= -a\dot{E}_1'\left\{1 + Z_0 Y_0 + \frac{Z_0 Y}{a^2}\right\}. \end{aligned}\right\} \tag{8}$$

135. We thus have,

primary E.M.F., $$\dot{E}_0 = -a\dot{E}_1'\left\{1 + Z_0 Y_0 + \frac{Z_0 Y}{a^2}\right\}, \tag{8}$$

secondary E.M.F., $$\dot{E}_1 = \dot{E}_1'\{1 - Z_1 Y\}, \tag{7}$$

primary current, $$\dot{I}_0 = -\frac{\dot{E}_1'}{a}\{Y + a^2 Y_0\}, \tag{6}$$

secondary current, $$\dot{I}_1 = Y\dot{E}_1', \tag{2}$$

as functions of the secondary induced E.M.F., $\dot{E}_1'$, as parameter.

From the above we derive

Ratio of transformation of E.M.Fs.:

$$\frac{\dot{E}_0}{\dot{E}_1} = -a \frac{1 + Z_0 Y_0 + \frac{Z_0 Y}{a^2}}{1 - Z_1 Y}. \tag{9}$$

Ratio of transformations of currents:

$$\frac{\dot{I}_0}{\dot{I}_1} = -\frac{1}{a}\left\{1 + a^2 \frac{Y_0}{Y}\right\}. \tag{10}$$

From this we get, at constant primary impressed E.M.F.,

$$\dot{E}_0 = \text{constant};$$

secondary induced E.M.F.,

$$\dot{E}_1{}' = -\frac{\dot{E}_0}{a} \frac{1}{1 + Z_0 Y_0 + \frac{Z_0 Y}{a^2}};$$

E.M.F. induced per turn,

$$\delta \dot{E} = -\frac{\dot{E}_0}{n_0} \frac{1}{1 + Z_0 Y_0 + \frac{Z_0 Y}{a^2}};$$

secondary terminal voltage,

$$\dot{E}_1 = -\frac{\dot{E}_0}{a} \frac{1 - Z_1 Y}{1 + Z_0 Y_0 + \frac{Z_0 Y}{a^2}};$$

primary current,

$$\dot{I}_0 = \frac{\dot{E}_0}{a^2} \frac{Y + a^2 Y_0}{1 + Z_0 Y_0 + \frac{Z_0 Y}{a^2}} = \dot{E}_0 \frac{\frac{Y}{a^2} + Y_0}{1 + Z_0 Y_0 + \frac{Z_0 Y}{a^2}};$$

secondary current,

$$\dot{I}_1 = -\frac{\dot{E}_0}{a} \frac{Y}{1 + Z_0 Y_0 + \frac{Z_0 Y}{a^2}}. \tag{11}$$

At constant secondary terminal voltage,

$$\dot{E}_1 = \text{const.};$$

secondary induced E.M.F.,

$$\dot{E}_1' = \frac{\dot{E}_1}{1 - Z_1 Y};$$

E.M.F. induced per turn,

$$\delta \dot{E} = \frac{\dot{E}_1}{n_1}\frac{1}{1 - Z_1 Y};$$

primary impressed E.M.F.,

$$\dot{E}_0 = -a\dot{E}_1 \frac{1 + Z_0 Y_0 + \frac{Z_0 Y}{a^2}}{1 - Z_1 Y};$$

primary current,

$$\dot{I}_0 = -\frac{\dot{E}_1}{a}\frac{Y + a^2 Y_0}{1 - Z_1 Y};$$

secondary current,

$$\dot{I}_1 = \dot{E}_1 \frac{Y}{1 - Z_1 Y}. \qquad (12)$$

136. Some interesting conclusions can be drawn from these equations.

The apparent impedance of the total transformer is

$$Z_t = \frac{\dot{E}_0}{\dot{I}_0} = a^2 \frac{1 + Z_0 Y_0 + \frac{Z_0 Y}{a^2}}{Y + a^2 Y_0} \qquad (13)$$

$$= \frac{1 + Z_0\left(Y_0 + \frac{Y}{a^2}\right)}{Y_0 + \frac{Y}{a^2}};$$

$$Z_t = \frac{1}{Y_0 + \frac{Y}{a^2}} + Z_0. \qquad (14)$$

Substituting now, $\frac{Y}{a^2} = Y'$, the total secondary admittance, reduced to the primary circuit by the ratio of turns, it is

$$Z_t = \frac{1}{Y_0 + Y'} + Z_0. \qquad (15)$$

$Y_0 + Y'$ is the total admittance of a divided circuit with the exciting current, of admittance Y_0, and the secondary current, of admittance Y' (reduced to primary), as branches. Thus :

$$\frac{1}{Y_0 + Y'} = Z_0' \tag{16}$$

is the impedance of this divided circuit, and

$$Z_t = Z_0' + Z_0. \tag{17}$$

That is :

The alternate-current transformer, of primary admittance Y_0, total secondary admittance Y, and primary impedance Z_0, is equivalent to, and can be replaced by, a divided circuit with the branches of admittance Y_0, the exciting current, and admittance $Y' = Y/a^2$, the secondary current, fed over mains of the impedance Z_0, the internal primary impedance.

This is shown diagrammatically in Fig. 106.

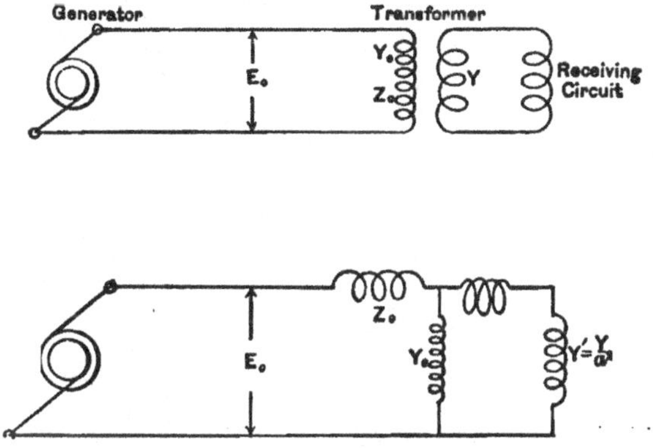

Fig. 106.

137. Separating now the internal secondary impedance from the external secondary impedance, or the impedance of the consumer circuit, it is

$$\frac{1}{Y} = Z_1 + Z; \tag{18}$$

where Z = external secondary impedance,

$$Z = \frac{\dot{E}_1}{I_1}. \tag{19}$$

Reduced to primary circuit, it is

$$\frac{1}{Y'} = \frac{a^2}{Y} = a^2 Z_1 + a^2 Z$$
$$= Z_1' + Z'. \tag{20}$$

That is:

An alternate-current transformer, of primary admittance Y_0, primary impedance Z_0, secondary impedance Z_1, and ratio of turns a, can, when the secondary circuit is closed by an impedance Z (the impedance of the receiver circuit), be replaced, and is equivalent to a circuit of impedance $Z' = a^2 Z$, fed over mains of the impedance $Z_0 + Z_1'$, where $Z_1' = a^2 Z_1$, shunted by a circuit of admittance Y_0, which latter circuit branches off at the points $a - b$, between the impedances Z_0 and Z_1'.

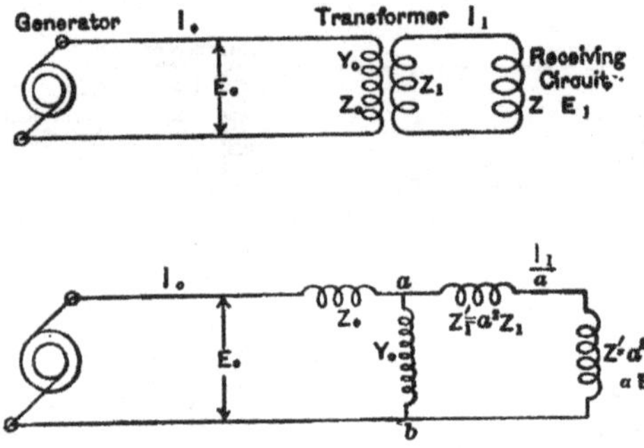

Fig. 107.

This is represented diagrammatically in Fig. 107.

It is obvious therefore, that if the transformer contains several independent secondary circuits they are to be considered as branched off at the points *a*, *b*, in diagram Fig. 107, as shown in diagram Fig. 108.

It therefore follows:

An alternate-current transformer, of x secondary coils, of the internal impedances Z_1^I, Z_1^{II}, . . . Z_1^x, closed by external secondary circuits of the impedances Z^I, Z^{II}, . . . Z^x, is equivalent to a divided circuit of $x + 1$ branches, one branch of

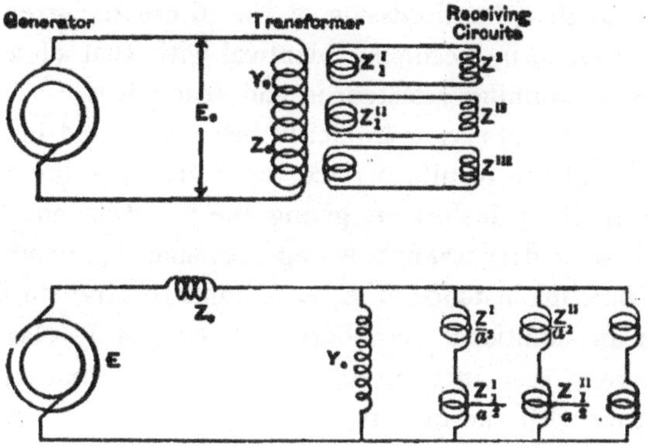

Fig. 108.

admittance Y_0, *the exciting current, the other branches of the impedances* $Z_1^I + Z^I$, $Z_1^{II} + Z^{II}$, ... $Z_1^x + Z^x$, *the latter impedances being reduced to the primary circuit by the ratio of turns, and the whole divided circuit being fed by the primary impressed E.M.F.* E_0, *over mains of the impedance* Z_0.

Consequently, transformation of a circuit merely changes all the quantities proportionally, introduces in the mains the impedance $Z_0 + Z_1'$, and a branch circuit between Z_0 and Z_1', of admittance Y_0.

Thus, double transformation will be represented by diagram, Fig. 109.

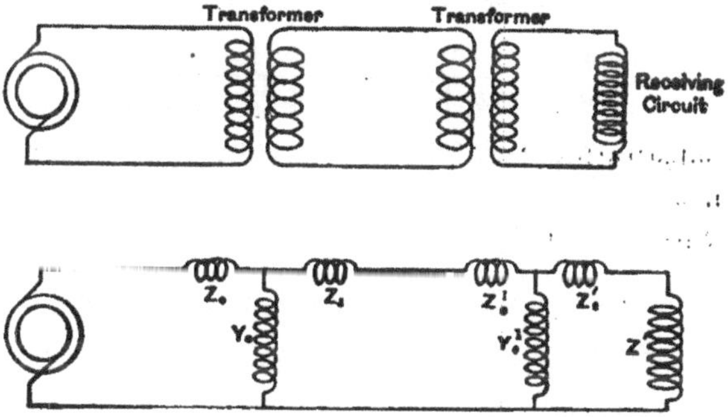

Fig. 109.

With this the discussion of the alternate-current transformer ends, by becoming identical with that of a divided circuit containing resistances and reactances.

Such circuits have explicitly been discussed in Chapter VIII., and the results derived there are now directly applicable to the transformer, giving the variation and the control of secondary terminal voltage, resonance phenomena, etc.

Thus, for instance, if $Z_1' = Z_0$, and the transformer contains an additional secondary coil, constantly closed by a condenser reactance of such size that this auxiliary circuit, together with the exciting circuit, gives the reactance $-x_0$, with a non-inductive secondary circuit $Z_1 = r_1$, we get the condition of transformation from constant primary potential to constant secondary current, and inversely, as previously discussed.

Non-inductive Secondary Circuit.

138. In a non-inductive secondary circuit, the external secondary impedance is,

$$Z = R_1,$$

or, reduced to primary circuit,

$$a^2 Z = a^2 R_1 = R$$

Assuming the secondary impedance, reduced to primary circuit, as equal to the primary impedance,

$$a^2 Z_1 = Z_0 = r_0 - j x_0,$$

it is,

$$\frac{Y}{a^2} = \frac{1}{a^2(Z + Z_1)} = \frac{1}{R + r_0 - j x_0}.$$

Substituting these values in Equations (9), (10), and (13), we have

Ratio of E.M.Fs. :

$$\frac{\dot{E}_0}{E_1} = -a \frac{1 + \dfrac{r_0 - j x_0}{R + r_0 - j x_0} + (r_0 - j x_0)(g_0 + j b_0)}{1 - \dfrac{(r_0 - j x_0)}{R + r_0 - j x_0}}$$

$$= -a\left\{1 + \frac{r_0 - jx_0}{R + r_0 - jx_0} + (r_0 - jx_0)(g_0 + jb_0)\right\}$$

$$\left\{1 + \frac{r_0 - jx_0}{R + r_0 - jx_0} + \left(\frac{r_0 - jx_0}{R + r_0 - jx_0}\right)^2 + \ldots\right\};$$

or, expanding, and neglecting terms of higher than third order,

$$\frac{\dot{E}_0}{\dot{E}_1} = -a\left\{1 + 2\frac{r_0 - jx_0}{R + r_0 - jx_0} + 2\left(\frac{r_0 - jx_0}{R + r_0 - jx_0}\right)^2 + (r_0 - jx_0)(g_0 + jb_0)\right\};$$

or, expanded,

$$\frac{\dot{E}_0}{\dot{E}_1} = -a\left\{1 + 2\frac{r_0 - jx_0}{R} + (r_0 - jx_0)(g_0 + jb_0)\right\}$$

Neglecting terms of tertiary order also,

$$\frac{\dot{E}_0}{\dot{E}_1} = -a\left\{1 + 2\frac{r_0 - jx_0}{R}\right\}.$$

Ratio of currents:

$$\frac{\dot{I}_0}{\dot{I}_1} = -\frac{1}{a}\{1 + (g_0 + jb_0)(R + r_0 - jx_0)\};$$

or, expanded,

$$\frac{\dot{I}_0}{\dot{I}_1} = -\frac{1}{a}\{1 + R(g_0 + jb_0) + (r_0 - jx_0)(g_0 + jb_0)\}.$$

Neglecting terms of tertiary order also,

$$\frac{\dot{I}_0}{\dot{I}_1} = -\frac{1}{a}\{1 + R(g_0 + jb_0)\}.$$

Total apparent primary admittance:

$$Z_t = \frac{\dot{E}_0}{\dot{I}_0} = \frac{1 + \dfrac{r_0 - jx_0}{R + r_0 - jx_0} + (r_0 - jx_0)(g_0 + jb_0)}{\dfrac{1}{R + r_0 - jx_0} + g_0 + jb_0}$$

$$= \{R + (r_0 - jx_0) + R(r_0 - jx_0)(g_0 + jb_0)\}\{1 - (g_0 + jb_0)(R + r_0 - jb_0) + (g_0 + jb_0)^2(R + r_0 - jb_0)^2 + \ldots\}$$

$$= \{R + 2(r_0 - jx_0) - R^2(g_0 + jb_0) - 2R(r_0 - jx_0)(g_0 + jb_0) + R^3(g_0 + jb_0)^2\};$$

or,

$$Z_t = R\left\{1 + 2\frac{r_o - jx_o}{R} - R(g_o + jb_o) - 2(r_o - jx_o)(g_o + jb_o) + R^2(g_o + jb_o)^2\right\}.$$

Neglecting terms of tertiary order also:

$$Z_t = R\left\{1 + 2\frac{r_o - jx_o}{R} - R(g_o + jb_o)\right\}.$$

Angle of lag in primary circuit:

$\tan \hat{\omega}_o = \frac{x_t}{r_t}$, hence,

$$\tan \hat{\omega}_o = \frac{2\frac{x_o}{R} + Rb_o + 2r_o b_o - 2x_o g_o - 2R^2 g_o b_o}{1 + \frac{2r_o}{R} - Rg_o - 2r_o g_o - 2x_o b_o + R^2 g_o^2 + R_1^2 b_o^2}.$$

Neglecting terms of tertiary order also:

$$\tan \hat{\omega}_o = \frac{2\frac{x_o}{R} + Rb_o}{1 + 2\frac{r_o}{R} - Rg_o}.$$

139. If, now, we represent the external resistance of the secondary circuit at full load (reduced to the primary circuit) by R_o, and denote,

$\frac{2r_o}{R_o} = \text{p} = \text{ratio}\ \frac{\text{Internal resistance of transformer}}{\text{External resistance of secondary circuit}} =$ percentage internal resistance,

$\frac{2x_o}{R_o} = \text{q} = \text{ratio}\ \frac{\text{Internal reactance of transformer}}{\text{External resistance of secondary circuit}} =$ percentage internal reactance,

$R_o g_o = \text{h} = \text{ratio}\ \frac{\text{Hysteretic energy current}}{\text{Total secondary current}} =$ percentage hysteresis,

$R_o b_o = \text{g} = \text{ratio}\ \frac{\text{Magnetizing current}}{\text{Total secondary current}} =$ percentage magnetizing current,

and if d represents the load of the transformer, as fraction of full load, we have

$$R = \frac{R_o}{d},$$

and,
$$\frac{2\,r_0}{R} = \text{p}\,d,$$
$$\frac{2\,x_0}{R} = \text{q}\,d,$$
$$R\,g_0 = \frac{\text{h}}{d},$$
$$R\,b_0 = \frac{\text{g}}{d}.$$

Substituting these values we get, as the equations of the transformer on non-inductive load,

Ratio of E.M.Fs. :
$$\frac{\dot{E}_0}{\dot{E}_1} = -\,a\left\{1 + d\,(\text{p} - j\text{q}) + \frac{(\text{p} - j\text{q})\,(\text{h} + j\text{g})}{2}\right\}$$
$$\doteq -\,a\{1 + d\,(\text{p} - j\text{q})\}$$

or, eliminating imaginary quantities,
$$\frac{e_0}{e_1} = a\sqrt{\left(1 + d\text{p} + \frac{\text{ph} + \text{qg}}{2}\right)^2 + \left(d\text{q} - \frac{\text{pg} - \text{qh}}{2}\right)^2}$$
$$\doteq a\,\sqrt{(1 + d\text{p})^2 + d^2\text{q}^2}$$
$$\doteq a\left\{1 + d\text{p} + \frac{\text{ph} + \text{qg} + d^2\text{q}^2}{2}\right\}$$
$$\doteq a\,\{1 + d\text{p}\}.$$

Ratio of currents :
$$\frac{\dot{I}_0}{\dot{I}_1} = -\,\frac{1}{a}\left\{1 + \frac{(\text{h} + j\text{g})}{d} + \frac{(\text{p} - j\text{q})(\text{h} + j\text{g})}{2}\right\}$$
$$\doteq -\,\frac{1}{a}\left\{1 + \left(\frac{\text{h} + j\text{g}}{d}\right)\right\};$$

or, eliminating imaginary quantities,
$$\frac{i_0}{i_1} = \frac{1}{a}\sqrt{\left(1 + \frac{\text{h}}{d} + \frac{\text{ph} + \text{qg}}{2}\right)^2 + \left(\frac{\text{g}}{d} + \frac{\text{pg} - \text{qh}}{2}\right)^2}$$
$$\doteq \frac{1}{a}\sqrt{\left(1 + \frac{\text{h}}{d}\right)^2 + \left(\frac{\text{g}}{d}\right)^2}$$
$$\doteq \frac{1}{a}\left\{1 + \frac{\text{h}}{d} + \frac{\text{ph} + \text{qg} + \text{g}^2}{2\,d^2}\right\}$$
$$\doteq \frac{1}{a}\left\{1 + \frac{\text{h}}{d}\right\}.$$

Total apparent primary impedance:

$$Z_t = \frac{R_0}{d}\left\{1 + d(p - jg) - \frac{h + jg}{d} - (p - jg)(h + jg) + \left(\frac{h + jg}{d^2}\right)^2\right\}$$

$$\cong \frac{R_0}{d}\left\{1 + d(p - jg) - \frac{h + jg}{d}\right\}$$

or, eliminating imaginary quantities,

$$z_t = \frac{R_0}{d}\sqrt{\left(1 + dp - \frac{h}{d} - ph - qg + \frac{h^2 - g^2}{d^2}\right)^2 + \left(dq - \frac{g}{d} - pg + gh + 2\frac{hg}{d^2}\right)^2}$$

$$\cong \frac{R_0}{d}\sqrt{\left(1 + dp - \frac{h}{d}\right)^2 + \left(dq - \frac{g}{d}\right)^2}$$

$$\cong \frac{R_0}{d}\left\{1 + dp - \frac{h}{d} - ph - 2qg + \frac{h^2 - g^2}{d^2} + \frac{d^2 q^2}{2} + \frac{g^2}{2d^2} + \frac{d^2 p^2}{2} + \frac{h^2}{2d^2}\right\}$$

$$\cong \frac{R_0}{d}\left\{1 + dp - \frac{h}{d}\right\}.$$

Angle of lag in primary circuit:

$$\tan \hat{\omega}_0 = \frac{dq + \frac{g}{d} + pg - gh - 2\frac{hg}{d^2}}{1 + dp - \frac{h}{d} - ph - qg + \frac{hg}{d}\frac{h^2 + g^2}{d^2}}$$

$$\cong \frac{dg + \frac{g}{d}}{1 + dp - \frac{h}{d}}.$$

That is,

An alternate-current transformer, feeding into a non-inductive secondary circuit, is represented by the constants:

R_0 = secondary external resistance at full load;
p = percentage resistance;
q = percentage reactance;
h = percentage hysteresis;
g = percentage magnetizing current;
d = secondary percentage load.

All these qualities being considered as reduced to the primary circuit by the square of the ratio of turns, a^2.

140. As an instance, a transformer of the following constants may be given:

$$e_0 = 1{,}000; \quad R_0 = 120; \quad q = .06;$$
$$a = 10; \quad p = .02. \quad h = .02;$$
$$g = .04.$$

Substituting these values, gives:

$$e_1 = \frac{100}{\sqrt{(1.0014 + .02\,d)^2 + (.0002 + .06\,d)^2}};$$

$$i_1 = \frac{e_1 d a^2}{R_0} = \frac{e_1 d}{1.2};$$

$$i_0 = .1\, i_1 \sqrt{\left(1.0014 + \frac{.02}{d}\right)^2 + \left(\frac{.04}{d} - .0002\right)^2};$$

$$\tan \hat{\omega}_0 = \frac{.06\,d + \frac{.04}{d} - .0004 - \frac{.0016}{d^2}}{1.9972 + .02\,d + \frac{.002}{d^2} - \frac{.02}{d}}.$$

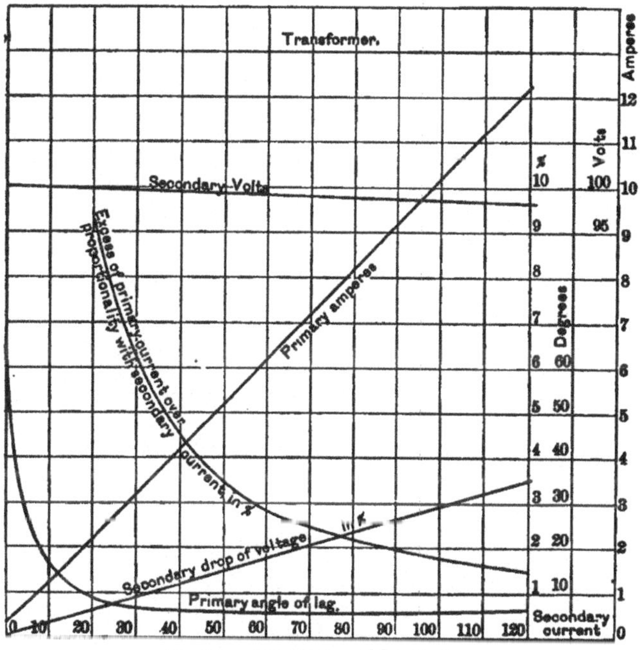

Fig. 110. Load Diagram of Transformer.

In diagram Fig. 110 are shown, for the values from $d = 0$ to $d = 1.5$, with the secondary current i_1 as abscissæ, the values:

secondary terminal voltage, in volts,
secondary drop of voltage, in per cent,
primary current, in amps,
excess of primary current over proportionality with secondary, in per cent,
primary angle of lag.

The power-factor of the transformer, cos $\hat{\omega}_0$, is .45 at open secondary circuit, and is above .99 from 25 amperes, upwards, with a maximum of .995 at full load.

CHAPTER XV.

THE GENERAL ALTERNATING-CURRENT TRANSFORMER OR FREQUENCY CONVERTER.

141. The simplest alternating-current apparatus is the alternating-current transformer. It consists of a magnetic circuit, interlinked with two electric circuits or sets of electric circuits. The one, the primary circuit, is excited by an impressed E.M.F., while in the other, the secondary circuit, an E.M.F. is induced. Thus, in the primary circuit, power is consumed, in the secondary circuit a corresponding amount of power produced; or in other words, power is transferred through space, from primary to secondary circuit. This transfer of power finds its mechanical equivalent in a repulsive thrust acting between primary and secondary. Thus, if the secondary coil is not held rigidly as in the stationary transformer, it will be repelled and move away from the primary. This mechanical effect is made use of in the induction motor, which represents a transformer whose secondary is mounted movably with regard to the primary in such a way that, while set in rotation, it still remains in the primary field of force. The condition that the secondary circuit, while revolving with regard to the primary, does not leave the primary field of magnetic force, requires that this field is not undirectional, but that an active field exists in every direction. One way of producing such a magnetic field is by exciting different primary circuits angularly displaced in space with each other by currents of different phase. Another way is to excite the primary field in one direction only, and get the cross magnetization, or the angularly displaced magnetic field, by the reaction of the secondary current.

We see, consequently, that the stationary transformer and the induction motor are merely different applications of the same apparatus, comprising a magnetic circuit interlinked with two electric circuits. Such an apparatus can properly be called a "*general alternating-current transformer.*" The equations of the stationary transformer and those of the induction motor are merely specializations of the general alternating-current transformer equations.

Quantitatively the main differences between induction motor and stationary transformer are those produced by the air-gap between primary and secondary, which is required to give the secondary mechanical movability. This air-gap greatly increases the magnetizing current over that in the closed magnetic circuit transformer, and requires an ironclad construction of primary and secondary to keep the magnetizing current within reasonable limits. An ironclad construction again greatly increases the self-induction of primary and secondary circuit. Thus the induction motor is a transformer of large magnetizing current and large self-induction; that is, comparatively large primary exciting susceptance and large reactance.

The general alternating-current transformer transforms between electrical and mechanical power, and changes not only E.M.Fs. and currents, but frequencies also, and may therefore be called a "frequency converter." Obviously, it also may change the number of phases.

142. Besides the magnetic flux interlinked with both primary and secondary electric circuit, a magnetic cross-flux passes in the transformer between primary and secondary, surrounding one coil only, without being interlinked with the other. This magnetic cross-flux is proportional to the current flowing in the electric circuit, and constitutes what is called the self-induction of the transformer. As seen, as self-induction of a transformer circuit, not the total flux produced by and interlinked with this circuit is understood, but only that — usually small — part of the flux which surrounds the one circuit without interlinking with the other, and is thus produced by the M.M.F. of one circuit only.

143. The mutual magnetic flux of the transformer is produced by the resultant M.M.F. of both electric circuits. It is determined by the counter E.M.F., the number of turns, and the frequency of the electric circuit, by the equation:

$$E = \sqrt{2}\,\pi N n \Phi 10^{-8}.$$

Where E = effective E.M.F.
N = frequency.
n = number of turns.
Φ = maximum magnetic flux.

The M.M.F. producing this flux, or the resultant M.M.F. of primary and secondary circuit, is determined by shape and magnetic characteristic of the material composing the magnetic circuit, and by the magnetic induction. At open secondary circuit, this M.M.F. is the M.M.F. of the primary current, which in this case is called the exciting current, and consists of an energy component, the magnetic energy current, and a reactive component, the magnetizing current.

144. In the general alternating-current transformer, where the secondary is movable with regard to the primary, the rate of cutting of the secondary electric circuit with the mutual magnetic flux is different from that of the primary. Thus, the frequencies of both circuits are different, and the induced E.M.Fs. are not proportional to the number of turns as in the stationary transformer, but to the product of number of turns into frequency.

145. Let, in a general alternating-current transformer:

s = ratio $\frac{\text{secondary}}{\text{primary}}$ frequency, or "slip";

thus, if

N = primary frequency, or frequency of impressed E.M.F.,

sN = secondary frequency;

and the E.M.F. induced per secondary turn by the mutual flux has to the E.M.F. induced per primary turn the ratio s,

$s = 0$ represents synchronous motion of the secondary;
$s < 0$ represents motion above synchronism—driven by external mechanical power, as will be seen;
$s = 1$ represents standstill;
$s > 1$ represents backward motion of the secondary

that is, motion against the mechanical force acting between primary and secondary (thus representing driving by external mechanical power).

Let

n_0 = number of primary turns in series per circuit;
n_1 = number of secondary turns in series per circuit;
$a = \frac{n_0}{n_1}$ = ratio of turns;
$Y_0 = g_0 + jb_0$ = primary exciting admittance per circuit;

where

g_0 = effective conductance;
b_0 = susceptance;
$Z_0 = r_0 - jx_0$ = internal primary self-inductive impedance per circuit,

where

r_0 = effective resistance of primary circuit;
x_0 = reactance of primary circuit;
$Z_{11} = r_1 - jx_1$ = internal secondary self-inductive impedance per circuit at standstill, or for $s = 1$,

where

r_1 = effective resistance of secondary coil;
x_1 = reactance of secondary coil at standstill, or full frequency, $s = 1$.

Since the reactance is proportional to the frequency, at the slip s, or the secondary frequency sN, the secondary impedance is:

$$Z_1 = r_1 - jsx_1.$$

Let the secondary circuit be closed by an external resistance r, and an external reactance, and denote the latter by x at frequency N, then at frequency sN, or slip s, it will be $= sx$, and thus:

$$Z = r - jsx = \text{external secondary impedance.*}$$

Let

$\dot{E}_0$ = primary impressed E.M.F. per circuit,
$\dot{E}'$ = E.M.F. consumed by primary counter E.M.F.,
$\dot{E}_1$ = secondary terminal E.M.F.,
$\dot{E}_1'$ = secondary induced E.M.F.,
$\dot{e}$ = E.M.F. induced per turn by the mutual magnetic flux, at full frequency N,
$\dot{I}_0$ = primary current,
$\dot{I}_{00}$ = primary exciting current,
$\dot{I}_1$ = secondary current.

It is then:

Secondary induced E.M.F.

$$\dot{E}_1' = sn_1e.$$

Total secondary impedance

$$Z_1 + Z = (r_1 + r) - js(x_1 + x);$$

hence, secondary current

$$\dot{I}_1 = \frac{\dot{E}_1'}{Z_1 + Z} = \frac{sn_1e}{(r_1 + r) - js(x_1 + x)};$$

Secondary terminal voltage

$$\dot{E}_1 = \dot{E}_1' - \dot{I}_1 Z_1 = \dot{I}_1 Z$$

$$= sn_1e\left\{1 - \frac{r_1 - jsx_1}{(r_1 + r) - js(x_1 + x)}\right\} = \frac{sn_1e(r - jsx)}{(r_1 + r) - js(x_1 + x)}.$$

---

* This applies to the case where the secondary contains inductive reactance only; or, rather, that kind of reactance which is proportional to the frequency. In a condenser the reactance is inversely proportional to the frequency, in a synchronous motor under circumstances independent of the frequency. Thus, in general, we have to set, $x = x' + x'' + x'''$, where x' is that part of the reactance which is proportional to the frequency, x'' that part of the reactance independent of the frequency, and x''' that part of the reactance which is inversely proportional to the frequency; and have thus, at slip s, or frequency sN, the external secondary reactance $sx' + x'' + \frac{x'''}{s}$.

E.M.F. consumed by primary counter E.M.F.

$$E' = -n_0 e;$$

hence, primary exciting current:

$$I_{00} = E' Y_0 = -n_0 e (g_0 + j b_0).$$

Component of primary current corresponding to secondary current I_1:

$$I_1' = -\frac{I_1}{a}$$

$$= -\frac{n_0 s e}{a^2\{(r_1 + r) - js(x_1 + x)\}},$$

hence, total primary current,

$$I_0 = I_{00} + I_1'$$

$$= -s n_0 e \left\{ \frac{1}{a_2} \frac{1}{(r_1 + r) - js(x_1 + x)} + \frac{g_0 + j b_0}{s} \right\}.$$

Primary impressed E.M.F.,

$$E_0 = E' + I_0 Z_0$$

$$= -n_0 e \left\{ 1 + \frac{s}{a^2} \frac{r_0 - j x_0}{(r_1 + r) - js(x_1 + x)} + (r_0 - j x_0)(g_0 + j b_0) \right\}$$

We get thus, as the

Equations of the General Alternating-Current Transformer:

Of ratio of turns, a; and ratio of frequencies, s; with the E.M.F. induced per turn at full frequency, e, as parameter, the values:

Primary impressed E.M.F.,

$$E_0 = -n_0 e \left\{ 1 + \frac{s}{a^2} \frac{r_0 - j x_0}{(r_1 + r) - js(x_1 + x)} + (r_0 - j x_0)(g_0 + j b_0) \right\}$$

Secondary terminal voltage,

$$E_1 = s n_1 e \left\{ 1 - \frac{r_1 - j s x_1}{(r_1 + r) - js(x_1 + x)} \right\} = s n^1 e \frac{r - j s x}{(r_1 + r) - js(x_1 + x)}.$$

Primary current,

$$I_0 = -s n_0 e \left\{ \frac{1}{a^2} \frac{1}{(r_1 + r) - js(x_1 + x)} + \frac{g_0 + j b_0}{s} \right\}.$$

Secondary current,

$$I_1 = \frac{s\, n_1 e}{(r_1 + r) - js\,(x_1 + x)}.$$

Therefrom, we get:

Ratio of currents,

$$\frac{I_0}{I_1} = -\frac{1}{a}\left\{1 + \frac{a^2}{s}(g_0 + j\,b_0)\,[(r_1 + r) - js(x_1 + x)]\right\}.$$

Ratio of E.M.Fs.,

$$\frac{E_0}{E_1} = -\frac{a}{s}\left\{\frac{1 + \dfrac{s}{a^2}\dfrac{r_0 - j\,x_0}{(r_1 + r) - js\,(x_1 + x)} + (r_0 - j\,x_0)(g_0 + j\,b_0)}{1 - \dfrac{r_1 - js\,x_1}{(r_1 + r) - js\,(x_1 + x)}}\right\}$$

Total apparent primary impedance,

$$Z_t = \frac{E_0}{I_0} = \frac{a^2}{s}\{(r_1 + r) - js(x_1 + x)\}$$

$$\left\{\frac{1 + \dfrac{s}{a^2}\dfrac{r_0 - j\,x_0}{(r_1 + r) + js\,(x_1 + x)} + (r_0 - j\,x_0)(g_0 + j\,b_0)}{1 + \dfrac{a^2}{s}(g_0 + j\,b_0)\,[(r_1 + r) - js\,(x_1 + x)]}\right\}$$

where $$x = x' + \frac{x''}{s} + \frac{x'''}{s^2}$$

in the general secondary circuit as discussed in foot-note, page 221.

Substituting in these equations:

$$s = 1,$$

gives the

General Equations of the Stationary Alternating-Current Transformer:

$$E_0 = -n_0 e\left\{1 + \frac{1}{a^2}\frac{Z_0}{Z_1 + Z} + Z_0 Y_0\right\}.$$

$$E_1 = n_1 e\left\{1 - \frac{Z_1}{Z_1 + Z}\right\} = n_1 e\,\frac{Z}{Z_1 + Z}.$$

$$I_0 = -n_0 e\left\{\frac{1}{a^2(Z_1 + Z)} + Y_0\right\}.$$

$$\dot{I}_1 = \frac{n_1 e}{Z_1 + Z}$$

$$\frac{\dot{I}_0}{\dot{I}_1} = -\frac{1}{a}\{1 + a^2 Y_0 (Z_1 + Z)\}.$$

$$\frac{\dot{E}_0}{\dot{E}_1} = -a\left\{\frac{1 + \dfrac{Z_0}{a^2(Z_1 + Z)} + Z_0 Y_0}{1 - \dfrac{Z_1}{Z_1 + Z}}\right\}$$

$$Z_t = \frac{\dot{E}_0}{\dot{I}_0} = a^2 (Z_1 + Z)\left\{\frac{1 + \dfrac{Z_0}{a^2(Z_1 + Z)} + Z_0 Y_0}{1 + a^2 Y_0 (Z_1 + Z)}\right\}$$

Substituting in the equations of the general alternating-current transformer,

$$Z = 0,$$

gives the

General Equations of the Induction Motor:

$$\dot{E}_0 = -n_0 e\left\{1 + \frac{s}{a^2}\frac{r_0 - j x_0}{r_1 - j s x_1} + (r_0 - j x_0)(g_0 + j b_0)\right\}$$

$$\dot{E}_1 = 0.$$

$$\dot{I}_0 = -s n_0 e\left\{\frac{1}{a^2(r_1 - j s x_1)} + \frac{g_0 + j b_0}{s}\right\}.$$

$$\dot{I}_1 = \frac{s n_1 e}{r_1 - j s x_1}.$$

$$\frac{\dot{I}_0}{\dot{I}_1} = -\frac{1}{a}\left\{1 + \frac{a^2}{s}(g_0 + j b_0)(r_1 - j s x_1)\right\}.$$

$$Z_t = \frac{a^2}{s}(r_1 - j s x_1)\left\{\frac{1 + \dfrac{s}{a^2}\dfrac{r_0 - j x_0}{r_1 - j s x_1} + (r_0 - j x_0)(g_0 + j b_0)}{1 + \dfrac{a^2}{s}(r_1 - j s x_1)(g_0 + j b_0)}\right\}.$$

Returning now to the general alternating-current transformer, we have, by substituting

$$(r_1 + r)^2 + s^2 (x_1 + x)^2 = z_k^2,$$

and separating the real and imaginary quantities,

$$\dot{E}_0 = -n_0 e\left\{\left[1 + \frac{s}{a^2 z_k^2}(r_0 (r_1 + r) + s x_0 (x_1 + x))\right.\right.$$

$$+ (r_0 g_0 + x_0 b_0)\Big] + j\Big[\frac{s}{a^2 z_k^2}(s r_0 (x_1 + x) - x_0 (r_1 + r))$$

$$+ (r_0 b_0 - x_0 g_0)\Big]\Big\}.$$

$$\dot{I}_0 = - s n_0 e \left\{ \left[\frac{r_1 + r}{a^2 z_k^2} + \frac{g_0}{s}\right] + j\left[\frac{s(x_1 + x)}{a^2 z_k^2} + \frac{b_0}{s}\right]\right\}$$

$$\dot{I}_1 = \frac{s n_1 e}{z_k^2}\left\{(r_1 + r) + js(x_1 + x)\right\}.$$

Neglecting the exciting current, or rather considering it as a separate and independent shunt circuit outside of the transformer, as can approximately be done, and assuming the primary impedance reduced to the secondary circuit as equal to the secondary impedance,

$$Y_0 = 0,$$

$$\frac{Z_0}{a^2} = Z_1.$$

Substituting this in the equations of the general transformer, we get,

$$\dot{E}_0 = - n_0 e \left\{ 1 + \frac{s}{z_k^2}[r_1(r_1 + r) + s x_1 (x_1 + x)]\right.$$

$$\left. + \frac{js}{z_k^2}[s r_1 (x_1 + x) - x_1 (r_1 + r)]\right\}.$$

$$\dot{E}_1 = \frac{s n_1 e}{z_k^2}\{[r(r_1 + r) + s^2 x (x_1 + x)] + js[r x_1 - x r_1]\}.$$

$$\dot{I}_0 = - \frac{s n_1 e}{a z_k^2}\{(r_1 + r) + js(x_1 + x)\}.$$

$$\dot{I}_1 = \frac{s n_1 e}{z_k^2}\{(r_1 + r) + js(x_1 + x)\}.$$

146. The true power is, in symbolic representation (see Chapter XII.):

$$P = [\dot{E}, \dot{I}]^1$$

denoting,

$$\frac{s\,n_1^2 e^2}{z_k^2} = w$$

gives:

Secondary output of the transformer

$$P_1 = [E_1, I_1]^1 = \left(\frac{s\,n_1 e}{z_k}\right)^2 r = s\,r\,w.$$

Internal loss in secondary circuit,

$$P_1^1 = i_1^2 r_1 = \left(\frac{s\,n_1 e}{z_k}\right)^2 r_1 = s\,r_1 w.$$

Total secondary power,

$$P_1 + P_1^1 = \left(\frac{s\,n_1 e}{z_k}\right)^2 (r + r_1) = s\,w\,(r + r_1).$$

Internal loss in primary circuit,

$$P_0^1 = i_0^2 r_0 = i_0^2 r_1 a^2 = \left(\frac{s n_1 e}{z_k}\right)^2 r_1 = s\,r_1 w$$

Total electrical output, plus loss,

$$P^1 = P_1 + P_1^1 + P_0^1 = \left(\frac{s n_1 e}{z_k}\right)^2 (r + 2\,r_1) = sw\,(r + 2\,r_1).$$

Total electrical input of primary,

$$P_0 = [E_0 I_0]^1 = s\left(\frac{n_1 e}{z_k}\right)^2 (r + r_1 + s r_1) = w\,(r + r_1 + s r_1).$$

Hence, mechanical output of transformer,

$$P = P_0 - P^1 = w\,(1 - s)\,(r + r_1).$$

Ratio,

$$\frac{\text{mechanical output}}{\text{total secondary power}} = \frac{P}{P_1 + P_1^1} = \frac{1 - s}{s} = \frac{\text{speed}}{\text{slip}}.$$

147. Thus,

In a general alternating transformer of ratio of turns, a, and ratio of frequencies, s, neglecting exciting current, it is:

Electrical input in primary,

$$P_0 = \frac{s\,n_1^2 e^2 (r + r_1 + s r_1)}{(r_1 + r)^2 + s^2 (x_1 + x)^2}$$

Mechanical output,

$$P = \frac{s(1-s)\, n_1^2 e^2 (r + r_1)}{(r_1 + r)^2 + s^2 (x_1 + x)^2}.$$

Electrical output of secondary,

$$P_1 = \frac{s^2 n_1^2 e^2 r}{(r_1 + r)^2 + s^2 (x_1 + x)^2}.$$

Losses in transformer,

$$P_0^1 + P_1^1 = P^1 = \frac{2 s^2 n_1^2 e^2 r_1}{(r_1 + r)^2 + s^2 (x_1 + x)^2}.$$

Of these quantities, P^1 and P_1 are always positive; P_0 and P can be positive or negative, according to the value of s. Thus the apparatus can either produce mechanical power, acting as a motor, or consume mechanical power; and it can either consume electrical power or produce electrical power, as a generator.

148. At

$s = 0$, synchronism, $P_0 = 0$, $P = 0$, $P_1 = 0$.

At $0 < s < 1$, between synchronism and standstill.

P_1, P and P_0 are positive; that is, the apparatus consumes electrical power P_0 in the primary, and produces mechanical power P and electrical power $P_1 + P_1^1$ in the secondary, which is partly, P_1^1, consumed by the internal secondary resistance, partly, P_1, available at the secondary terminals.

In this case it is:

$$\frac{P_1 + P_1^1}{P} = \frac{s}{1 - s};$$

that is, of the electrical power consumed in the primary circuit, P_0, a part P_0^1 is consumed by the internal primary resistance, the remainder transmitted to the secondary, and divides between electrical power, $P_1 + P_1^1$, and mechanical power, P, in the proportion of the slip, or drop below synchronism, s, to the speed: $1 - s$.

In this range, the apparatus is a motor.

At $s > 1$; or, backwards driving,

$P < 0$, or negative; that is, the apparatus requires mechanical power for driving.

It is then: $P_0 - P_0^1 - P_1^1 < P_1$;

that is: the secondary electrical power is produced partly by the primary electrical power, partly by the mechanical power, and the apparatus acts simultaneously as transformer and as alternating-current generator, with the secondary as armature.

The ratio of mechanical input to electrical input is the ratio of speed to synchronism.

In this case, the secondary frequency is higher than the primary.

At $s < 0$, beyond synchronism,

$P < 0$; that is, the apparatus has to be driven by mechanical power.

$P_0 < 0$; that is, the primary circuit produces electrical power from the mechanical input.

At $r + r_1 + s r_1 = 0$, or, $s < -\frac{r + r_1}{r_1}$;

the electrical power produced in the primary becomes less than required to cover the losses of power, and P_0 becomes positive again.

We have thus:

$$s < -\frac{r + r_1}{r_1}$$

consumes mechanical and primary electric power; produces secondary electric power.

$$-\frac{r + r_1}{r_1} < s < 0$$

consumes mechanical, and produces electrical power in primary and in secondary circuit.

$$0 < s < 1$$

consumes primary electric power, and produces mechanical and secondary electrical power.

$$1 < s$$

consumes mechanical and primary electrical power; produces secondary electrical power.

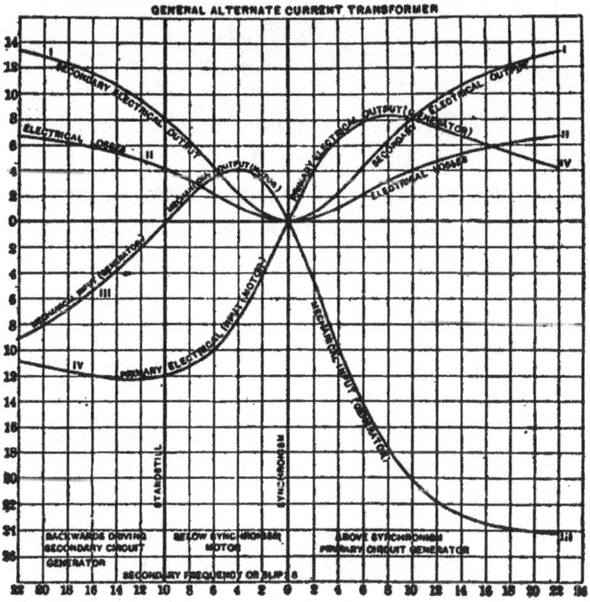

Fig. 111.

149. As an instance, in Fig. 111 are plotted, with the slip s as abscissæ, the values of:

Secondary electrical output	as Curve I.;
Total internal loss	as Curve II.;
Mechanical output	as Curve III.;
Primary electrical input	as Curve IV.;

for the values:

$$n_1 e = 100.0; \qquad r = .4;$$
$$r_1 = .1; \qquad x = .3;$$
$$x_1 = .2;$$

hence,
$$P_1 = \frac{16,000 s^2}{1 + s^2};$$

$$P_0^1 + P_1^1 = \frac{8,000 s^2}{1 + s^2};$$

$$P_0 = \frac{4,000 s + (5 + s)}{1 + s^2};$$

$$P = \frac{20,000 s (1 - s)}{1 + s^2}.$$

150. Since the most common practical application of the general alternating current transformer is that of frequency converter, that is to change from one frequency to another, either with or without change of the number of phases, the following characteristic curves of this apparatus are of great interest.

1. The regulation curve; that is, the change of secondary terminal voltage as function of the load at constant impressed primary voltage.

2. The compounding curve; that is, the change of primary impressed voltage required to maintain constant secondary terminal voltage.

In this case the impressed frequency and the speed are constant, and consequently the secondary frequency. Generally the frequency converter is used to change from a low frequency, as 25 cycles, to a higher frequency, as 62.5 cycles, and is then driven backward, that is, against its torque, by mechanical power. Mostly a synchronous motor is employed, connected to the primary mains, which by over-excitation compensates also for the lagging current of the frequency converter.

Let,

$Y_0 = g_0 + jb_0$ = primary exciting admittance per circuit of the frequency converter.

$Z_1 = r_1 - jx_1$ = internal self inductive impedance per secondary circuit, at the secondary frequency

$Z_0 = r_0 - jx_0$ = internal self inductive impedance per primary circuit at the primary frequency.

a = ratio of secondary to primary turns per circuit.

b = ratio of number of secondary to number of primary circuits.

c = ratio of secondary to primary frequencies.

Let,

e = induced E.M.F. per secondary circuit at secondary frequency.

$Z = r - jx$ = external impedance per secondary circuit at secondary frequency, that is load on secondary system, where $x = 0$ for noninductive lead.

We then have,

total secondary impedance,

$$Z + Z_1 = (r + r_1) - j(x + x_1)$$

secondary current,

$$I_1 = \frac{e}{Z + Z_1} = e(a_1 + ja_2),$$

where,

$$a_1 = \frac{r + r_1}{(r + r_1)^2 + (x + x_1)^2} \qquad a_2 = \frac{x + x_1}{(r + r_1)^2 + (x + x_1)^2}$$

secondary terminal voltage,

$$E_1 = I_1 Z = e\frac{Z}{Z + Z_1}$$

$$= e(r - jx)(a_1 + ja_2) = e(b_1 + jb_2).$$

where,

$$b_1 = (ra_1 + xa_2) \qquad b_2 = (ra_2 - xa_1)$$

primary induced E.M.F. per circuit,

$$E^1 = \frac{e}{ac}$$

primary load current per circuit,

$$I^1 = abI_1 = abe(a_1 + ja_2)$$

primary exciting current per circuit,

$$I_{00} = \frac{Y_0 e}{ac} = (g_0 + jb_0)\frac{e}{ac}$$

thus, total primary current,

$$\dot{I}_0 = \dot{I}^1 + \dot{I}_{00}$$
$$= e\,(c_1 + jc_2)$$

where,

$$c_1 = aba_1 + \frac{g_0}{ac} \qquad c_2 = aba_2 + \frac{b_0}{ac}$$

primary terminal voltage:

$$\dot{E}_0 = \dot{E}^1 + \dot{I}_0 Z_0$$
$$= e\,(d_1 + jd_2)$$

where,

$$d_1 = \frac{1}{ac} + r_0c_1 + x_0c_2 \qquad d_2 = r_0c_2 - x_0c_1$$

or absolute,

$$e_0 = e\sqrt{d_1^2 + d_2^2}$$
$$e = \frac{e_0}{\sqrt{d_1^2 + d_2^2}}$$

substituting this value of e in the preceding equations, gives, as function of the primary impressed E.M.F., e_0

secondary current,

$$\dot{I}_1 = \frac{e_0\,(a_1 + ja_2)}{\sqrt{d_1^2 + d_2^2}} \quad \text{or, absolute,} \quad I_1 = e_0\sqrt{\frac{a_1^2 + a_2^2}{d_1^2 + d_2^2}}$$

secondary terminal voltage,

$$\dot{E}_1 = \frac{e_0\,(b_1 + jb_2)}{\sqrt{d_1^2 + d_2^2}} \qquad E_1 = e_0\sqrt{\frac{b_1^2 + b_2^2}{d_1^2 + d_2^2}}$$

primary current,

$$\dot{I}_0 = \frac{e_0\,(c_1 + jc_2)}{\sqrt{d_1^2 + d_2^2}} \qquad I_0 = e_0\sqrt{\frac{c_1^2 + c_2^2}{d_1^2 + d_2^2}}$$

primary impressed E.M.F.

$$\dot{E}_0 = \frac{e_0\,(d_1 + jd_2)}{\sqrt{d_1^2 + d_2^2}}$$

secondary output,

$$P_1 = [\dot{E}_1\dot{I}_1]^1$$
$$= \frac{e_0^2\,(a_1b_1 + a_2b_2)}{d_1^2 + d_2^2}$$

primary electrical input,

$$P_0 = [E_0 I_0]^1 = \frac{e_0^2 (c_1 d_1 + c_2 d_2)}{d_1^2 + d_2^2}$$

primary apparent input, voltamperes,

$$Q_0 = e_0 I_0$$

Substituting thus different values for the secondary internal impedance Z gives the regulation curve of the frequency converter.

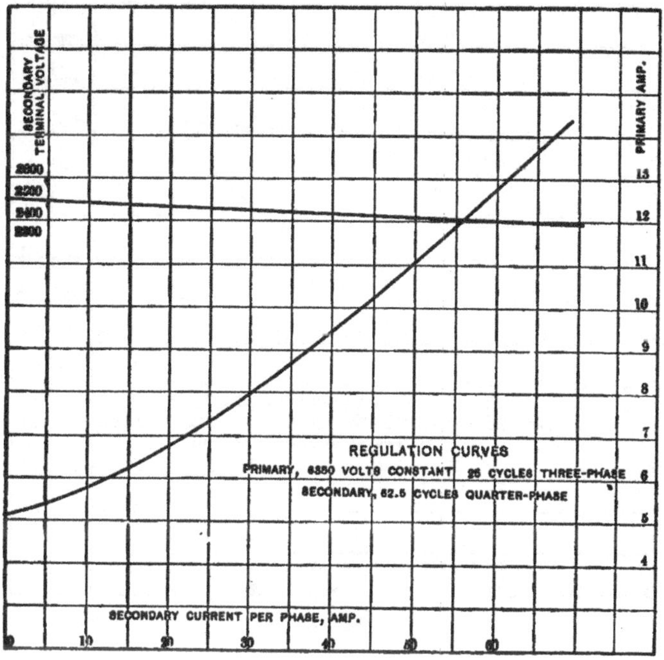

Fig. 112.

Such a curve, taken from tests of a 200 KW frequency converter changing from 6300 volts 25 cycles three-phase, to 2500 volts 62.5 cycles quarter-phase, is given in Fig. 112.

From the secondary terminal voltage,

$$\dot{E}_1 = e\,(b_1 + jb_2)$$

it follows, absolute,

$$e_1 = e\,\sqrt{b_1^2 + b_2^2}$$

$$e = \frac{e_1}{\sqrt{b_1^2 + b_2^2}}$$

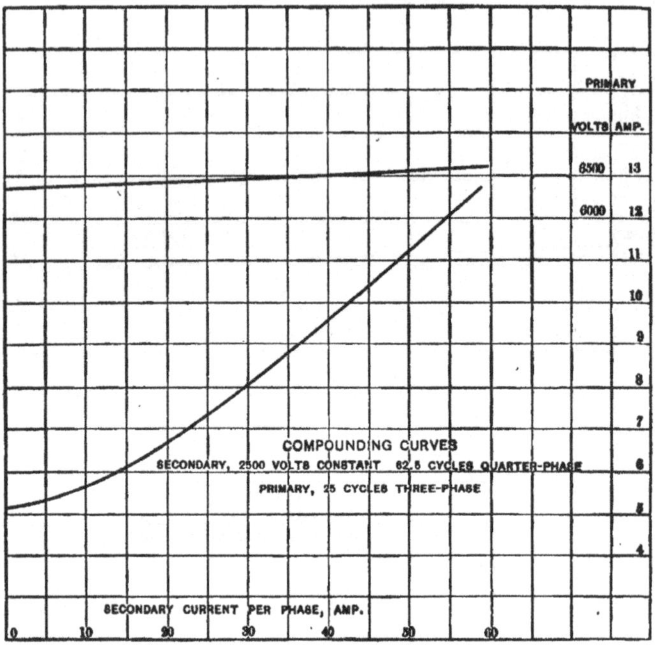

Fig. 113.

Substituting these values in the above equation gives the quantities as functions of the secondary terminal voltage, that is at constant e_1, or the compounding curve.

The compounding curve of the frequency converter above mentioned is given in Fig. 113.

CHAPTER XVI.

INDUCTION MOTOR.

151. A specialization of the general alternating-current transformer is the induction motor. It differs from the stationary alternating-current transformer, which is also a specialization of the general transformer, in so far as in the stationary transformer only the transfer of electrical energy from primary to secondary is used, but not the mechanical force acting between the two, and therefore primary and secondary coils are held rigidly in position with regard to each other. In the induction motor, only the mechanical force between primary and secondary is used, but not the transfer of electrical energy, and thus the secondary circuits closed upon themselves. Transformer and induction motor thus are the two limiting cases of the general alternating-current transformer. Hence the induction motor consists of a magnetic circuit interlinked with two electric circuits or sets of circuits, the primary and the secondary circuit, which are movable with regard to each other. In general a number of primary and a number of secondary circuits are used, angularly displaced around the periphery of the motor, and containing E.M.Fs. displaced in phase by the same angle. This multi-circuit arrangement has the object always to retain secondary circuits in inductive relation to primary circuits and *vice versa,* in spite of their relative motion.

The result of the relative motion between primary and secondary is, that the E.M.Fs. induced in the secondary or the motor armature are not of the same frequency as the E.M.Fs. impressed upon the primary, but of a frequency which is the difference between the impressed frequency and the frequency of rotation, or equal to the "slip," that is, the difference between synchronism and speed (in cycles).

Hence, if

N = frequency of main or primary E.M.F.,
and s = percentage slip;
sN = frequency of armature or secondary E.M.F.,
and $(1 - s) N$ = frequency of rotation of armature.

In its reaction upon the primary circuit, however, the armature current is of the same frequency as the primary current, since it is carried around mechanically, with a frequency equal to the difference between its own frequency and that of the primary. Or rather, since the reaction of the secondary on the primary must be of primary frequency — whatever the speed of rotation — the secondary frequency is always such as to give at the existing speed of rotation a reaction of primary frequency.

152. Let the primary system consist of p_0 equal circuits, displaced angulary in space by $1/p_0$ of a period, that is, $1/p_0$ of the width of two poles, and excited by p_0 E.M.Fs. displaced in phase by $1/p_0$ of a period; that is, in other words, let the field circuits consist of a symmetrical p_0-phase system. Analogously, let the armature or secondary circuits consist of a symmetrical p_1-phase system.

Let

n_0 = number of primary turns per circuit or phase;
n_1 = number of secondary turns per circuit or phase;
$a = \frac{n_0 p_0}{n_1 p_1}$ = ratio of total primary turns to total secondary turns or ratio of transformation.

Since the number of secondary circuits and number of turns of the secondary circuits, in the induction motor — as in the stationary transformer — is entirely unessential, it is preferable to reduce all secondary quantities to the primary system, by the ratio of transformation, a; thus

if $\dot{E}_1'$ = secondary E.M.F. per circuit, $\dot{E}_1 = a\dot{E}_1'$
= secondary E.M.F. per circuit reduced to primary system;

if $\dot{I}_1'$ = secondary current per circuit, $\dot{I}_1 = \frac{\dot{I}_1'}{a}$
= secondary current per circuit reduced to primary system;

if r_1' = secondary resistance per circuit, $r_1 = a^2 r_1'$
= secondary resistance per circuit reduced to primary system;

if x_1' = secondary reactance per circuit, $x_1 = a^2 x_1'$
= secondary reactance per circuit reduced to primary system;

if z_1' = secondary impedance per circuit, $z_1 = a^2 z_1'$
= secondary impedance per circuit reduced to primary system;

that is, the number of secondary circuits and of turns per secondary circuit is assumed the same as in the primary system.

In the following discussion, as secondary quantities, the values reduced to the primary system shall be exclusively used, so that, to derive the true secondary values, these quantities have to be reduced backwards again by the factor

$$a = \frac{n_0 p}{n_1 p_1}.$$

153. Let

Φ = total maximum flux of the magnetic field per motor pole,

We then have

$E = \sqrt{2}\,\pi n N \Phi 10^{-8}$ = effective E.M.F. induced by the magnetic field per primary circuit.

Counting the time from the moment where the rising magnetic flux of mutual induction Φ (flux interlinked with both electric circuits, primary and secondary) passes through zero, in complex quantities, the magnetic flux is denoted by

$$\dot{\Phi} = j\Phi,$$

and the primary induced E.M.F.,

$$\dot{E} = -e;$$

where

$e = \sqrt{2}\pi n N \Phi 10^{-8}$ may be considered as the "Active E.M.F. of the motor," or "Counter E.M.F."

Since the secondary frequency is sN, the secondary induced E.M.F. (reduced to primary system) is $\dot{E}_1 = -se$.

Let

$\dot{I}_0$ = exciting current, or current passing through the motor, per primary circuit, when doing no work (at synchronism),

and

$$Y = g + jb = \text{primary admittance per circuit} = \frac{\dot{I}_0}{e}.$$

We thus have,

ge = magnetic energy current, ge^2 = loss of power by hysteresis (and eddy currents) per primary coil.

Hence

$p_0 ge^2$ = total loss of energy by hysteresis and eddys as calculated according to Chapter X.
be = magnetizing current, and
$n_0 be$ = effective M.M.F. per primary circuit;

hence $\frac{p_0}{2} n_0 be$ = total effective M.M.F.;

and

$\frac{p_0}{\sqrt{2}} n_0 be$ = total maximum M.M.F., as resultant of the M.M.Fs. of the p_0-phases, combined by the parallelogram of M.M.Fs.*

If $\mathcal{R}$ = reluctance of magnetic circuit per pole, as discussed in Chapter X., it is

$$\frac{p_0}{\sqrt{2}} n_0 be = \mathcal{R}\Phi.$$

* Complete discussion hereof, see Chapter XXV.

Thus, from the hysteretic loss, and the reluctance, the constants, g and b, and thus the admittance, Y are derived.

Let r_0 = resistance per primary circuit;
x_0 = reactance per primary circuit;

thus,

$Z_0 = r_0 - j x_0$ = impedance per primary circuit;
r_1 = resistance per secondary circuit reduced to primary system;
x_1 = reactance per secondary circuit reduced to primary system, at full frequency, N;

hence,

$s x_1$ = reactance per secondary circuit at slip s;

and

$Z_1 = r_1 - j s x_1$ = secondary internal impedance.

154. We now have,

Primary induced E.M.F.,

$$\dot{E} = - e.$$

Secondary induced E.M.F.,

$$\dot{E}_1 = - s e.$$

Hence,

Secondary current,

$$\dot{I}_1 = \frac{\dot{E}_1}{Z_1} = - \frac{s e}{r_1 - j s x_1}$$

Component of primary current, corresponding thereto, or primary load current,

$$\dot{I}' = - \dot{I}_1 = \frac{s e}{r_1 - j s x_1};$$

Primary exciting current,

$$\dot{I}_0 = e Y = e (g + j b); \text{ hence,}$$

Total primary current,

$$I = I' + I_0$$
$$= e\left\{\frac{s}{r_1 - jsx_1} + (g + jb)\right\};$$

E.M.F. consumed by primary impedance,

$$E_z = Z_0 I$$
$$= e\,(r_0 - jx_0)\left\{\frac{s}{r_1 - jsx_1} + (g + jb)\right\};$$

E.M.F. required to overcome the primary induced E.M.F.,

$$-E = e;$$

hence,

Primary terminal voltage,

$$E_0 = e + E_z$$
$$= e\left\{1 + \frac{s\,(r_0 - jx_0)}{r_1 - jsx_1} + (r_0 - jx_0)\,(g + jb)\right\}.$$

We get thus, in an induction motor, at slip s and active E.M.F. e,

Primary terminal voltage,

$$E_0 = e\left\{1 + \frac{s\,(r_0 - jx_0)}{r_1 - jsx_1} + (r_0 - jx_0)\,(g + jb)\right\};$$

Primary current,

$$I = e\left\{\frac{s}{r_1 - jsx_1} + (g + jb)\right\};$$

or, in complex expression,

Primary terminal voltage,

$$E_0 = e\left\{1 + s\frac{Z_0}{Z_1} + Z_0 Y\right\};$$

Primary current,

$$I = e\left\{\frac{s}{Z_1} + Y\right\}.$$

To eliminate e, we divide, and get,

Primary current, at slip s, and impressed E.M.F., $\dot{E}_0$;

$$\dot{I} = \frac{s + Z_1 Y}{Z_1 + s Z_0 + Z_0 Z_1 Y} \dot{E}_0;$$

or,

$$\dot{I} = \frac{s + (r_1 - jsx_1)(g + jb)}{(r_1 - jsx_1) + s(r_0 - jx_0) + (r_0 - jx_0)(r_1 - jsx_1)(g + jb)} E_0.$$

Neglecting, in the denominator, the small quantity $Z_0 Z_1 Y$, it is

$$\dot{I} = \frac{s + Z_1 Y}{Z_1 + s Z_0} E_0$$

$$= \frac{s + (r_1 - jsx_1)(g + jb)}{(r_1 - jsx_1) + s(r_0 - jx_0)} E_0$$

$$= \frac{(s + r_1 g + s x_1 b) + j(r_1 b - s x_1 g)}{(r_1 + s r_0) - js(x_1 + x_0)} E_0,$$

or, expanded,

$$\dot{I} = \frac{[(sr_1 + s^2 r_0) + r_1^2 g + sr_1(r_0 g - x_0 b) + s^2 x_1(x_0 g + x_1 g + r_0 b)] + j[s^2(x_0 + x_1) + r_1^2 b + sr_1(x_0 g + r_0 b) + s^2 x_1(x_0 b + x_1 b - r_0 g)]}{(r_1 + sr_0)^2 + s^2(x_1 + x_0)^2} E_0.$$

Hence, displacement of phase between current and E.M.F.,

$$\tan \hat{\omega}_0 = \frac{s^2(x_0 + x_1) + r_1^2 b + sr_1(x_0 g + r_0 b) + s^2 x_1(x_0 b + x_1 b - r_0 g)}{(sr_1 + s^2 r_0) + r_1^2 g + sr_1(r_0 g - x_0 b) + s^2 x_1(x_0 g + x_1 g + r_0 b)}.$$

Neglecting the exciting current, I_0, altogether, that is, setting $Y = 0$,

We have

$$\dot{I} = s \dot{E}_0 \frac{(r_1 + s r_0) + js(x_0 + x_1)}{(r_1 + s r_0)^2 + s^2(x_0 + x_1)^2}$$

$$= \frac{s \dot{E}_0}{(r_1 + sr_0) - js(x_0 + x_1)};$$

$$\tan \hat{\omega}_0 = \frac{s(x_0 + x_1)}{r_1 + s r_0}.$$

155. In graphic representation, the induction motor diagram appears as follows:—

Denoting the magnetism by the vertical vector $\overline{O\Phi}$ in Fig. 114, the M.M.F. in ampere-turns per circuit is represented by vector $\overline{OF}$, leading the magnetism $\overline{O\Phi}$ by the angle of hysteretic advance α. The E.M.F. induced in the secondary is proportional to the slip s, and represented by $\overline{OE_1}$ at the amplitude of 180°. Dividing $\overline{OE_1}$ by a in the proportion of $r_1 \div s x_1$, and connecting a with the middle b of the upper arc of the circle $\overline{OE_1}$, this line intersects the lower arc of the circle at the point $\overline{I_1 r_1}$. Thus, $\overline{OI_1r_1}$ is the E.M.F. consumed by the secondary resistance, and $\overline{OI_1x_1}$ equal and parallel to $\overline{E_1I_1r_1}$ is the E.M.F. consumed by the secondary reactance. The angle, $E_1OI_1r_1 = \hat{\omega}_1$ is the angle of secondary lag.

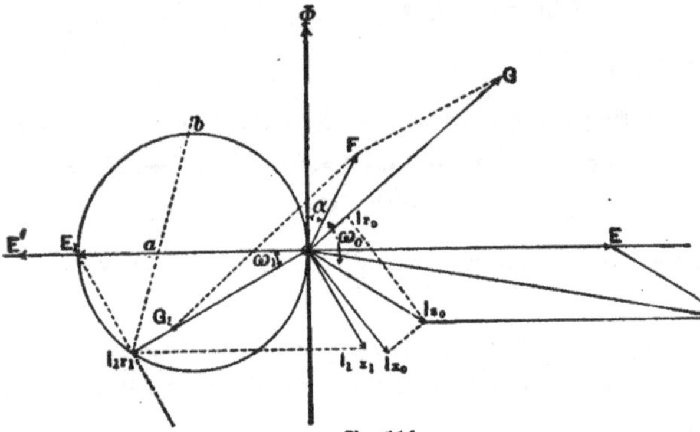

Fig. 114.

The secondary M.M.F. $\overline{OG_1}$ is in the direction of the vector $\overline{OI_1r_1}$. Completing the parallelogram of M.M.Fs. with $\overline{OF}$ as diagonal and $\overline{OG_1}$ as one side, gives the primary M.M.F. $\overline{OG}$ as other side. The primary current and the E.M.F. consumed by the primary resistance, represented by $\overline{OIr_0}$, is in line with $\overline{OG}$, the E.M.F. consumed by the primary reactance 90° ahead of $\overline{OG}$, and represented by $\overline{OIx_0}$, and their resultant $\overline{OIz_0}$ is the E.M.F. consumed by the primary impedance. The E.M.F. induced in the primary circuit is $\overline{OE'}$, and the E.M.F. required to overcome this counter E.M.F. is $\overline{OE}$ equal and opposite to $\overline{OE'}$. Combining $\overline{OE}$ with $\overline{OIz_0}$ gives the primary terminal voltage represented by vector $\overline{OE_0}$, and the angle of primary lag, $E_0OG = \hat{\omega}_0$.

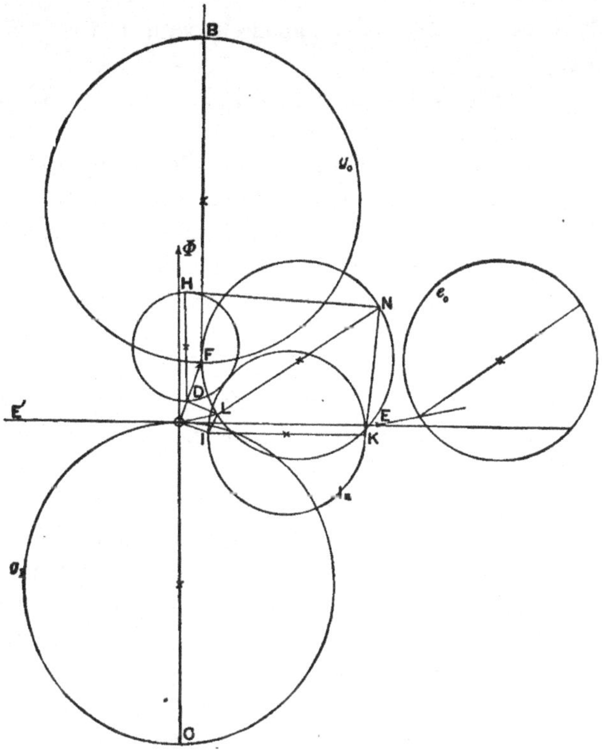

Fig. 115.

156. Thus far the diagram is essentially the same as the diagram of the stationary alternating-current transformer. Regarding dependence upon the slip of the motor, the locus of the different quantities for different values of the slip s is determined thus,

Let $E_1 = s E$

Assume in opposition to $\overline{O\Phi}$, a point A, such that

$$OA + I_1 r_1 = E_1 + I_1 s x_1, \text{ then}$$

$$OA = \frac{I_1 r_1 \times E_1}{I_1 s x_1} = \frac{I_1 r_1 \times sE}{I_1 s x_1} = \frac{r_1}{x_1} E = \text{constant.}$$

That is, $I_1 r_1$ lies on a half-circle with $OA = \frac{r_1}{x_1} E'$ as diameter.

That means G_1 lies on a half-circle g_1 in Fig. 115 with $\overline{OC}$ as diameter. In consequence hereof, G_0 lies on half-circle g_0 with $\overline{FB}$ equal and parallel to $\overline{OC}$ as diameter.

Thus Ir_0 lies on a half-circle with $\overline{DH}$ as diameter, which circle is perspective to the circle $\overline{FB}$, and Ix_0 lies on a half-circle with $\overline{IK}$ as diameter, and Iz_0 on a half-circle with $\overline{LN}$ as diameter, which circle is derived by the combination of the circles Ir_0 and Ix_0.

The primary terminal voltage E_0 lies thus on a half-circle e_0 equal to the half-circle Iz_0, and having to point E the same relative position as the half-circle Iz_0 has to point 0.

This diagram corresponds to constant intensity of the maximum magnetism, $\overline{O\Phi}$. If the primary impressed voltage E_0 is kept constant, the circle e_0 of the primary impressed voltage changes to an arc with O as center, and all the corresponding points of the other circles have to be reduced in accordance herewith, thus giving as locus of the other quantities curves of higher order which most conveniently are constructed point for point by reduction from the cirçle of the loci in Fig. 115.

Torque and Power.

157. The torque developed per pole by an electric motoı equals the product of effective magnetism, $\Phi / \sqrt{2}$, times effective armature M.M.F., $F / \sqrt{2}$, times the sine of the angle between both,

$$T' = \frac{\Phi F}{2} \sin(\Phi F).$$

If n_1 = number of turns, I_1 = current, per circuit, with p_1-armature circuits, the total maximum current polarization, or M.M.F. of the armature, is

$$F_1 = \frac{p_1 n_1 I_1}{\sqrt{2}}.$$

Hence the torque per pole,

$$T' = \frac{p_1 n_1 \Phi I_1}{2\sqrt{2}} \sin(\Phi I_1).$$

If q = the number of poles of the motor, the total torque of the motor is,

$$T = \frac{q p_1 n_1 \Phi I_1}{2\sqrt{2}} \sin(\Phi I_1).$$

The secondary induced E.M.F., E_1, lags 90° behind the inducing magnetism, hence reaches a maximum displaced in space by 90° from the position of maximum magnetization. Thus, if the secondary current, I_1, lags behind its E.M.F., E_1, by angle, $\hat{\omega}_1$, the space displacement between armature current and field magnetism is

$$\measuredangle (I_1 \Phi) = 90° + \hat{\omega}_1$$

hence

$$\sin(\Phi I_1) = \cos \hat{\omega}_1$$

We have, however,

$$\cos \hat{\omega}_1 = \frac{r_1}{\sqrt{r_1^2 + s^2 x_1^2}},$$

$$I_1 = \frac{e s}{\sqrt{r_1^2 + s^2 x_1^2}},$$

$$e = \sqrt{2}\pi n_1 \Phi N 10^{-8},$$

thus,

$$n_1 \Phi = \frac{e 10^8}{\sqrt{2}\pi N};$$

substituting these values in the equation of the torque, it is

$$T = \frac{q p_1 s r_1 e^2 10^8}{4\pi N (r_1^2 + s^2 x_1^2)};$$

or, in practical (C.G.S.) units,

$$T = \frac{q p_1 s r_1 e^2}{4 \pi N (r_1^2 + s^2 x_1^2)},$$

is the Torque of the Induction Motor.

At the slip s, the frequency N, and the number of poles q, the linear speed at unit radius is

$$v = \frac{4 \pi N}{q} (1 - s);$$

hence the output of the motor,

$$P = T v,$$

or, substituted,

$$P = \frac{p_1 r_1 e^2 s (1 - s)}{r_1^2 + s^2 x_1^2},$$

is the Power of the Induction Motor.

158. We can arrive at the same results in a different way:

By the counter E.M.F. e of the primary circuit with current $\dot{I} = \dot{I}_0 + \dot{I}_1$ the power is consumed, $e \dot{I} = e \dot{I}_0 + e \dot{I}_1$. The power $e \dot{I}_0$ is that consumed by the primary hysteresis and eddys. The power $e \dot{I}_1$ disappears in the primary circuit by being transmitted to the secondary system.

Thus the total power impressed upon the secondary system, per circuit, is

$$P_1 = e \dot{I}_1.$$

Of this power a part, $\dot{E}_1 \dot{I}_1$, is consumed in the secondary circuit by resistance. The remainder,

$$P' = \dot{I}_1 (e - \dot{E}_1),$$

disappears as electrical power altogether; hence, by the law of conservation of energy, must reappear as some other form of energy, in this case as mechanical power, or as the output of the motor (including friction).

Thus the mechanical output per motor circuit is

$$P' = \dot{I}_1 (e - \dot{E}_1).$$

Substituting,

$$E_1 = s e;$$

$$I_1 = \frac{s e}{r_1 - j s x_1};$$

it is

$$P' = \frac{e^2 s (1 - s)}{r_1 - j s x_1}$$

$$= \frac{e^2 s (1 - s)(r_1 + j s x_1)}{r_1^2 + s^2 x_1^2};$$

hence, since the imaginary part has no meaning as power,

$$P' = \frac{r_1 e^2 s (1 - s)}{r_1^2 + s^2 x_1^2};$$

and the total power of the motor,

$$P = \frac{p_1 r_1 e^2 s (1 - s)}{r_1^2 + s^2 x_1^2}.$$

At the linear speed,

$$v = \frac{4 \pi N}{q}(1 - s)$$

at unit radius the torque is

$$T = \frac{q p_1 r_1 e^2 s}{4 \pi N (r_1^2 + s^2 x_1^2)}.$$

In the foregoing, we found

$$E_0 = e \left\{ 1 + s \frac{Z_0}{Z_1} + Z_0 Y \right\}$$

or, approximately,

$$E_0 = e \left\{ 1 + s \frac{Z_0}{Z_1} \right\};$$

or,

$$e = \frac{E_0 Z_1}{s Z_0 + Z_1};$$

expanded,

$$e = E_0 \frac{r_1 - j s x_1}{(r_1 + s r_0) - j s (x_1 + x_0)};$$

or, eliminating imaginary quantities,

$$e = E_0 \sqrt{\frac{r_1^2 + s^2 x_1^2}{(r_1 + s r_0)^2 + s^2 (x_1 + x_0)^2}}.$$

Substituting this value in the equations of torque and of power, they become,

torque, $$T = \frac{q p_1 r_1 E_0^2 s}{4 \pi N \{ (r_1 + s r_0)^2 + s^2 (x_1 + x_0)^2 \}};$$

power, $$P = \frac{p_1 r_1 E_0^2 s (1 - s)}{(r_1 + s r_0)^2 + s^2 (x_1 + x_0)^2}.$$

Maximum Torque.

159. The torque of the induction motor is a maximum for that value of slip s, where

$$\frac{dT}{ds} = 0,$$

or, since $$T = \frac{q p_1 r_1 E_0^2 s}{4 \pi N \{ (r_1 + s r_0)^2 + s^2 (x_1 + x_0)^2 \}},$$

for, $$\frac{d}{ds} \left\{ \frac{(r_1 + s r_0)^2 + s^2 (x_1 + x_0)^2}{s} \right\} = 0;$$

expanded, this gives,

$$-\frac{r_1^2}{s^2} + r_0^2 + (x_1 + x_0)^2 = 0,$$

or, $$s_t = \frac{r_1}{\sqrt{r_0^2 + (x_1 + x_0)^2}}.$$

Substituting this in the equation of torque, we get the value of maximum torque,

$$T_t = \frac{q p_1 E_0^2}{8 \pi N \{ r_0 + \sqrt{r_0^2 + (x_1 + x_0)^2} \}}.$$

That is, independent of the secondary resistance, r_1.

The power corresponding hereto is, by substitution of s_t in P,

$$P_t = \frac{p_1 E_0^2 \{ \sqrt{r_0^2 + (x_1 + x_0)^2} - r_1 \}}{2 \sqrt{r_0^2 + (x_1 + x_0)^2} \{ \sqrt{r_0^2 + (x_1 + x_0)^2} + r_0 \}}.$$

This power is not the maximum output of the motor, but already below the maximum output. The maximum output is found at a lesser slip, or higher speed, while at the maximum torque point the output is already on the decrease, due to the decrease of speed.

With increasing slip, or decreasing speed, the torque of the induction motor increases; or inversely, with increasing load, the speed of the motor decreases, and thereby the torque increases, so as to carry the load down to the slip s_t, corresponding to the maximum torque. At this point of load and slip the torque begins to decrease again; that is, as soon as with increasing load, and thus increasing slip, the motor passes the maximum torque point s_t, it "falls out of step," and comes to a standstill.

Inversely, the torque of the motor, when starting from rest, will increase with increasing speed, until the maximum torque point is reached. From there towards synchronism the torque decreases again.

In consequence hereof, the part of the torque-speed curve below the maximum torque point is in general unstable, and can be observed only by loading the motor with an apparatus, whose countertorque increases with the speed faster than the torque of the induction motor.

In general, the maximum torque point, s_t, is between synchronism and standstill, rather nearer to synchronism. Only in motors of very large armature resistance, that is low efficiency, $s_t > 1$, that is, the maximum torque falls below standstill, and the torque constantly increases from synchronism down to standstill.

It is evident that the position of the maximum torque point, s_t can be varied by varying the resistance of the secondary circuit, or the motor armature. Since the slip of the maximum torque point, s_t, is directly proportional to the armature resistance, r_1, it follows that very constant speed and high efficiency will bring the maximum torque point near synchronism, and give small starting torque, while good starting torque means a maximum torque point at low speed; that is, a motor with poor speed regulation and low efficiency.

Thus, to combine high efficiency and close speed regulation with large starting torque, the armature resistance has to be varied during the operation of the motor, and the motor started with high armature resistance, and with increasing speed this armature resistance cut out as far as possible.

160. If $s_t = 1,$
it is $r_1 = \sqrt{r_0^2 + (x_1 + x_0)^2}.$

In this case the motor starts with maximum torque, and when overloaded does not drop out of step, but gradually slows down more and more, until it comes to rest.

If, $s_t > 1,$
then $r_1 > \sqrt{r_0^2 + (x_1 + x_0)^2}.$

In this case, the maximum torque point is reached only by driving the motor backwards, as countertorque.

As seen above, the maximum torque T_t, is entirely independent of the armature resistance, and likewise is the current corresponding thereto, independent of the armature resistance. Only the speed of the motor depends upon the armature resistance.

Hence the insertion of resistance into the motor armature does not change the maximum torque, and the current corresponding thereto, but merely lowers the speed at which the maximum torque is reached.

The effect of resistance inserted into the induction motor is merely to consume the E.M.F., which otherwise would find its mechanical equivalent in an increased speed, analogous as resistance in the armature circuit of a continuous-current shunt motor.

Further discussion on the effect of armature resistance is found under "Starting Torque."

Maximum Power.

161. The power of an induction motor is a maximum for that slip, s_p, where

$$\frac{dP}{ds} = 0;$$

or, since $$P = \frac{p_1 r_1 E_0^2 s(1-s)}{(r_1 + s r_0)^2 + s^2 (x_1 + x_0)^2},$$

$$\frac{d}{ds}\left\{\frac{(r_1 + s r_0)^2 + s^2 (x_1 + x_0)^2}{s(1-s)}\right\} = 0;$$

expanded, this gives

$$s_p = \frac{r_1}{r_1 + \sqrt{(r_1 + r_0)^2 + (x_1 + x_0)^2}};$$

substituted in P, we get the maximum power,

$$P_p = \frac{p_1 E_0^2}{2\left\{(r_1 + r_0) + \sqrt{(r_1 + r_0)^2 + (x_1 + x_0)^2}\right\}}.$$

This result has a simple physical meaning: $(r_1 + r_0) = r$ is the total resistance of the motor, primary plus secondary (the latter reduced to the primary). $(x_1 + x_0)$ is the total reactance, and thus $\sqrt{r_1 + r_0)^2 + (x_1 + x_0)^2} = z$ is the total impedance of the motor. Hence

$$P_p = \frac{p_1 E_0^2}{2\{r + z\}},$$

is the maximum output of the induction motor, at the slip

$$s_p = \frac{r_1}{r_1 + z}.$$

The same value has been derived in Chapter IX., as the maximum power which can be transmitted into a non-inductive receiver circuit over a line of resistance r, and impedance z, or as the maximum output of a generator, or of a stationary transformer. Hence:

The maximum output of an induction motor is expressed by the same formula as the maximum output of a generator, or of a stationary transformer, or the maximum output which can be transmitted over an inductive line into a non-inductive receiver circuit.

The torque corresponding to the maximum output P_p is,

$$T_p = \frac{q p_1 E_0^2 (r_1 + z)}{8 \pi N z (r + z)}.$$

This is not the maximum torque; but the maximum torque, T_t, takes place at a lower speed, that is, greater slip,

$$s_t = \frac{r_1}{\sqrt{r_0^2 + (x_1 + x_0)^2}},$$

since,
$$\frac{r_1}{\sqrt{r_0^2 + (x_1 + x_0)^2}} > \frac{r_1}{r_1 + \sqrt{(r_1 + r_0) + (x_1 + x_0)^2}};$$

that is,
$$s_t > s_p.$$

It is obvious from these equations, that, to reach as large an output as possible, r and z should be as small as possible; that is, the resistances $r_1 + r_0$, and the impedances, z, and thus the reactances, $x_1 + x_0$, should be small. Since $r_1 + r_0$ is usually small compared with $x_1 + x_0$ it follows, that the problem of induction motor design consists in constructing the motor so as to give the minimum possible reactances, $x_1 + x_0$.

Starting Torque.

162. In the moment of starting an induction motor, the slip is

$$s = 1;$$

hence, starting current,

$$I = \frac{1 + (r_1 - jx_1)(g + jb)}{(r_1 - jx_1) + (r_0 - jx_0) + (r_1 - jx_1)(r_0 - jx_0)(g + jb)} E_0;$$

or, expanded, with the rejection of the last term in the denominator, as insignificant,

$$I = \frac{[(r_1 + r_0) + g(r_1[r_1 + r_0] + x_1[x_1 + x_0]) + b(r_0x_1 - x_0r_1)] - j[(x_1 + x_0) + b(r_1[r_1 + r_0] + x_1[x_1 + x_0]) - g(r_0x_1 - x_0r_1)]}{(r_1 + r_0]^2 + (x_1 + x_0)^2} E_0;$$

and, displacement of phase, or angle of lag,

$$\tan \hat{\omega}_0 = \frac{(x_1 + x_0) + b(r_1[r_1 + r_0] + x_1[x_1 + x_0]) - g(r_0x_1 - x_0r_1)}{(r_1 + r_0) + g(r_1[r_1 + r_0] + x_1[x_1 + x_0]) + b(r_0x_1 - x_0r_1)}.$$

Neglecting the exciting current, $g = 0 = b$, these equations assume the form,

$$I = \frac{(r_1 + r_0) + j(x_1 + x_0)}{(r_1 + r_0)^2 + (x_1 + x_0)^2} E_0 = \frac{E_0}{(r_1 + r_0) - j(x_1 + x_0)};$$

or, eliminating imaginary quantities,

$$I = \frac{E_0}{\sqrt{(r_1 + r_0)^2 + (x_1 + x_0)^2}} = \frac{E_0}{z};$$

and

$$\tan \hat{\omega}_0 = \frac{x_1 + x_0}{r_1 + r_0}.$$

That means, that in starting the induction motor without additional resistance in the armature circuit, — in which case $x_1 + x_0$ is large compared with $r_1 + r_0$, and the total impedance, z, small, — the motor takes excessive and greatly lagging currents.

The starting torque is

$$T_0 = \frac{q p_1 r_1 E_0^2}{4 \pi N \{(r_1 + r_0)^2 + (x_1 + x_0)^2\}}$$

$$= \frac{q p_1 E_0^2}{4 \pi N} \frac{r_1}{z^2}.$$

That is, the starting torque is proportional to the armature resistance, and inversely proportional to the square of the total impedance of the motor.

It is obvious thus, that, to secure large starting torque, the impedance should be as small, and the armature resistance as large, as possible. The former condition is the condition of large maximum output and good efficiency and speed regulation; the latter condition, however, means inefficiency and poor regulation, and thus cannot properly be fulfilled by the internal resistance of the motor, but only by an additional resistance which is short-circuited while the motor is in operation.

Since, necessarily,

$$r_1 < z,$$

we have,
$$T_0 < \frac{q p_1 E_0^2}{4 \pi N z};$$

and since the starting current is, approximately,

$$I = \frac{E_0}{z},$$

we have,
$$T_0 < \frac{q p_1}{4 \pi N} E_0 I.$$

$$T_{00} = \frac{q p_1}{4 \pi N} E_0 I$$

would be the theoretical torque developed at 100 per cent efficiency and power factor, by E.M.F., E_0, and current, I, at synchronous speed.

Thus,
$$T_0 < T_{00},$$

and the ratio between the starting torque T_0, and the theoretical maximum torque, T_{00}, gives a means to judge the perfection of a motor regarding its starting torque.

This ratio, T_0 / T_{00}, exceeds .9 in the best motors.

Substituting $I = E_0 / z$ in the equation of starting torque, it assumes the form,

$$T_0 = \frac{q p_1}{4 \pi N} I^2 r_1.$$

Since $4 \pi N / q$ = synchronous speed, it is :

The starting torque of the induction motor is equal to the resistance loss in the motor armature, divided by the synchronous speed.

The armature resistance which gives maximum starting torque is

$$\frac{d T_0}{d r_1} = 0;$$

or since, $$T_0 = \frac{q p_1 E_0^2}{4 \pi N} \frac{r_1}{(r_1 + r_0)^2 + (x_1 + x_0)^2}$$

$$\frac{d}{d r_1} \left\{ \frac{(r_1 + r_0)^2 + (x_1 + x_0)^2}{r_1} \right\} = 0;$$

expanded, this gives,

$$r_1 = \sqrt{r_0^2 + (x_1 + x_0)^2},$$

the same value as derived in the paragraph on "maximum torque."

Thus, adding to the internal armature resistance, r_1' in starting the additional resistance,

$$r_1'' = \sqrt{r_0^2 + (x_1 + x_0)^2} - r_1',$$

makes the motor start with maximum torque, while with increasing speed the torque constantly decreases, and reaches zero at synchronism. Under these conditions, the induction motor behaves similarly to the continuous-current series motor, varying in the speed with the load, the difference being, however, that the induction motor approaches a definite speed at no load, while with the series motor the speed indefinitely increases with decreasing load.

The additional armature resistance, r_1'', required to give a certain starting torque, if found from the equation of starting torque:

Denoting the internal armature resistance by r_1', the total armature resistance is $r_1 = r_1' + r_1''$.

and thus, $$T_0 = \frac{q p_1 E_0^2}{4 \pi N} \frac{r_1' + r_1''}{(r_1' + r_1'' + r_0)^2 + (x_1 + x_0)^2};$$

hence,

$$r_1'' = -r_1' - r_0 + \frac{q p_1 E_0^2}{8 \pi N T_0} \pm \sqrt{\left(\frac{q p_1 E_0^2}{8 \pi N T_0}\right)^2 - \frac{q p_1 E_0^2 r_0}{4 \pi N T_0} - (x_1 + x_0)^2}.$$

This gives two values, one above, the other below, the maximum torque point.

Choosing the positive sign of the root, we get a larger armature resistance, a small current in starting, but the torque constantly decreases with the speed.

Choosing the negative sign, we get a smaller resistance, a large starting current, and with increasing speed the torque first increases, reaches a maximum, and then decreases again towards synchronism.

These two points correspond to the two points of the speed-torque curve of the induction motor, in Fig. 116, giving the desired torque T_o.

The smaller value of r_1'' will give fairly good speed regulation, and thus in small motors, where the comparatively large starting current is no objection, the permanent armature resistance may be chosen to represent this value.

The larger value of r_1'' allows to start with minimum current, but requires cutting out of the resistance after the start, to secure speed regulation and efficiency.

Synchronism.

163. At synchronism, $s = 0$, we have,

$$I_s = E_0 (g + jb);$$

or,

$$I_s = E_0 \sqrt{g^2 + b^2};$$

$$P = 0, \ T = 0;$$

that is, power and torque are zero. Hence, the induction motor can never reach complete synchronism, but must slip sufficiently to give the torque consumed by friction.

Running near Synchronism.

164. When running near synchronism, at a slip s above the maximum output point, where s is small, from .02 to .05 at full load, the equations can be simplified by neglecting terms with s, as of higher order.

We then have, current,

$$I = \frac{s + r_1(g + jb)}{r_1} E_0;$$

or, eliminating imaginary quantities,

$$I = \sqrt{\left(\frac{s}{r_1} + g\right)^2 + b^2}\, E_0;$$

angle of lag,

$$\tan \hat{\omega}_0 = \frac{s^2(x_1 + x_0) + r_1^2 b}{s r_1 + r_1^2 g} = \frac{s^2 \frac{x_1 + x_0}{r_1} + r_1 b}{s + r_1 g};$$

$$P = \frac{p_1 E_0^2 s}{r_1};$$

$$T = \frac{q p_1 E_0^2 s}{4 \pi N r_1};$$

or, inversely,

$$s = \frac{r_1 P}{p_1 E_0^2};$$

$$s = \frac{4 \pi N r_1 T}{q p_1 E_0^2}$$

that is,

Near sychronism, the slip, s, of an induction motor, or its drop in speed, is proportional to the armature resistance, r_1 and to the power, P, or torque, T.

Example.

165. As an instance are shown, in Fig. 116, characteristic curves of a 20 horse-power three-phase induction motor, of 900 revolutions synchronous speed, 8 poles, frequency of 60 cycles.

The impressed E.M.F. is 110 volts between lines, and the motor star connected, hence the E.M.F. impressed per circuit :

$$\frac{110}{\sqrt{3}} = 63.5; \text{ or } E_0 = 63.5.$$

The constants of the motor are:

Primary admittance, $Y = .1 + .4\ j.$
Primary impedance, $Z = .03 - .09\ j.$
Secondary impedance, $Z_1 = .02 - .085\ j.$

In Fig. 116 is shown, with the speed in per cent of synchronism, as abscissæ, the torque in kilogrammetres, as ordinates, in drawn lines, for the values of armature resistance:

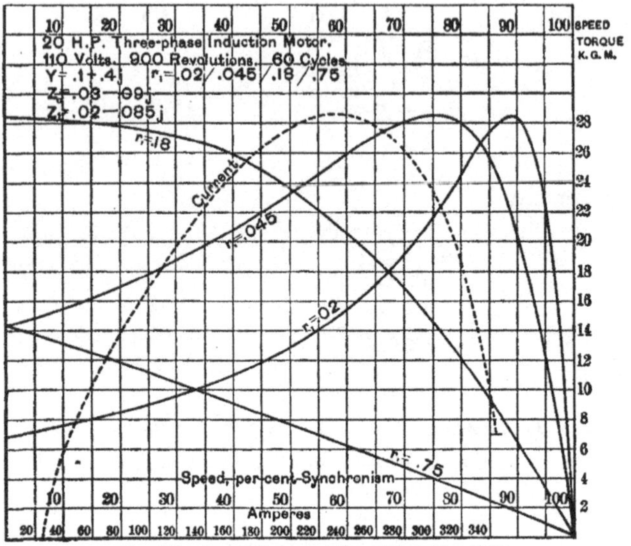

Fig. 116. Speed Characteristics of Induction Motor.

$r_1 = .02$: short circuit of armature, full speed.
$r_1 = .045$: .025 ohms additional resistance.
$r_1 = .18$: .16 ohms additional, maximum starting torque.
$r_1 = .75$: .73 ohms additional, same starting torque as $r_1 = .045$.

On the same Figure is shown the current per line, in dotted lines, with the verticals or torque as abscissæ, and the horizontals or amperes as ordinates. To the same torque always corresponds the same current, no matter what the speed be.

On Fig. 117 is shown, with the current input per line as abscissæ, the torque in kilogrammetres and the output in horse-power as ordinates in drawn lines, and the speed and the magnetism, in per cent of their synchronous values, as ordinates in dotted lines, for the armature resistance $r_1 = .02$ or short circuit.

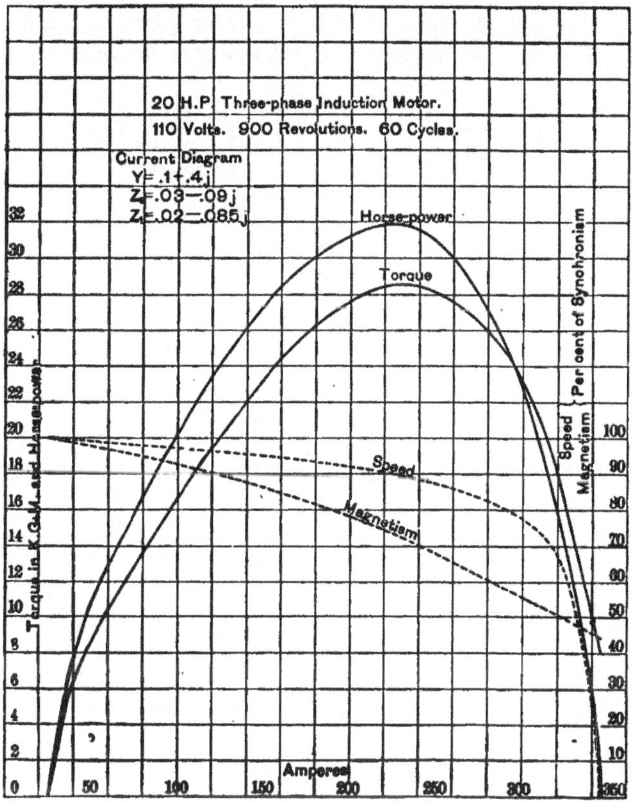

Fig. 117. Current Characteristics of Induction Motor.

In Fig. 118 is shown, with the speed, in per cent of synchronism, as abscissæ, the torque in drawn line, and the output in dotted line, for the value of armature resistance $r_1 = .045$, for the whole range of speed from 120 per cent backwards speed to 220 per cent beyond synchronism, showing the two maxima, the motor maximum at $s = .25$, and the generator maximum at $s = -.25$.

166. As seen in the preceding, the induction motor is characterized by the three complex imaginary constants,

$Y_0 = g_0 + jb_0$, the primary exciting admittance,
$Z_0 = r_0 - jx_0$, the primary self-inductive impedance, and
$Z_1 = r_1 - jx_1$, the secondary self-inductive impedance,

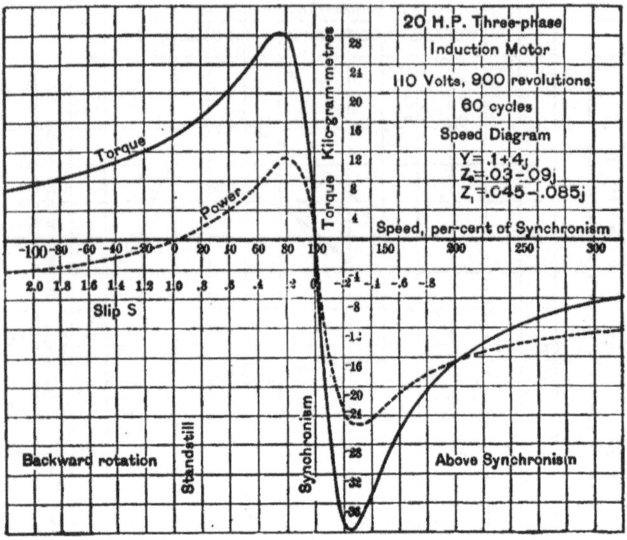

Fig. 118. Speed Characteristics of Induction Motor.

reduced to the primary by the ratio of secondary to primary turns.

From these constants and the impressed E.M.F. e_0, the motor can be calculated as follows:

Let,

e = counter E.M.F. of motor, that is E.M.F. induced in the primary by the mutual magnetic flux.

At the slip s the E.M.F. induced in the secondary circuit is,

se

Thus the secondary current,

$$I_1 = \frac{se}{r_1 - jsx_1} = e\,(a_1 + ja_2)$$

where,

$$a_1 = \frac{sr_1}{r_1^2 + s^2x_1^2} \qquad a_2 = \frac{s^2x_1}{r_1^2 + s^2x_1^2}$$

The primary exciting current is,

$$I_{00} = eY_0 = e\,(g_0 + jb_0)$$

thus, the total primary current,

$$I_0 = I_1 + I_{00} = e\,(b_1 + jb_2)$$

where,

$$b_1 = a_1 + g_0, \qquad b_2 = a_2 + b_0$$

The E.M.F. consumed by the primary impedance is,

$$E^1 = I_0Z_0 = e\,(r_0 - jx_0)\,(b_1 + jb_2)$$

the primary counter E.M.F. is e, thus the primary impressed E.M.F.,

$$E_0 = e + E^1 = e\,(c_1 + jc_2)$$

where,

$$c_1 = 1 + r_0b_1 + x_0b_2 \qquad c_2 = r_0b_2 - x_0b_1$$

or, absolute,

$$e_0 = e\,\sqrt{c_1^2 + c_2^2}$$

hence,

$$e = \frac{e_0}{\sqrt{c_1^2 + c_2^2}}$$

This value substituted gives,

Secondary current,

$$I_1 = e_0\frac{a_1 + ja_2}{\sqrt{c_1^2 + c_2^2}} \qquad I_1 = e_0\sqrt{\frac{a_1^2 + a_2^2}{c_1^2 + c_2^2}}$$

Primary current,

$$I_0 = e_0\frac{b_1 + jb_2}{\sqrt{c_1^2 + c_2^2}} \qquad I_0 = e_0\sqrt{\frac{b_1^2 + b_2^2}{c_1^2 + c_2^2}}$$

Impressed E.M.F.,

$$E_0 = e_0\,\frac{c_1 + jc_2}{\sqrt{c_1^2 + c_2^2}}$$

Thus torque, in synchronous watts (that is, the watts output the torque would produce at synchronous speed),

$$T = [e\dot{I}_1]^1$$

$$= \frac{e_0^2 a_1}{c_1^2 + c_2^2}$$

hence, the torque in absolute units,

$$T_0 = \frac{T}{N} = \frac{e_0^2 a_1}{(c_1^2 + c_2^2)\, N}$$

where $N =$ frequency.

The power output is torque times speed, thus:

$$P_1 = T(1 - s) = \frac{e_0^2 a_1 (1 - s)}{c_1^2 + c_2^2}$$

The power input is,

$$P_0 = [\dot{E}_0 \dot{I}_0] = [\dot{E}_0 \dot{I}_0]^1 + j\, [\dot{E}_0 \dot{I}_0]^j = P_0^1 + jP_0^j$$

$$= \frac{e_0^2 (b_1 c_1 + b_2 c_2)}{c_1^2 + c_2^2} + j \frac{e_0^2 (b_2 c_1 - b_1 c_2)}{c_1^2 + c_2^2}$$

The voltampere input,

$$Q_0 = e_0 I_0 = \frac{e_0^2 \sqrt{b_1^2 + b_2^2}}{\sqrt{c_1^2 + c_2^2}}$$

hence,

efficiency,

$$\frac{P_1}{P_0^1} = \frac{a_1 (1 - s)}{b_1 c_1 + b_2 c_2}$$

power factor,

$$\frac{P_0^1}{Q_0} = \frac{b_1 c_1 + b_2 c_2}{\sqrt{(b_1^2 + b_2^2)(c_1^2 + c_2^2)}}$$

apparent efficiency,

$$\frac{P_1}{Q_0} = \frac{a_1 (1 - s)}{\sqrt{(b_1^2 + b_2^2)(c_1^2 + c_2^2)}}$$

torque efficiency, *

$$\frac{T}{P_0^1} = \frac{a_1}{b_1 c_1 + b_2 c_2}$$

* That is the ratio of actual torque to torque which would be produced, if there were no losses of energy in the motor, at the same power input.

apparent torque efficiency,*

$$\frac{T}{Q_0} = \frac{a_1}{\sqrt{(b_1^2 + b_2^2)(c_1^2 + c_2^2)}}$$

167. Most instructive in showing the behavior of an induction motor are the load curves and the speed curves.

The load curves are curves giving, with the power output as abscissae, the current imput, speed, torque, power factor, efficiency, and apparent efficiency, as ordinates.

The speed curves give, with the speed as abscissae, the torque, current input, power factor, torque efficiency, and apparent torque efficiency, as ordinates.

The load curves characterize the motor especially at its normal running speeds near synchronism, the speed curves over the whole range of speed.

In Fig. 119 are shown the load curves, and in Fig. 120 the speed curves of a motor of the constants,

$$Y_0 = .01 + .1j$$
$$Z_0 = .1 \quad - .3j$$
$$Z_1 = .1 \quad - .3j$$

INDUCTION GENERATOR.

168. In the foregoing, the range of speed from $s = 1$, standstill, to $s = 0$, synchronism, has been discussed. In this range the motor does mechanical work.

It consumes mechanical power, that is, acts as generator or as brake outside of this range.

For, $s > 1$, backwards driving, P becomes negative, representing consumption of power, while T remains positive; hence, since the direction of rotation has changed, represents consumption of power also. All this power is consumed in the motor, which thus acts as brake.

For, $s < 0$, or negative, P and T become negative, and the machine becomes an electric generator, converting mechanical into electric energy.

* That is the ratio of actual torque to torque which would be produced if there were neither losses of energy nor phase displacement in the motor, at the same voltampere input.

The calculation of the induction generator at constant frequency, that is, at a speed increasing with the load by the negative slip, s_1, is the same as that of the induction motor except that s_1 has negative values, and the load curves for the machine shown as motor in Fig. 119 are shown in Fig. 121 for negative slip s_1 as induction generator.

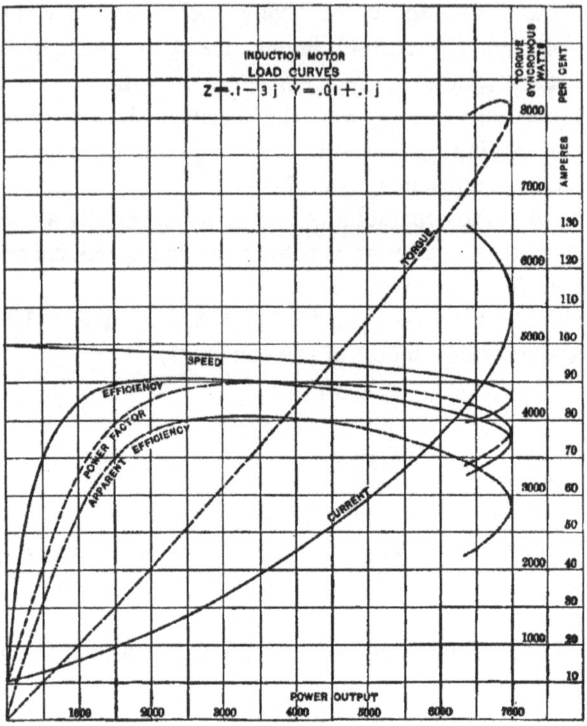

Fig. 119.

Again, a maximum torque point and a maximum output point are found, and the torque and power increase from zero at synchronism up to a maximum point, and then decrease again, while the current constantly increases.

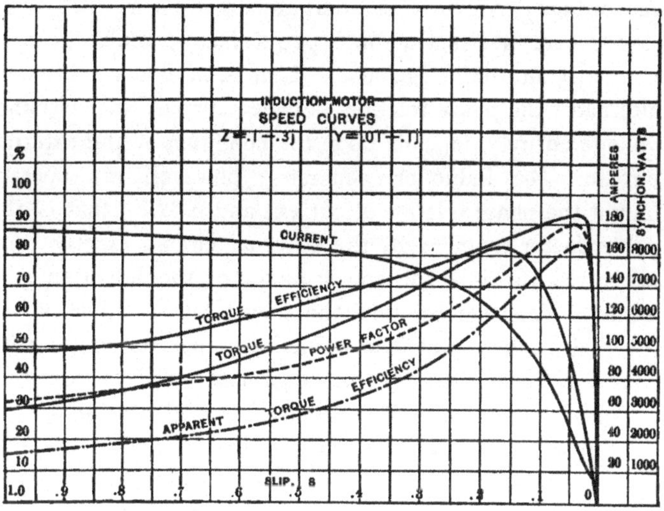

Fig. 120.

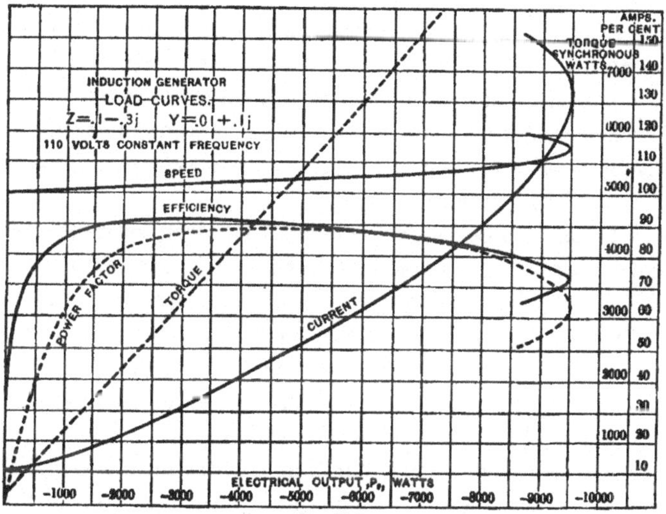

Fig. 121.

169. The induction generator differs essentially from the ordinary synchronous alternator in so far as the induction generator has a definite power factor, while the synchronous alternator has not. That is, in the synchronous alternator the phase relation between current and terminal voltage entirely depends upon the condition of the external circuit. The induction generator, however, can operate only if the phase relation of current and E.M.F., that is, the power factor required by the external circuit, exactly coincides with the internal power factor of the induction generator. This requires that the power factor either of the external circuit or of the induction generator varies with the voltage, so as to permit the generator and the external circuit to adjust themselves to equality of power factor.

Beyond magnetic saturation the power factor decreases; that is, the lead of current increases in the induction machine. Thus, when connected to an external circuit of constant power factor the induction generator will either not generate at all, if its power factor is lower than that of the external circuit, or, if its power factor is higher than that of the external circuit, the voltage will rise until by magnetic saturation in the induction generator its power factor has fallen to equality with that of the external circuit. This, however, requires magnetic saturation in the induction generator, which is objectionable, due to excessive hysteresis losses in the alternating field.

To operate below saturation,— that is, at constant internal power factor,—the induction generator requires an external circuit with leading current, whose power factor varies with the voltage, as a circuit containing synchronous motors or synchronous converters. In such a circuit, the voltage of the induction generator remains just as much below the counter E.M.F. of the synchronous motor as necessary to give the required leading exciting current of the induction generator, and the synchronous motor can thus to a certain extent be called the exciter of the induction generator.

When operating self-exciting, that is shunt-wound, converters from the induction generator, below saturation of both the converter and the induction generator, the conditions are unstable also, and the voltage of one of the two machines must rise beyond saturation of its magnetic field.

When operating in parallel with synchronous alternating generators, the induction generator obviously takes its leading exciting current from the synchronous alternator, which thus carries a lagging wattless current.

170. To generate constant frequency, the speed of the induction generator must increase with the load. Inversely, when driven at constant speed, with increasing load on the induction generator, the frequency of the current generated thereby decreases. Thus, when calculating the characteristic curves of the constant speed induction generator, due regard has to be taken of the decrease of frequency with increase of load, or what may be called the slip of frequency, s.

Let in an induction generator,

$$Y_0 = g_0 + jb_0 = \text{primary exciting admittance,}$$
$$Z_0 = r_0 - jx_0 = \text{primary self-inductive impedance,}$$
$$Z_1 = r_1 - jx_1 = \text{secondary self-inductive impedance,}$$

reduced to primary, all these quantities being reduced to the frequency of synchronism with the speed of the machine, N.

Let e = induced E.M.F., reduced to full frequency.

s = slip of frequency, thus: $(1-s)\,N$ = frequency generated by machine.

We then have

Secondary induced E.M.F.

$$se$$

thus, secondary current,

$$I_1 = \frac{se}{r_1 - jsx_1} - e\,(a_1 + ja_2)$$

where,

$$a_1 = \frac{sr_1}{r_1^2 + s^2x_1^2} \qquad a_2 = \frac{s^2x_1}{r_1^2 + s^2x_1^2}$$

primary exciting current,

$$\dot{I}_{00} = EY_0 = e\,(g_0 + jb_0)$$

thus, total primary current,

$$\dot{I}_0 = \dot{I}_1 + \dot{I}_{00} = e\,(b_1 + jb_2)$$

where,

$$b_1 = a_1 + g_0 \qquad b_2 = a_2 + b_0$$

primary impedance voltage,

$$\dot{E}^1 = \dot{I}_0\,(r_0 - j[1 - s]\,x_0)$$

primary induced E.M.F.,

$$e\,(1 - s)$$

thus, primary terminal voltage,

$$\dot{E}_0 = e\,(1 - s) - \dot{I}_0\,(r_0 - j\,[1 - s]\,x_0) = e\,(c_1 + jc_2)$$

where,

$$c_1 = 1 - s - r_0b_1 - (1 - s)\,x_0b_2 \qquad c_2 = (1 - s)\,x_0b_1 - r_0b_2$$

hence, absolute,

$$e_0 = e\,\sqrt{c_1^2 + c_2^2}$$

and,

$$e = \frac{e_0}{\sqrt{c_1^2 + c_2^2}}$$

Thus,

Secondary current,

$$\dot{I}_1 = \frac{e_0\,(a_1 + ja_2)}{\sqrt{c_1^2 + c_2^2}} \qquad I_1 = e_0\sqrt{\frac{a_1^2 + a_2^2}{c_1^2 + c_2^2}}$$

Primary current,

$$\dot{I}_0 = \frac{e_0\,(b_1 + jb_2)}{\sqrt{c_1^2 + c_2^2}} \qquad I_0 = e_0\sqrt{\frac{b_1^2 + b_2^2}{c_1^2 + c_2^2}}$$

Primary terminal voltage,

$$\dot{E}_0 = \frac{e_0\,(c_1 + jc_2)}{\sqrt{c_1^2 + c_2^2}}$$

Torque and mechanical power input,

$$T = P_1 = [e\ I_1]^1 = \frac{e_0^2 a_1}{c_1^2 + c_2^2}$$

Electrical output,

$$P_0 = P_0^1 + jP_0^j = [E_0 I_0] = [E_0 I_0]^1 + j[E_0 I_0]^j$$

$$= \frac{e_0^2}{c_1^2 + c_2^2}\left\{(b_1 c_1 + b_2 c_2) + j(b_2 c_1 - b_1 c_2)\right\}$$

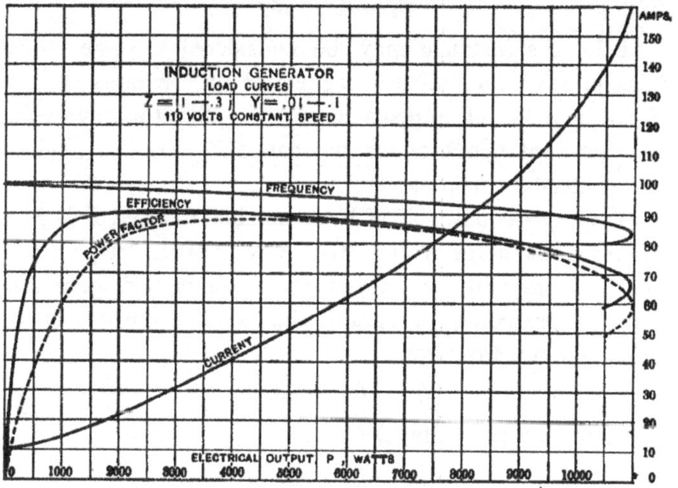

Fig. 122.

Voltampere output,

$$Q_0 = e_0 I_0$$

$$= e_0^2 \frac{\sqrt{b_1^2 + b_2^2}}{c_1^2 + c_2^2}$$

Efficiency,

$$\frac{P_0^1}{P_1} = \frac{b_1 c_1 + b_2 c_2}{a_1}$$

power factor,

$$\cos \hat{\omega} = \frac{P_0^1}{Q_0} = \frac{b_1 c_1 + b_2 + b_2 c_2}{\sqrt{(b_1^2 + b_2^2)(c_1^2 + c_2^2)}}$$

or,

$$\tan \hat{\omega} = \frac{P_0^j}{P_0^1} = \frac{b_2 c_1 - b_1 c_2}{b_1 c_1 + b_2 c_2}$$

In Fig. 122 is plotted the load characteristic of a constant speed induction generator, at constant terminal voltage $e_0 = 110$, and the constants,

$$Y_0 = .01 + .1j$$
$$Z_0 = .1 \quad - .3j$$
$$Z_1 = .1 \quad - .3j$$

171. As instance may be considered a power transmission from an induction generator of constants Y_0, Z_0, Z_1, over a line of impedance $Z = r - jx$, into a synchronous motor of synchronous impedance $Z_2 = r_2 - jx_2$, operating at constant field excitation.

Let, e_0 = counter E.M.F. or nominal induced E.M.F. of synchronous motor at full frequency; that is, frequency of synchronism with the speed of the induction generator. By the preceding paragraph the primary current of the induction generator was,

$$\dot{I}_0 = e\,(b_1 + jb_2)$$

primary terminal voltage,

$$\dot{E}_0 = e\,(c_1 + jc_2)$$

thus, terminal voltage at synchronous motor terminals,

$$\dot{E}_0' = \dot{E}_0 - \dot{I}_0\,(r - j[1 - s]\,x)$$
$$= e\,(d_1 + jd_2)$$

where,

$$d_1 = c_1 - r_1 b_1 - (1 - s)\,x_1 b_2 \qquad d_2 = (1 - s)\,x_1 b_1 - r_1 b_2$$

Counter E.M.F. of synchronous motor,

$$\dot{E}_2 = \dot{E}_0' - \dot{I}_0\,(r_2 - j[1 - s]\,x_2)$$
$$= e\,(f_1 + jf_2)$$

where,

$$f_1 = d_1 - r_2 b_1 - (1 - s)\,x_2 b_2 \qquad f_2 = (1 - s)\,x_2 b_1 - r_2 b_2$$

or absolute,

$$E_2 = e\,\sqrt{f_1^2 + f_2^2}$$

since, however,

$$E_2 = e_2(1 - s)$$

we have,

$$e = \frac{e_2(1 - s)}{\sqrt{f_1^2 + f_2^2}}$$

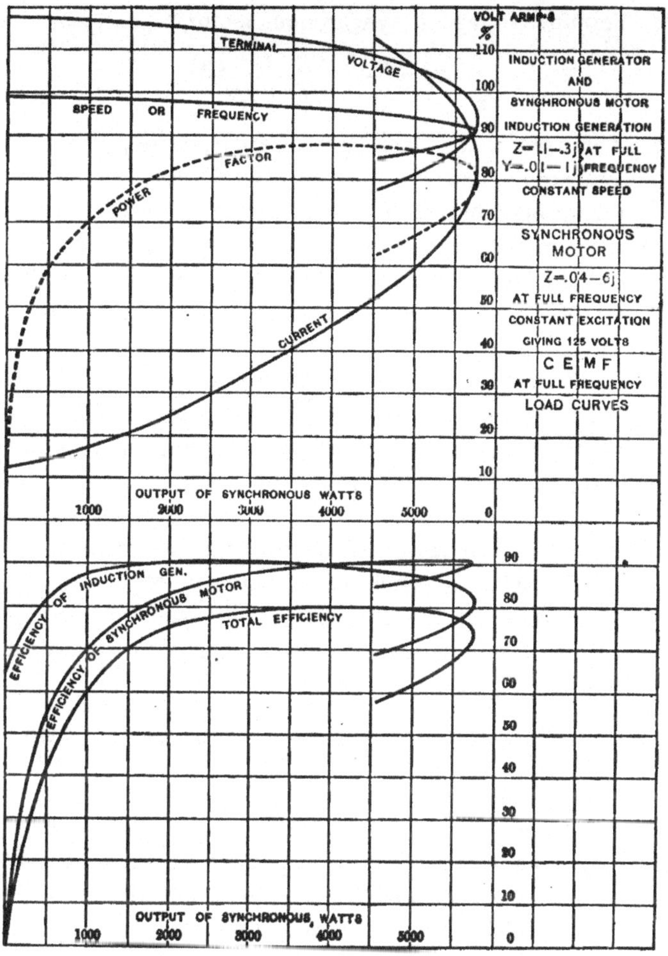

Fig. 123.

Thus,

Current, $$I_0 = \frac{e_2 (1 - s)(b_1 + jb_2)}{\sqrt{f_1^2 + f_2^2}}$$

Terminal voltage at induction generator,

$$E_3 = \frac{e_2 (1 - s)(c_1 + jc_2)}{\sqrt{f_1^2 + f_2^2}}$$

Terminal voltage at synchronous motor,

$$E_0' = \frac{e_2 (1 - s)(d_1 + jd_2)}{\sqrt{f_1^2 + f_2^2}}$$

and herefrom in the usual way the efficiencies, power factor, etc. are derived.

When operated from an induction generator, a synchronous motor gives a load characteristic very similar to that of an induction motor operated from a synchronous generator, but in the former case the current is leading, in the latter lagging.

In either case, the speed gradually falls off with increasing load (in the synchronous motor, due to the falling off of the frequency of the induction generator), up to a maximum output point, where the motor drops out of step and comes to standstill.

Such a load characteristic of the induction generator in Fig. 121, feeding a synchronous motor of counter E.M.F. $e_0 = 125$ volts (at full frequency) and synchronous impedance $Z_2 = .04 - 6j$, over a line of negligible impedance is shown in Fig. 123.

CONCATENATION, OR TANDEM CONTROL OF INDUCTION MOTORS.

172. If of two induction motors the secondary of the first motor is connected to the primary of the second motor, the second machine operates as motor with the E.M.F. and frequency impressed upon it by the secondary of the first machine, which acts as general alternating-current transformer, converting a part of the primary impressed power into secondary electrical power for the supply of the second machine, and a part into mechanical work.

The frequency of the secondary E.M.F. of the first motor, and thus the frequency impressed upon the second motor, is the frequency of slip below complete synchronism, s. The frequency of the secondary induced E.M.F. of the second motor is the difference between its impressed frequency, s, and its speed; thus, if both motors are connected together mechanically to turn at the same speed, $1 - s$, the secondary frequency of the second motor is $2s - 1$, hence equal to zero at $s = .5$. That is, the second motor reaches its synchronism at half speed. At this speed its torque becomes equal to zero, the energy current flowing into it, and consequently the energy component of the secondary current of the first motor, and thus the torque of the first motor becomes equal to zero also, when neglecting the hysteresis energy current of the second motor. That is, a system of concatenated motors with short-circuited secondary of the second motor approaches half synchronism, in the same manner as the ordinary induction motor approaches synchronism. With increasing load, its slip below half synchronism increases.

More generally, any pair of induction motors connected in concatenation divide the speed so that the sum of their two respective speeds approaches synchronism at no load; or, still more generally, any number of concatenated motors run at such speeds that the sum of the speeds approaches synchronism at no load.

With mechanical connection between the two motors, concatenation thus offers a means to operate a pair of induction motors at full efficiency at half speed in tandem, as well as at full speed in parallel, and thus gives the same advantage as the series-parallel control of the continuous-current motor.

In starting, a concatenated system is controlled by resistance in the armature of the second motor.

Since, with increasing speed, the frequency impressed upon the second motor decreases proportionally to the decrease of voltage, when neglecting internal losses in the first motor, the magnetic density of the second motor remains practically constant, and thus its torque the same as when operated at full voltage and full frequency under the same conditions.

At half synchronism the torque of the concatenated couple becomes zero, and above half synchronism the second motor runs beyond its impressed frequency; that is, becomes generator. In this case, due to the reversal of current in the secondary of the first motor, its torque becomes negative also, that is the concatenated couple becomes induction generator above half synchronism. At about two-thirds synchronism, with low resistance armature, the torque of the couple becomes zero again, and once more positive between about two-thirds synchronism and full synchronism, and negative once more beyond full synchronism. With high resistance in the secondary of the second motor, the second range of positive torque, below full synchronism, disappears, more or less.

173. The calculation of a concatenated couple of induction motors is as follows,

Let

N = frequency of main circuit,
s = slip of the first motor from synchronism.

the frequency induced in the secondary of the first motor and thus impressed upon the primary of the second motor is, sN.

The speed of the first motor is $(1-s)N$, thus the slip of the second motor, or the frequency induced in its secondary, is

$$sN - (1-s)N = (2s-1)N.$$

Let

e = counter E.M.F. induced in the secondary of the second motor, reduced to full frequency.

$Z_0 = r_0 - jx_0$ = primary self-inductive impedance.

$Z_1 = r_1 - jx_1$ = secondary self-inductance impedance.

$Y = g + jb$ = primary exciting admittance of each motor, all reduced to full frequency and to the primary by the ratio of turns.

We then have,

Second motor,

secondary induced E.M.F.,

$$e\,(2s - 1)$$

secondary current,

$$I_1 = \frac{e\,(2s-1)}{r_1 - j\,(2s-1)\,x_1} = e\,(a_1 + ja_2)$$

where,

$$a_1 = \frac{(2s-1)\,r_1}{r_1^2 + (2s-1)^2 x_1^2} \qquad a_2 = \frac{(2s-1)^2 x_1}{r_1^2 + (2s-1)^2 x_1^2}$$

primary exciting current,

$$I_0 = e\,(g + jb)$$

thus, total primary current,

$$I_2 = I_1 + I_0 = e\,(b_1 + jb_2)$$

where,

$$b_1 = a_1 + g \qquad b_2 = a_2 + b$$

primary induced E.M.F.,

$$se$$

primary impedance voltage,

$$I_2\,(r_0 - jsx_0)$$

thus, primary impressed E.M.F.,

$$E_2 = se + I_2\,(r_0 - jsx_0) = e\,(c_1 + jc_2)$$

where,

$$c_1 = s + r_0 b_1 + sx_0 b_2 \qquad c_2 = r_0 b_2 - sx_0 b_1$$

First motor,

secondary current,

$$I_2 = e\,(b_1 + jb_2)$$

secondary induced E.M.F.,

$$\dot{E}_3 = \dot{E}_2 + \dot{I}_2(r_1 - jsx_1) = e(d_1 + jd_2)$$

where,

$$d_1 = c_1 + r_1 b_1 + sx_1 b_2 \qquad d_2 = c_2 + r_1 b_2 - sx_1 b_1$$

primary induced E.M.F.,

$$\dot{E}_4 = \frac{\dot{E}_3}{s} = e(f_1 + jf_2)$$

where,

$$f_1 = \frac{d_1}{s} \qquad f_2 = \frac{d_2}{s}^*$$

primary exciting current,

$$\dot{I}_4 = \dot{E}_4 (g + jb)$$

total primary current,

$$\dot{I} = \dot{I}_2 + \dot{I}_4 = e(g_1 + jg_2)$$

where,

$$g_1 = b_1 + gf_1 - bf_2 \qquad g_2 = b_2 + gf_2 + bf_1$$

primary impedance voltage,

$$\dot{I}(r_0 - jx_0)$$

thus, primary impressed E.M.F.,

$$\dot{E}_0 = \dot{E}_4 + \dot{I}(r_0 - jx_0) = e(h_1 + jh_2)$$

where,

$$h_1 = f_1 + r_0 g_1 + x_0 g_2 \qquad h_2 = f_2 + r_0 g_2 - x_0 g_1$$

or, absolute,

$$e_0 = e\sqrt{h_1^2 + h_2^2}$$

and,

$$e = \frac{e_0}{\sqrt{h_1^2 + h_2^2}}$$

Substituting now this value of e in the preceding gives the values of the currents and E.M.F.'s in the different circuits of the motor series.

* At $s = 0$ these terms f_1 and f_2 become indefinite, and thus at and very near synchronism have to be derived by substituting the complete expressions for f_1 and f_2.

In the second motor, the torque is,

$$T_2 = [e\dot{I}_1]^1 = e^2 a_1$$

hence, its power output,

$$P_3 = (1 - s)\ T_2 = (1 - s)\ e^2 a_1$$

The power input is,

$$\begin{aligned} P_2 &= [\dot{E}_2 \dot{I}_2] = [\dot{E}_2 \dot{I}_2]^1 + j[\dot{E}_2 \dot{I}_2]^j \\ &= e^2 [(c_1 + jc_2)\ (b_1 + jb_2)] \end{aligned}$$

hence, the efficiency,

$$\frac{P_3}{P_2^1} = \frac{(1 - s)\ e^2 a_1}{[\dot{E}_2 \dot{I}_2]^1} = \frac{(1 - s)\, a_1}{c_1 b_1 + c_2 b_2}$$

the power factor,

$$\frac{P_2^1}{Q_2} = \frac{[\dot{E}_2 \dot{I}_2]^1}{E_2 I_2} = \frac{c_1 b_1 + c_2 b_2}{\sqrt{(c_1^2 + c_2^2)\ (b_1^2 + b_2^2)}}$$

etc.

In the first motor,

the torque is,

$$\begin{aligned} T_1 &= [\dot{E}_1 \dot{I}_2]^1 = e^2 [(f_1 + jf_2)\ (b_1 + jb_2)]^1 \\ &= e^2\ (f_1 b_1 + f_2 b_2) \end{aligned}$$

the power output,

$$\begin{aligned} P_4 &= T_1\ (1 - s) \\ &= e^2\ (1 - s)\ (f_1 b_1 + f_2 b_2) \end{aligned}$$

the power input,

$$\begin{aligned} P_1 &= [\dot{E}_0 \dot{I}] = e^2 [(h_1 + jh_2)\ (g_1 + jg_2)] \\ &= [\dot{E}_0 \dot{I}]^1 + j[\dot{E}_0 \dot{I}]^j \end{aligned}$$

Thus, the efficiency,

$$\frac{P_4}{[\dot{E}_0 \dot{I}]^1 - [\dot{E}_2 \dot{I}_2]^1} = \frac{(1 - s)\ (f_1 b_1 + f_2 b_2)}{(h_1 g_1 + h_2 g_2) - (c_1 b_1 + c_2 b_2)}$$

the power factor of the whole system,

$$\frac{P_1}{E_0 I} = \frac{h_1 g_1 + h_2 g_2}{\sqrt{(h_1^2 + h_2^2)\ (g_1^2 + g_2^2)}}$$

the power factor of the first motor,

$$\frac{P_1 - P_2}{E_0 I - E_2 I_2} = \frac{(h_1 g_1 + h_2 g_2) - (c_1 b_1 + c_2 b_2)}{\sqrt{(h_1^2 + h_2^2)(g_1^2 + g_2^2)} - \sqrt{(c_1^2 + c_2^2)(b_1^2 + b_2^2)}}$$

the total efficiency of the system,

$$\frac{P_4 + P_3}{[E_0 I]^1} = \frac{(1 - s)(f_1 b_1 + f_2 b_2 + a_1)}{h_2 g_1 + h_2 g_2}$$

etc.

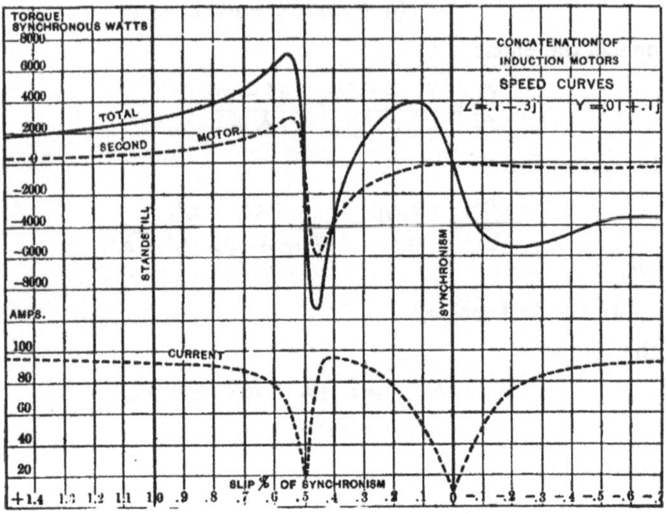

Fig. 124. Concatenation of Induction Motors. Speed Curves.
$Z = .1 - .3j \quad Y = .01 + .1j$

174. As instance are given in Fig. 124, the curves of total torque, of torque of the second motor, and of current, for the range of slip from $s = +1.5$ to $s = -.7$ for a pair of induction motors in concatenation, of the constants:

$$Z_0 = Z_1 = .1 - .3j$$
$$Y = .01 + .1j.$$

As seen, there are two ranges of positive torque for the whole system, one below half synchronism, and one from about two-thirds to full synchronism, and two ranges of negative torque, or generator action of the motor, from half to two-third synchronism, and above full synchronism.

With higher resistance in the secondary of the second motor, the second range of positive torque of the system disappears more or less, and the torque curves become as shown in Fig. 125.

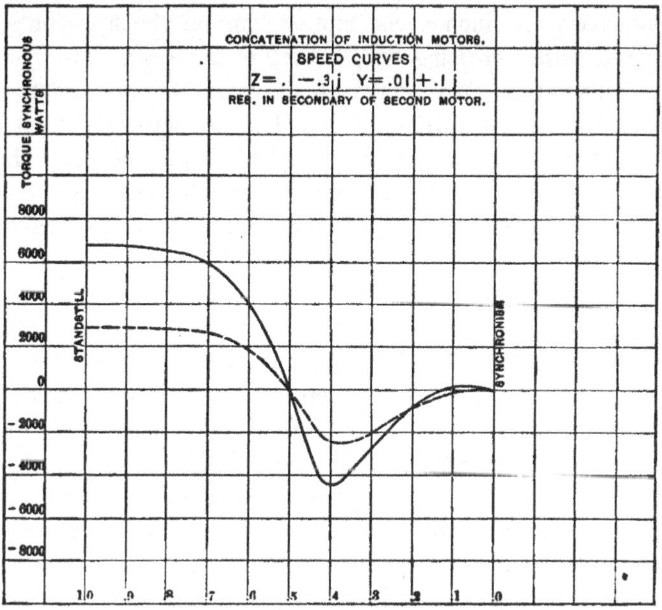

Fig. 125. Concatenation of Induction Motors. Speed Curves.

$Z = .1 - .3j \qquad Y = .01 + .1j$

SINGLE-PHASE INDUCTION MOTOR.

175. The magnetic circuit of the induction motor at or near synchronism consists of two magnetic fluxes superimposed upon each other in quadrature, in time, and in position. In the polyphase motor these fluxes are produced by E.M.Fs. displaced in phase. In the monocyclic motor one of the fluxes is due to the primary energy circuit, the other to the primary exciting circuit. In the single-phase motor the one flux is produced by the primary circuit, the other by the currents induced in the secondary or armature, which are carried into quadrature position by the rotation of the armature. In consequence thereof, while in all these motors the magnetic distribution is the same at or near synchronism, and can be represented by a rotating field of uniform intensity and uniform velocity, it remains such in polyphase and monocyclic motors; but in the single-phase motor, with increasing slip, — that is, decreasing speed, — the quadrature field decreases, since the induced armature currents are not carried to complete quadrature position; and thus only a component available for producing the quadrature flux. Hence, approximately, the quadrature flux of a single-phase motor can be considered as proportional to its speed; that is, it is zero at standstill.

Since the torque of the motor is proportional to the product of secondary current times magnetic flux in quadrature, it follows that the torque of the single-phase motor is equal to that of the same motor under the same condition of operation on a polyphase circuit, multiplied with the speed; hence equal to zero at standstill.

Thus, while single-phase induction motors are quite satisfactory at or near synchronism, their torque decreases proportionally to the speed, and becomes zero at standstill. That is, they are not self-starting, but some starting device has to be used.

Such a starting device may either be mechanical or electrical. All the electrical starting devices essentially consist in impressing upon the motor at standstill a magnetic quadrature flux. This may be produced either by some outside E.M.F., as in the monocyclic starting device, or by displacing the circuits of two or more primary coils from each other, either by mutual induction between the coils, — that is, by using one as secondary to the other, — or by impedances of different inductance factors connected with the different primary coils.

176. The starting-devices of the single-phase induction motor by producing a quadrature magnetic flux can be subdivided into three classes:

1. Phase-Splitting Devices. Two or more primary circuits are used, displaced in position from each other, and either in series or in shunt with each other, or in any other way related, as by transformation. The impedances of these circuits are made different from each other as much as possible, to produce a phase displacement between them. This can be done either by inserting external impedances into the circuits, as a condenser and a reactive coil, or by making the internal impedances of the motor circuits different, as by making one coil of high and the other of low resistance.

2. Inductive Devices. The different primary circuits of the motor are inductively related to each other in such a way as to produce a phase displacement between them. The inductive relation can be outside of the motor or inside, by having the one coil induced by the other; and in this latter case the current in the induced coil may be made leading, accelerating coil, or lagging, shading coil.

3. Monocyclic Devices. External to the motor an essentially wattless E.M.F. is produced in quadrature with the main E.M.F. and impressed upon the motor, either directly or after combination with the single-phase main E.M.F. Such wattless quadrature E.M.F. can be produced by the common connection of two impedances of different power factor, as an inductance and a resistance, or an inductance and a condensance connected in series across the mains.

The investigation of these starting-devices offers a very instructive application of the symbolic method of investigation of alternating-current phenomena, and a study thereof is thus recommended to the reader.*

* See paper on the Single-phase Induction Motor, A.I.E.E. Transactions, 1898.

177. As a rule, no special motors are built for single-phase operation, but polyphase motors used in single-phase circuits, since for starting the polyphase primary winding is required, the single primary coil motor obviously not allowing the application of phase-displacing devices for producing the starting quadrature flux.

Since at or near synchronism, at the same impressed E.M.F. — that is, the same magnetic density — the total voltamperes excitation of the single-phase induction motor must be the same as of the same motor on polyphase circuit, it follows that by operating a quarter-phase motor from single-phase circuit on one primary coil, its primary exciting admittance is doubled. Operating a three-phase motor single-phase on one circuit its primary exciting admittance is trebled. The self-inductive primary impedance is the same single-phase as polyphase, but the secondary impedance reduced to the primary is lowered, since in single-phase operation all secondary circuits correspond to the one primary circuit used. Thus the secondary impedance in a quarter-phase motor running single-phase is reduced to one-half, in a three-phase motor running single-phase reduced to one-third. In consequence thereof the slip of speed in a single-phase induction motor is usually less than in a polyphase motor; but the exciting current is considerably greater, and thus the power factor and the efficiency are lower.

The preceding considerations obviously apply only when running so near synchronism that the magnetic field of the single-phase motor can be assumed as uniform, that is the cross magnetizing flux produced by the armature as equal to the main magnetic flux.

When investigating the action of the single-phase motor at lower speeds and at standstill, the falling off of the magnetic quadrature flux produced by the armature current, the change of secondary impedance, and where a starting device is used the effect of the magnetic field produced by the starting device, have to be considered.

The exciting current of the single-phase motor consists of the primary exciting current or current producing the main magnetic flux, and represented by a constant admittance Y_0^1, the primary exciting admittance of the motor, and the secondary exciting current, that is that component of primary current corresponding to the secondary current which gives the excitation for the quadrature magnetic flux. This latter magnetic flux is equal to the main magnetic flux Φ_0 at synchronism, and falls off with decreasing speed to zero at standstill, if no starting device is used or to $\Phi_1 = t\Phi_0$ at standstill if by a starting device a quadrature magnetic flux is impressed upon the motor, and at standstill t = ratio of quadrature or starting magnetic flux to main magnetic flux.

Thus the secondary exciting current can be represented by an admittance Y_1^1 which changes from equality with the primary exciting admittance Y_0^1 at synchronism, to $Y_1^1 = 0$, respectively to $Y_1^1 = t\,Y_0^1$ at standstill. Assuming thus that the starting device is such that its action is not impaired by the change of speed, at slip s the secondary exciting admittance can be represented by:

$$Y_1^1 = [1 - (1 - t)\,s]\;Y_0^1$$

The secondary impedance of the motor at synchronism is the joint impedance of all the secondary circuits, since all secondary circuits correspond to the same primary circuit, hence $= \frac{Z_1}{3}$ with a three-phase secondary, and $= \frac{Z_1}{2}$ with a two-phase secondary with impedance Z_1 per circuit.

At standstill, however, the secondary circuits correspond to the primary circuit only with their projection in the direction of the primary flux, and thus as resultant only one-half of the secondary circuits are effective, so that the secondary impedance at standstill is equal to $2\,Z_1 / 3$ with a three-phase, and equal to Z_1 with a two-phase secondary. Thus the effective secondary impedance of the single-phase motor changes with the speed and can at the slip s be represented by $Z_1^1 = \frac{(1+s)Z_1}{3}$ in a three-phase motor, and $Z_1^1 = \frac{(1+s)Z_1}{2}$ in a two-phase motor, with the impedance Z_1 per secondary circuit.

In the single-phase motor without starting device, due to the falling off of the quadrature flux, the torque at slip s is:

$$T = a_1 e^2 (1 - s)$$

In a single-phase motor with a starting device which at standstill produces a ratio of magnetic fluxes t, the torque at standstill is;

$$T_0 = t T_1$$

where T_1 = total torque of the same motor on polyphase circuit.

Thus denoting the value $\frac{T_0}{a_1 e^2} = v$

the single-phase motor torque at standstill is:

$$T_0 = a_1 e^2 v$$

and the single-phase motor torque at slip s is:

$$T = a_1 e^2 [1 - (1 - v) s]$$

178. In the single-phase motor considerably more advantage is gained by compensating for the wattless magnetizing component of current by capacity than in the polyphase motor, where this wattless current is relatively small. The use of shunted capacity, however, has the disadvantage of requiring a wave of impressed E.M.F. very close to sine shape; since even with a moderate variation from sine shape the wattless charging current of the condenser of higher frequency may lower the power factor more than the compensation for the wattless component of the fundamental wave raises it, as will be seen in the chapter on General Alternating Current Waves.

Thus the most satisfactory application of the condenser in the single-phase motor is not in shunt to the primary circuit, but in a tertiary circuit; that is, in a circuit stationary with regard to the primary impressed circuit, but induced by the revolving secondary circuit.

In this case the condenser is supplied with an E.M.F. transformed twice, from primary to secondary, and from secondary to tertiary, through multitooth structures in a uniformly revolving field, and thus a very close approximation to sine wave produced at the condenser, irrespective of the wave shape of primary impressed E.M.F.

With the condenser connected into a tertiary circuit of a single-phase induction motor, the wattless magnetizing current of the motor is supplied by the condenser in a separate circuit, and the primary coil carries the energy current only, and thus the efficiency of the motor is essentially increased.

The tertiary circuit may be at right angles to the primary, or under any other angle. Usually it is applied on an angle of 60°, so as to secure a mutual induction between tertiary and primary for starting, which produces in starting in the condenser a leading current, and gives the quadrature magnetic flux required.

179. The most convenient way to secure this arrangement is the use of a three-phase motor which with two of its terminals 1–2, is connected to the single-phase mains, and with terminals 1 and 3 to a condenser.

Let $Y_0 = g_0 + jb_0$ = primary exciting admittance of the motor per delta circuit.

$Z_0 = r_0 - jx_0$ = primary self-inductive impedance per delta circuit.

$Z_1 = r_1 - jx_1$ = secondary self-inductive impedance per delta circuit reduced to primary.

Let

$Y_3 = g_3 - jb_3$ = admittance of the condenser connected between terminals 1 and 3.

If then, as single-phase motor,

t = ratio of auxiliary quadrature flux to main flux in starting,

h = ratio of E.M.F. induced in condenser circuit to E.M.F. induced in main circuit in starting,

$$v = \frac{\text{starting torque}}{a_1 e^2 \text{ in starting}}.$$

It is single-phase

$Y_0^1 = 1.5\ Y_0 = 1.5\ (g_0 + jb_0)$ = primary exciting admittance,

$Y_1^1 = 1.5\ Y_0\,[1 - (1 - t)\ s]$
$= 1.5\ (g_0 + jb_0)\ [1 - (1 - t)\ s]$ = secondary exciting admittance at slip s.

$Z_0^1 = \frac{2\,Z_0}{3} = \frac{2\,(r_0 - jx_0)}{3}$, = primary self-inductive impedance.

$Z_1^1 = \frac{(1 + s)}{3}\,Z_1 = \frac{(1 + s)}{3}\,(r_1 - jsx_1)$ = secondary self-inductive impedance.

$Z_2^1 = \frac{2\,Z_0}{3} = \frac{2\,(r_0 - jx_0)}{3}$ = tertiary self-inductive impedance of motor.

Thus,

$$Y_4 = \frac{1}{Z_2^1 + \frac{1}{Y_3}} = \text{total admittance of tertiary circuit}$$

Since the E.M.F. induced in the tertiary circuit decreases from e at synchronism to he at standstill, the effective tertiary admittance or admittance reduced to an induced E.M.F. e is at slip s

$$Y_4^1 = [1 - (1 - h)\ s]\ Y_4$$

Let then,

e = counter E.M.F. of primary circuit,
s = slip.

We have,

secondary load current

$$I_1 = \frac{se}{Z_1^1} = \frac{3\,se}{(1+s)\,(r_1 - jsx_1)} = e\,(a_1 + ja_2)$$

secondary exciting current

$$I_1^1 = eY_1^1 = 1.5\,eY_0\,[1 - (1 - t)\,s]$$

secondary condenser current

$$I_4 = eY_4^1 = eY_4\,[1 - (1 - h)\,s]$$

thus, total secondary current

$$I^1 = I_1 + I_1^1 + I_4$$

primary exciting current

$$I_0^1 = eY_0^1 = 1.5\,eY_0$$

thus, total primary current

$$\begin{aligned} I_0 &= I^1 + I_0^1 \\ &= I_1 + I_4 + I_1^1 + I_0^1 \\ &= e\,(b_1 + jb_2) \end{aligned}$$

primary impressed E.M.F.

$$\begin{aligned} E_0 &= e + Z_0^1 I_0 \\ &= e\,(c_1 + jc_2) \end{aligned}$$

thus, main counter E.M.F.

$$e = \frac{E_0}{c_1 + jc_2}$$

or,

$$E = \frac{e_0}{c_1 + jc_2}$$

and, absolute

$$e = \frac{e_0}{\sqrt{c_1^2 + c_2^2}}$$

hence, primary current

$$I_0 = \frac{e_0\,(b_1 + jb_2)}{c_1 + jc_2} \qquad I_0 = e_0\sqrt{\frac{b_1^2 + b_2^2}{c_1^2 + c_2^2}}$$

voltampere input,

$$Q_0 = e_0 I_0$$

power input

$$P_0 = [I_0 e_0]^1 = e_0^2 \frac{b_1 c_1 + b_2 c_2}{c_1^2 + c_2^2}$$

torque at slip s

$$T = T^1 [1 - (1 - v) s]$$
$$= \frac{e_0^2 a_1}{c_1^2 + c_2^2} [1 - (1 - v) s]$$

and, power output

$$P = T(1 - s)$$
$$= \frac{e_0^2 a_1}{c_1^2 + c_2^2} (1 - s) [1 - (1 - v) s]$$

and herefrom in the usual manner the efficiency, apparent efficiency, torque efficiency, apparent torque efficiency, and power factor.

The derivation of the constants t, h, v, which have to be determined before calculating the motor, is as follows:

Let e_0 = single-phase impressed E.M.F.,
Y = total stationary admittance of motor per delta circuit,
E_3 = E.M.F. at condenser terminals in starting.

In the circuit between the single-phase mains from terminal 1 over terminal 3 to 2, the admittances $Y + Y_3$, and Y, are connected in series, and have the respective E.M.Fs. E_3 and $e_0 - E_3$. It is thus,

$$Y + Y_3 \div Y = e_0 - E_3 \div E_3,$$

since with the same current passing through both circuits, the impressed E.M.Fs. are inverse proportional to the respective admittances.

Thus,

$$E_3 = \frac{e_0 Y}{2Y + Y_3} = e_0 (h_1 + j h_2)$$

and quadrature E.M.F.

$$e_0 h_2$$

hence

$$E_2 = e_0 \sqrt{h_1^2 + h_2^2}$$

thus

$$h = \sqrt{h_1^2 + h_2^2}$$

Since in the three-phase E.M.F. triangle, the altitude corresponding to the quadrature magnetic flux $= \frac{e_0}{2\sqrt{3}}$, and the quadrature and main fluxes are equal, in the single-phase motor the ratio of quadrature to main flux is

$$t = \frac{2 h_2}{\sqrt{3}} = 1.155\, h_2$$

From t, v is derived as shown in the preceding.

For further discussion on the Theory and Calculation of the Single-phase Induction Motor, see American Institute Electrical Engineers Transactions, January, 1900.

SYNCHRONOUS INDUCTION MOTOR.

180. The induction motor discussed in the foregoing consists of one or a number of primary circuits acting upon a movable armature which comprises a number of closed secondary circuits displaced from each other in space so as to offer a resultant circuit in any direction. In consequence thereof the motor can be considered as a transformer, having to each primary circuit a corresponding secondary circuit, — a secondary coil, moving out of the field of the primary coil, being replaced by another secondary coil moving into the field.

In such a motor the torque is zero at synchronism, positive below, and negative above, synchronism.

If, however, the movable armature contains one closed circuit only, it offers a closed secondary circuit only in the direction of the axis of the armature coil, but no secondary circuit at right angles therewith. That is, with the rotation of the armature the secondary circuit, corresponding to a primary circuit, varies from short circuit at coincidence of the axis of the armature coil with the axis of the primary coil, to open circuit in quadrature therewith, with the periodicity of the armature speed. That is, the apparent admittance of the primary circuit varies periodically from open-circuit admittance to the short-circuited transformer admittance.

At synchronism such a motor represents an electric circuit of an admittance varying with twice the periodicity of the primary frequency, since twice per period the axis of the armature coil and that of the primary coil coincide. A varying admittance is obviously identical in effect with a varying reluctance, which will be discussed in the chapter on reaction machines. That is, the induction motor with one closed armature circuit is, at synchronism, nothing but a reaction machine, and consequently gives zero torque at synchronism if the maxima and minima of the periodically varying admittance coincide with the maximum and zero values of the primary circuit, but gives a definite torque if they are displaced therefrom. This torque may be positive or negative according to the phase displacement between admittance and primary circuit; that is, the lag or lead of the maximum admittance with regard to the primary maximum. Hence an induction motor with single-armature circuit at synchronism acts either as motor or as alternating-current generator according to the relative position of the armature circuit to the primary circuit. Thus it can be called a synchronous induction motor or synchronous induction generator, since it is an induction machine giving torque at synchronism.

Power factor and apparent efficiency of the synchronous induction motor as reaction machine are very low. Hence it is of practical application only in cases where a small amount of power is required at synchronous rotation, and continuous current for field excitation is not available.

The current induced in the armature of the synchronous induction motor is of double the frequency impressed upon the primary.

Below and above synchronism the ordinary induction motor, or induction generator, torque is superimposed upon the synchronous induction machine torque. Since with the frequency of slip the relative position of primary and of secondary coil changes, the synchronous induction machine torque alternates periodically with the frequency of slip. That is, upon the constant positive or negative torque below or above synchronism an alternating torque of the frequency of slip is superimposed, and thus the resultant torque pulsating with a positive mean value below, a negative mean value above, synchronism.

When started from rest, a synchronous induction motor will accelerate like an ordinary single-phase induction motor, but not only approach synchronism, as the latter does, but run up to complete synchronism under load. When approaching synchronism it makes definite beats with the frequency of slip, which disappear when synchronism is reached.

THE HYSTERESIS MOTOR.

181. In a revolving magnetic field, a circular iron disk, or iron cylinder of uniform magnetic reluctance in the direction of the revolving field, is set in rotation, even if subdivided so as to preclude the induction of eddy currents. This rotation is due to the effect of hysteresis of the revolving disks or cyclinder, and such a motor may thus be called a hysteresis motor.

Let *I* be the iron disk exposed to a rotating magnetic field or resultant M.M.F. The axis of resultant magnetization in the disk *I* does not coincide with the axis of the rotating field, but lags behind the latter, thus producing a couple. That is, the component of magnetism in a direction of the rotating disk, *I*, ahead of the axis of rotating M.M.F., is rising, thus below, and in a direction behind the axis of rotating M.M.F. decreasing; that is, above proportionality with the M.M.F., in consequence of the lag of magnetism in the hysteresis loop, and thus the axis of resultant magnetism in the iron disk, I, does not coincide with the axis of rotating M.M.F., but is shifted backwards by an angle, α, which is the angle of hysteretic lead in Chapter X., § 79.

The induced magnetism gives with the resultant M.M.F. a mechanical couple, —

$$T = mF\Phi \sin \alpha,$$

where

F = resultant M.M.F.,
Φ = resultant magnetism,
α = angle of hysteretic advance of phase,
m = a constant.

The apparent or voltampere input of the motor is, —

$$Q = mF\Phi.$$

Thus the apparent torque efficiency, —

$$\frac{T}{Q} = \sin \alpha,$$

and the power of the motor is, —

$$P = (1 - s)\,T = (1 - s)\,mF\phi \sin \alpha,$$

where

s = slip as fraction of synchronism.

The apparent efficiency is, —

$$\frac{P}{Q} = (1 - s) \sin \alpha.$$

Since in a magnetic circuit containing an air gap the angle α is extremely small, a few degrees only, it follows that the apparent efficiency of the hysteresis motor is extremely low, the motor consequently unsuitable for producing larger amounts of mechanical work.

From the equation of torque it follows, however, that at constant impressed E.M.F., or current, — that is, constant F, – the torque is constant and independent of the speed; and therefore such a motor arrangement is suitable, and occasionally used as alternating-current meter.

The same result can be reached from a different point of view. In such a magnetic system, comprising a movable iron disk, I, of uniform magnetic reluctance in a revolving field, the magnetic reluctance — and thus the distribution of magnetism — is obviously independent of the speed, and consequently the current and energy expenditure of the impressed M.M.F. independent of the speed also. If, now, —

$V =$ volume of iron of the movable part,
$B =$ magnetic density,
and $\eta =$ coefficient of hysteresis,

the energy expended by hysteresis in the movable disk, I, is per cycle, —

$$W_0 = V \eta B^{1.6},$$

hence, if $N =$ frequency, the energy supplied by the M.M.F. to the rotating iron disk in the hysteretic loop of the M.M.F. is, —

$$P_0 = N V \eta B^{1.6}.$$

At the slip, $s N$, that is, the speed $(1 - s) N$, the energy expended by hysteresis in the rotating disk is, however, —

$$P_1 = s N V \eta B^{1.6}$$

Hence, in the transfer from the stationary to the revolving member the magnetic energy, —

$$P = P_0 - P_1 = (1 - s) N V \eta B^{1.6}$$

has disappeared, and thus reappears as mechanical work, and the torque is, —

$$F = \frac{P}{(1 - s) N} = V \eta B^{1.6},$$

that is, independent of the speed.

Since, as seen in Chapter X., sin α is the ratio of the energy of the hysteretic loop to the total apparent energy, in voltampere, of the magnetic cycle, it follows that the apparent efficiency of such a motor can never exceed the value $(1 - s) \sin \alpha$, or a fraction of the primary hysteretic energy.

The primary hysteretic energy of an induction motor, as represented by its conductance, g, being a part of the loss in the motor, and thus a very small part of its output only, it follows that the output of a hysteresis motor is a very small fraction only of the output which the same magnetic structure could give with secondary short-circuited winding, as regular induction motor.

As secondary effect, however, the rotary effort of the magnetic structure as hysteresis motor appears more or less in all induction motors, although usually it is so small as to be neglected.

If in the hysteresis motor the rotary iron structure has not uniform reluctance in all directions — but is, for instance, bar-shaped or shuttle-shaped — on the hysteresis motor effect is superimposed the effect of varying magnetic reluctance, which tends to accelerate the motor to synchronism, and maintain it therein, as shall be more fully investigated under "Reaction Machine" in Chapter XX.

CHAPTER XVII.

ALTERNATING-CURRENT GENERATOR.

182. In the alternating-current generator, E.M.F. is induced in the armature conductors by their relative motion through a constant or approximately constant magnetic field.

When yielding current, two distinctly different M.M.Fs. are acting upon the alternator armature — the M.M.F. of the field due to the field-exciting spools, and the M.M.F. of the armature current. The former is constant, or approximately so, while the latter is alternating, and in synchronous motion relatively to the former; hence, fixed in space relative to the field M.M.F., or uni-directional, but pulsating in a single-phase alternator. In the polyphase alternator, when evenly loaded or balanced, the resultant M.M.F. of the armature current is more or less constant.

The E.M.F. induced in the armature is due to the magnetic flux passing through and interlinked with the armature conductors. This flux is produced by the resultant of both M.M.Fs., that of the field, and that of the armature.

On open circuit, the M.M.F. of the armature is zero, and the E.M.F. of the armature is due to the M.M.F. of the field coils only. In this case the E.M.F. is, in general, a maximum at the moment when the armature coil faces the position midway between adjacent field coils, as shown in Fig. 126, and thus incloses no magnetism. The E.M.F. wave in this case is, in general, symmetrical.

An exception from this statement may take place only in those types of alternators where the magnetic reluctance of the armature is different in different directions; thereby, during the synchronous rotation of the armature, a pulsation of the magnetic flux passing through it is produced. This pulsation of the magnetic flux induces E.M.F. in the field spools, and thereby makes the field current pulsating also. Thus, we have, in this case, even on open circuit, no

Fig. 126.

rotation through a constant magnetic field, but rotation through a pulsating field, which makes the E.M.F. wave unsymmetrical, and shifts the maximum point from its theoretical position midway between the field poles. In general this secondary reaction can be neglected, and the field M.M.F. be assumed as constant.

The relative position of the armature M.M.F. with respect to the field M.M.F. depends upon the phase relation existing in the electric circuit. Thus, if there is no displacement of phase between current and E.M.F., the current reaches its maximum at the same moment as the E.M.F.; or, in the position of the armature shown in Fig. 126, midway between the field poles. In this case the armature current tends neither to magnetize nor demagnetize the field, but merely distorts it; that is, demagnetizes the trailing-pole corner, *a*, and magnetizes the leading-pole corner, *b*. A change of the total flux, and thereby of the resultant E.M.F., will take place in this case only when the magnetic densities are so near to saturation that the rise of density at the leading-pole corner will be less than the decrease of density at the trailing-pole corner. Since the internal self-inductance of the alternator itself causes a certain lag of the current behind the induced E.M.F., this condition of no displacement can exist only in a circuit with external negative reactance, as capacity, etc.

If the armature current lags, it reaches the maximum later than the E.M.F.; that is, in a position where the armature coil partly faces the following-field pole, as shown in diagram in Fig. 127. Since the armature current flows

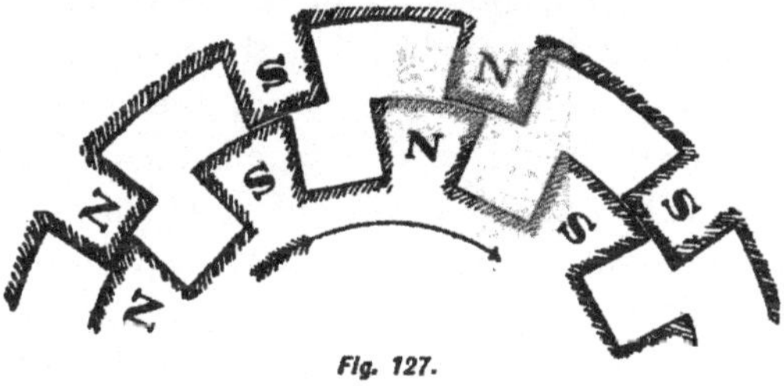

Fig. 127.

in opposite direction to the current in the following-field pole (in a generator), the armature in this case will tend to demagnetize the field.

If, however, the armature current leads,—that is, reaches its maximum while the armature coil still partly faces the

Fig. 128.

preceding-field pole, as shown in diagram Fig. 128,—it tends to magnetize this field coil, since the armature current flows in the same direction with the exciting current of the preceding-field spools.

Thus, with a leading current, the armature reaction of the alternator strengthens the field, and thereby, at constant-field excitation, increases the voltage; with lagging current it weakens the field, and thereby decreases the voltage in a generator. Obviously, the opposite holds for a synchronous motor, in which the armature current flows in the opposite direction; and thus a lagging current tends to magnetize, a leading current to demagnetize, the field.

183. The E.M.F. induced in the armature by the resultant magnetic flux, produced by the resultant M.M.F. of the field and of the armature, is not the terminal voltage of the machine; the terminal voltage is the resultant of this induced E.M.F. and the E.M.F. of self-inductance and the E.M.F. representing the energy loss by resistance in the alternator armature. That is, in other words, the armature current not only opposes or assists the field M.M.F. in creating the resultant magnetic flux, but sends a second magnetic flux in a local circuit through the armature, which flux does not pass through the field spools, and is called the magnetic flux of armature self-inductance.

Thus we have to distinguish in an alternator between armature reaction, or the magnetizing action of the armature upon the field, and armature self-inductance, or the E.M.F. induced in the armature conductors by the current flowing therein. This E.M.F. of self-inductance is (if the magnetic reluctance, and consequently the reactance, of the armature circuit is assumed as constant) in quadrature behind the armature current, and will thus combine with the induced E.M.F. in the proper phase relation. Obviously the E.M.F. of self-inductance and the induced E.M.F. do not in reality combine, but their respective magnetic fluxes combine in the armature core, where they pass through the same structure. These component E.M.Fs. are therefore mathematical fictions, but their resultant is real. This means that, if the armature current lags, the E.M.F. of self-inductance will be more than 90° behind the induced E.M.F., and therefore in partial opposition, and will tend to reduce the terminal voltage. On the other hand, if the armature current leads, the E.M.F. of self-inductance will be less than 90° behind the induced E.M.F., or in partial conjunction therewith, and increase the terminal voltage. This means that the E.M.F. of self-inductance increases the terminal voltage with a leading, and decreases it with a lagging current, or, in other words, acts in the same manner as the armature reaction. For this reason both actions can be combined in one, and represented by what is called the *synchronous reactance* of the alternator. In the following, we shall represent the total reaction of the armature of the alternator by the one term, *synchronous reactance.* While this is not exact, as stated above, since the reactance should be resolved into the magnetic reaction due to the magnetizing action of the armature current, and the electric reaction due to the self-induction of the armature current, it is in general sufficiently near for practical purposes, and well suited to explain the phenomena taking place under the various conditions of load. This synchronous reactance, x, is frequently not constant, but is pulsating, owing to the synchronously varying reluctance of the armature magnetic circuit, and the field magnetic circuit; it may, however, be considered in what follows as constant; that is, the E.M.Fs. induced thereby may be represented by their equivalent sine waves. A specific discussion of the distortions of the wave shape due to the pulsation of the synchronous reactance is found in Chapter XX. The synchronous reactance, x, is not a true reactance in the ordinary sense of the word, but an *equivalent* or *effective* reactance. Sometimes the total effects taking place in the alternator armature, are represented by a magnetic reaction, neglecting the self-inductance altogether, or rather replacing it by an increase of the armature reaction or armature M.M.F. to such a value as to include the self-inductance. This assumption is mostly made in the preliminary designs of alternators.

184. Let E_0 = induced E.M.F. of the alternator, or the E.M.F. induced in the armature coils by their rotation through the constant magnetic field produced by the current in the field spools, or the open circuit voltage, more properly called the "nominal induced E.M.F.," since in reality it does not exist, as before stated.

Then $$E_0 = \sqrt{2}\,\pi\, n N M 10^{-8};$$

where

n = total number of turns in series on the armature,
N = frequency,
M = total magnetic flux per field pole.

Let x_0 = synchronous reactance,
r_0 = internal resistance of alternator;
then $Z_0 = r_0 - j\,x_0$ = internal impedance.

If the circuit of the alternator is closed by the external impedance,

$$Z = r - j\,x,$$

the *current* is

$$\dot{I} = \frac{\dot{E}_0}{Z_0 + Z} = \frac{\dot{E}_0}{(r_0 + r) - j\,(x_0 + x)},$$

or,

$$I = \frac{E_0}{\sqrt{(r_0 + r)^2 + (x_0 + x)^2}};$$

and, *terminal voltage*,

$$\dot{E} = \dot{I}Z = \dot{E}_0 - \dot{I}Z_0 = \frac{\dot{E}_0\,(r - j\,x)}{(r_0 + r) - j\,(x_0 + x)},$$

or,

$$E = \frac{E_0\sqrt{r^2 + x^2}}{\sqrt{(r_0 + r)^2 + (x_0 + x)^2}}$$
$$= E_0 \frac{1}{\sqrt{1 + 2\,\frac{r_0 r + x_0 x}{r^2 + x^2} + \frac{r_0^2 + x_0^2}{r^2 + x^2}}};$$

or, expanded in a series,

$$E = E_o \left\{ 1 - \frac{r_o r + x_o x}{r^2 + x^2} + \frac{2\,(r_o r + x_o x) - (r_o x + x_o r)}{2\,(r^2 + x^2)} \pm \cdots \right\}.$$

As shown, the terminal voltage varies with the conditions of the external circuit.

185. As an instance, in Figs. 129–134, at constant induced E.M.F.,

$$E_o = 2500\,;$$

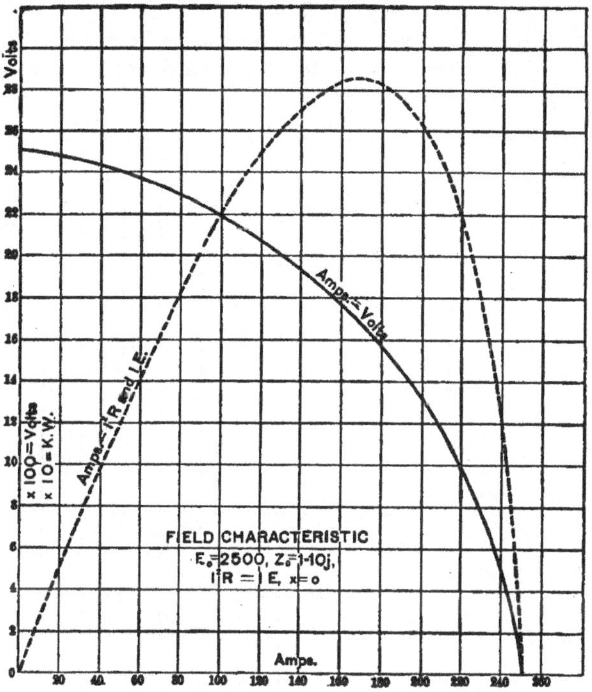

Fig. 129. Field Characteristic of Alternator on Non-Inductive Load.

and the values of the internal impedance,

$$Z_o = r_o - j\,x_o = 1 - 10\,j.$$

With the current I as abscissæ, the terminal voltages E as ordinates in drawn line, and the kilowatts output, $= I^2\,r$, in dotted lines, the kilovolt-amperes output, $= I\,E$, in dash-dotted lines, we have, for the following conditions of external circuit:

In Fig. 129, non-inductive external circuit, $x = 0$.

In Fig. 130, inductive external circuit, of the condition, $r / x = + .75$, with a power factor, .6.

In Fig. 131, inductive external circuit, of the condition, $r = 0$ with a power factor, 0.

In Fig. 132, external circuit with leading current, of the condition, $r / x = - .75$, with a power factor, .6.

In Fig. 133, external circuit with leading current, of the condition, $r = 0$, with a power factor, 0.

In Fig. 134, all the volt-ampere curves are shown together as complete ellipses, giving also the negative or synchronous motor part of the curves.

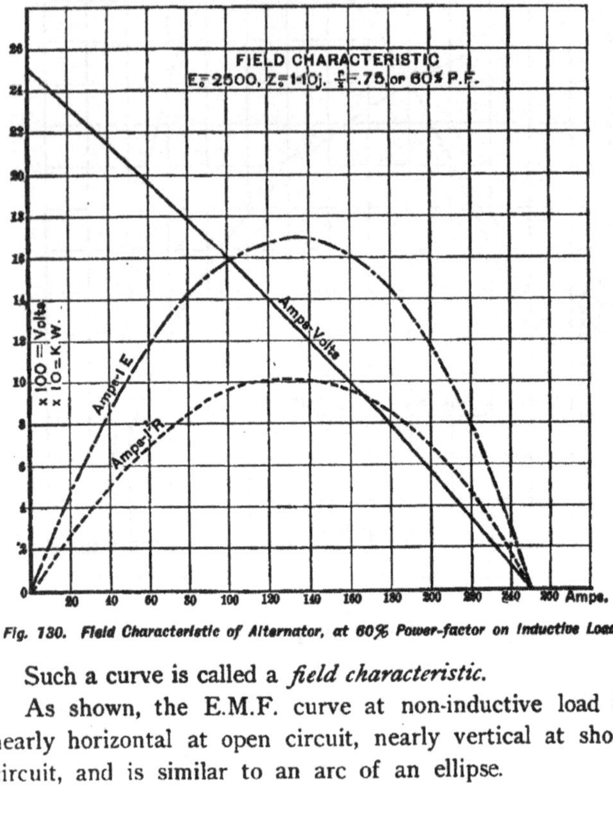

Fig. 130. Field Characteristic of Alternator, at 60% Power-factor on Inductive Load.

Such a curve is called a *field characteristic.*

As shown, the E.M.F. curve at non-inductive load is nearly horizontal at open circuit, nearly vertical at short circuit, and is similar to an arc of an ellipse.

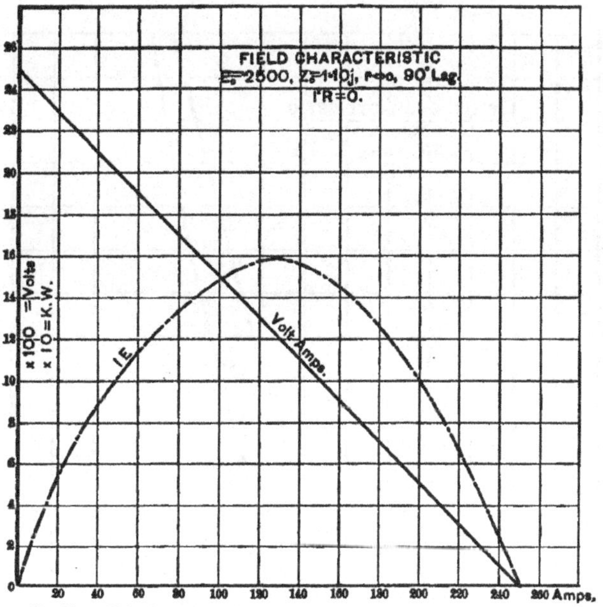

Fig. 131. Field Characteristic of Alternator, on Wattless Inductive Load.

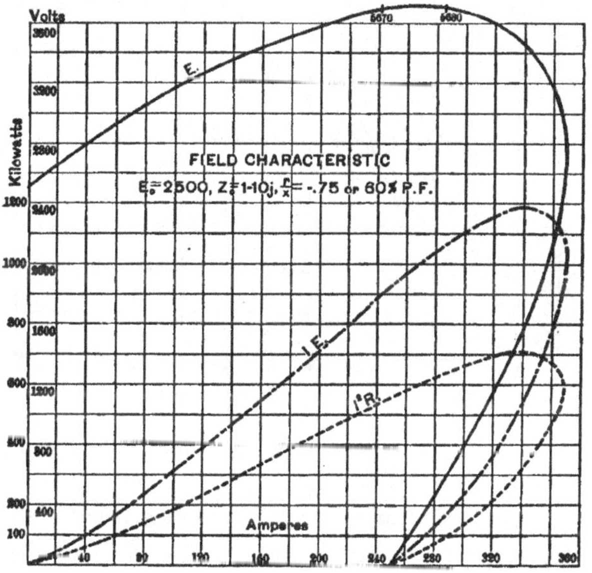

Fig. 132. Field Characteristic of Alternator, at 60% Power-factor on Condenser Load.

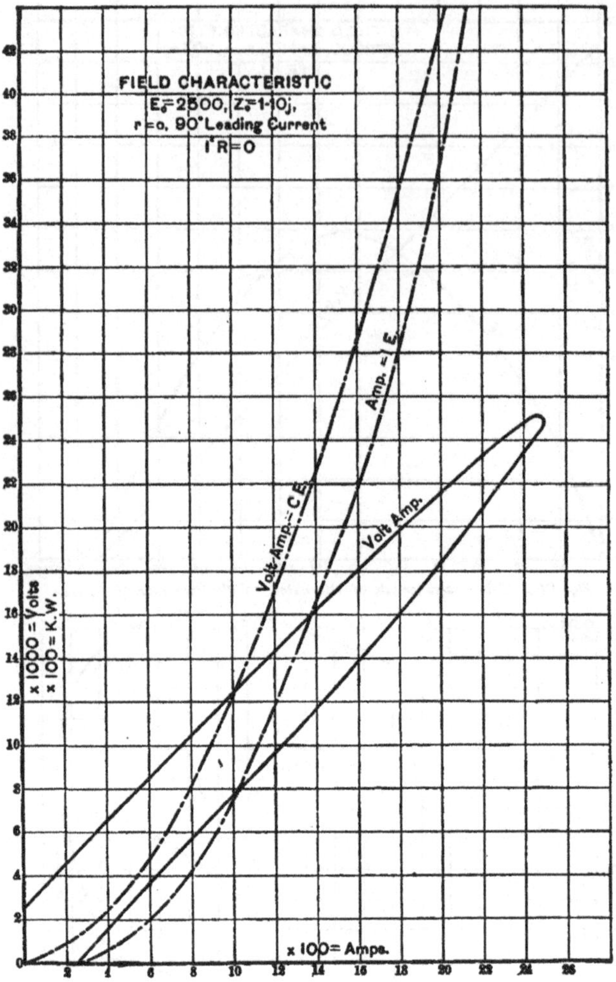

Fig. 133. Field Characteristic of Alternator, on Wattless Condenser Load.

With reactive load the curves are more nearly straight lines.

The voltage drops on inductive, rises on capacity load.

The output increases from zero at open circuit to a maximum, and then decreases again to zero at short circuit.

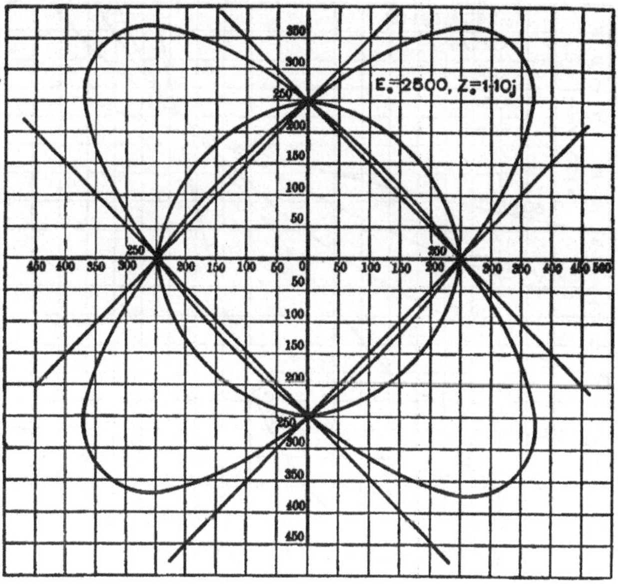

Fig. 134. Field Characteristic of Alternator.

186. The dependence of the terminal voltage, E, upon the phase relation of the external circuit is shown in Fig. 135, which gives, at impressed E.M.F.,

$$E_0 = 2,500 \text{ volts},$$

for the currents,

$$I = 50, 100, 150, 200, 250 \text{ amperes},$$

the terminal voltages, E, as ordinates, with the inductance factor of the external circuit,

$$\frac{x}{\sqrt{r^2 + x^2}},$$

as abscissæ.

187. If the internal impedance is negligible compared with the external impedance, then, approximately,

$$E = \frac{E_0 \sqrt{r^2 + x^2}}{\sqrt{(r_0 + r)^2 + (x_0 + x)^2}} = E_0;$$

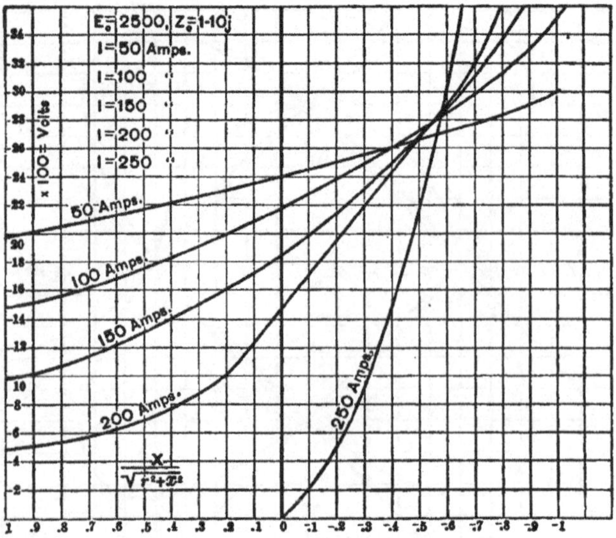

Fig. 135. Regulation of Alternator on Various Loads.

that is, *an alternator with small internal resistance and synchronous reactance tends to regulate for constant terminal voltage.*

Every alternator does this near open circuit, especially on non-inductive load.

Even if the synchronous reactance, x_0, is not quite negligible, this regulation takes place, to a certain extent, on non-inductive circuit, since for

$$x = 0, \quad E = \frac{E_0}{\sqrt{1 + 2\frac{r_0}{r} + \frac{x_0^2}{r^2}}};$$

and thus the expression of the terminal voltage, E, contains the synchronous reactance, x_0, only as a term of second order in the denominator.

On inductive circuit, however, x_0 appears in the denominator as a term of first order, and therefore constant potential regulation does not take place as well.

With a non-inductive external circuit, if the synchronous reactance, x_0, of the alternator is very large compared with the external resistance, r,

current $$I = \frac{E_0}{x_0} \frac{1}{\sqrt{1 + \left(\frac{r_0 + r}{x_0}\right)^2}} = \frac{E_0}{x_0},$$

approximately, or constant; or, if the external circuit contains the reactance, x,

$$I = \frac{E_0}{x_0 + x} \frac{1}{\sqrt{1 + \left(\frac{r_0 + r}{x_0 + x}\right)^2}} = \frac{E_0}{x_0 + x},$$

approximately, or constant.

The terminal voltage of a non-inductive circuit is

$$E = \frac{E_0}{x_0} r,$$

approximately, or proportional to the external resistance.

In an inductive circuit,

$$E = \frac{E_0}{x_0 + x} \sqrt{r^2 + x^2},$$

approximately, or proportional to the external impedance.

188. That is, on a non-inductive external circuit, an alternator with very low synchronous reactance regulates for constant terminal voltage, as a constant-potential machine; an alternator with a very high synchronous reactance regulates for a terminal voltage proportional to the external resistance, as a constant-current machine.

Thus, every alternator acts as a constant-potential machine near open circuit, and as a constant-current machine near short circuit. Between these conditions, there is a range where the alternator regulates approximately as a constant power machine, that is current and E.M.F. vary in inverse proportion, as between 130 and 200 amperes in Fig. 129.

The modern alternators are generally more or less machines of the first class; the old alternators, as built by Jablockkoff, Gramme, etc., were machines of the second class, used for arc lighting, where constant-current regulation is an advantage.

Obviously, large external reactances cause the same regulation for constant current independently of the resistance, r, as a large internal reactance, x_0.

On non-inductive circuit, if

$$I = \frac{E_0}{\sqrt{(r + r_0)^2 + x_0^2}},$$

$$E = \frac{E_0 r}{\sqrt{(r + r_0)^2 + x_0^2}},$$

the output is $$P = IE = \frac{E_0^2 r}{(r + r_0)^2 + x_0^2};$$

and $$\frac{dP}{dr} = \frac{x_0^2 - r^2 + r_0^2}{\{(r + r_0)^2 + x_0^2\}^2} E_0^2;$$

hence, if $$x_0 = \sqrt{r^2 - r_0^2},$$

or $$r = \sqrt{r_0^2 + x_0^2} = z_0;$$

then $$\frac{dP}{dr} = 0.$$

That is, the power is a maximum, and

$$P = \frac{E_0^2}{2\{z_0 + r_0\}},$$

$$E = \frac{E_0}{\sqrt{2\left\{1 + \frac{r_0}{z_0}\right\}}},$$

and $$I = \frac{E_0}{\sqrt{2 z_0 \{z_0 + r_0\}}}.$$

Therefore, with an external resistance equal to the internal impedance, or, $r = z_0 = \sqrt{r_0^2 + x_0^2}$, the output of an alternator is a maximum, and near this point it regulates for constant output; that is, an increase of current causes a proportional decrease of terminal voltage, and inversely.

The field characteristic of the alternator shows this effect plainly.

CHAPTER XVIII.

SYNCHRONIZING ALTERNATORS.

189. All alternators, when brought to synchronism with each other, will operate in parallel more or less satisfactorily. This is due to the reversibility of the alternating-current machine; that is, its ability to operate as synchronous motor. In consequence thereof, if the driving power of one of several parallel-operating generators is withdrawn, this generator will keep revolving in synchronism as a synchronous motor; and the power with which it tends to remain in synchronism is the maximum power which it can furnish as synchronous motor under the conditions of running.

190. The principal and foremost condition of parallel operation of alternators is equality of frequency; that is, the transmission of power from the prime movers to the alternators must be such as to allow them to run at the same frequency without slippage or excessive strains on the belts or transmission devices.

Rigid mechanical connection of the alternators cannot be considered as synchronizing; since it allows no flexibility or phase adjustment between the alternators, but makes them essentially one machine. If connected in parallel, a difference in the field excitation, and thus the induced E.M.F. of the machines, must cause large cross-current; since it cannot be taken care of by phase adjustment of the machines.

Thus rigid mechanical connection is not desirable for parallel operation of alternators.

191. The second important condition of parallel operation is uniformity of speed; that is, constancy of frequency.

If, for instance, two alternators are driven by independent single-cylinder engines, and the cranks of the engines happen to be crossed, the one engine will pull, while the other is near the dead-point, and conversely. Consequently, alternately the one alternator will tend to speed up and the other slow down, then the other speed up and the first slow down. This effect, if not taken care of by fly-wheel capacity, causes a "hunting" or pumping action; that is, a fluctuation of the lights with the period of the engine revolution, due to the alternating transfer of the load from one engine to the other, which may even become so excessive as to throw the machines out of step, especially when by an approximate coincidence of the period of engine impulses (or a multiple thereof), with the natural period of oscillation of the revolving structure, the effect is made cumulative. This difficulty as a rule does not exist with turbine or water-wheel driving.

192. In synchronizing alternators, we have to distinguish the phenomena taking place when throwing the machines in parallel or out of parallel, and the phenomena when running in synchronism.

When connecting alternators in parallel, they are first brought approximately to the same frequency and same voltage; and then, at the moment of approximate equality of phase, as shown by a phase-lamp or other device, they are thrown in parallel.

Equality of voltage is much less important with modern alternators than equality of frequency, and equality of phase is usually of importance only in avoiding an instantaneous flickering of the lights on the system. When two alternators are thrown together, currents pass between the machines, which accelerate the one and retard the other machine until equal frequency and proper phase relation are reached.

With modern ironclad alternators, this interchange of mechanical power is usually, even without very careful adjustment before synchronizing, sufficiently limited not to endanger the machines mechanically; since the cross-currents, and thus the interchange of power, are limited by self-induction and armature reaction.

In machines of very low armature reaction, that is, machines of "very good constant potential regulation," much greater care has to be exerted in the adjustment to equality of frequency, voltage, and phase, or the interchange of current may become so large as to destroy the machine by the mechanical shock; and sometimes the machines are so sensitive in this respect that it is preferable not to operate them in parallel. The same applies in getting out of step.

193. When running in synchronism, nearly all types of machines will operate satisfactorily; a medium amount of armature reaction is preferable, however, such as is given by modern alternators — not too high to reduce the synchronizing power too much, nor too low to make the machine unsafe in case of accident, such as falling out of step, etc.

If the armature reaction is very low, an accident, — such as a short circuit, falling out of step, opening of the field circuit, etc., — may destroy the machine. If the armature reaction is very high, the driving-power has to be adjusted very carefully to constancy; since the synchronizing power of the alternators is too weak to hold them in step, and carry them over irregularities of the driving-power.

194. Series operation of alternators is possible only by rigid mechanical connection, or by some means whereby the machines, with regard to their synchronizing power, act essentially in parallel; as, for instance, by the arrangement shown in Fig. 120, where the two alternators, A_1, A_2, are connected in series, but interlinked by the two coils of a large transformer, T, of which the one is connected across the terminals of one alternator, and the other across the terminals of the other alternator in such a way that, when operating in series, the coils of the transformer will

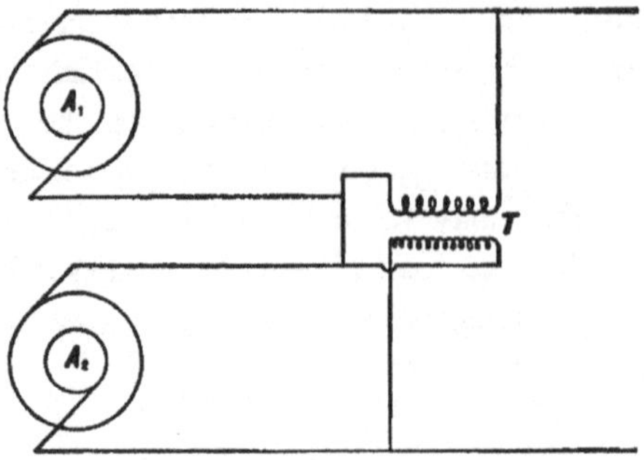

Fig. 136.

be without current. In this case, by interchange of power through the transformers, the series connection will be maintained stable.

195. In two parallel operating alternators, as shown in Fig. 137, let the voltage at the common bus bars be assumed

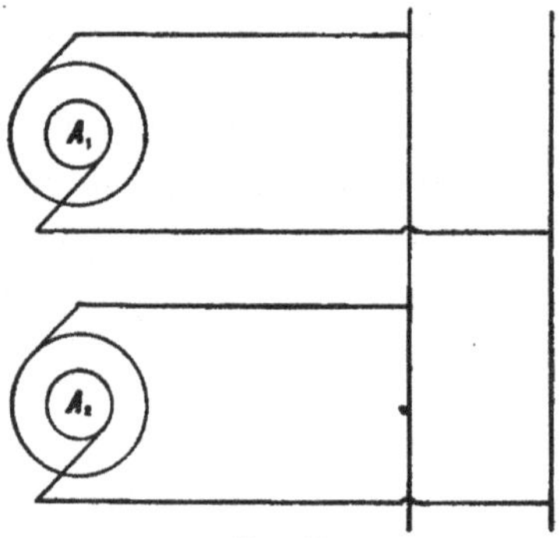

Fig. 137.

as zero line, or real axis of coördinates of the complex representation; and let —

e = difference of potential at the common bus bars of the two alternators,
$Z = r - jx$ = impedance of external circuit,
$Y = g + jb$ = admittance of external circuit;

hence, the current in external circuit is

$$\dot{I} = \frac{e}{r - jx} = e(g + jb).$$

Let

$\dot{E}_1 = e_1 - je_1' = a_1(\cos \hat{\omega}_1 - j \sin \hat{\omega}_1)$ = induced E.M.F. of first machine;
$\dot{E}_2 = e_2 - je_2' = a_2(\cos \hat{\omega}_2 - j \sin \hat{\omega}_2)$ = induced E.M.F. of second machine;
$\dot{I}_1 = i_1 + ji_1'$ = current of first machine;
$\dot{I}_2 = i_2 + ji_2'$ = current of second machine;
$Z_1 = r_1 - jx_1$ = internal impedance, and $Y_1 = g_1 + jb_1$ = internal admittance, of first machine;
$Z_2 = r_2 - jx_2$ = internal impedance, and $Y_2 = g_2 + jb_2$ = internal admittance, of second machine.

Then,

$$e_1^2 + e_1'^2 = a_1^2;$$
$$e_2^2 + e_2'^2 = a_2^2;$$
$\dot{E}_1 = e + \dot{I}_1 Z_1$, or $e_1 - je_1' = (e + i_1 r_1 + i_1' x_1) - j(i_1 x_1 - i_1' r_1)$;
$\dot{E}_2 = e + \dot{I}_2 Z_2$, or $e_2 - je_2' = (e + i_2 r_2 + i_2' x_2) - j(i_2 x_2 - i_2' r_2)$;
$\dot{I} = \dot{I}_1 + \dot{I}_2$, or $eg + jeb = (i_1 + i_2) + j(i_1' + i_2')$.

This gives the equations —

$$e_1 = e + i_1 r_1 + i_1' x_1;$$
$$e_2 = e + i_2 r_2 + i_2' x_2;$$
$$e_1' = i_1 x_1 - i_1' r_1;$$
$$e_2' = i_2 x_2 - i_2' r_2;$$
$$eg = i_1 + i_2;$$
$$eb = i_1' + i_2';$$
$$e_1^2 + e_1'^2 = a_1^2;$$
$$e_2^2 + e_2'^2 = a_2^2;$$

or eight equations with nine variables: e_1, e_1', e_2, e_2', i_1, i_1', i_2, i_2', e.

Combining these equations by twos,

$$e_1 r_1 + e_1' x_1 = e r_1 + i_1 z_1^2;$$
$$e_2 r_2 + e_2' x_2 = e r_2 + i_2 z_2^2;$$

substituted in

$$i_1 + i_2 = e g,$$

we have

$$e_1 g_1 + e_1' b_1 + e_2 g_2 + e_2' b_2 = e(g_1 + g_2 + g);$$

and analogously,

$$e_1 b_1 - e_1' g_1 + e_2 b_2 - e_2' g_2 = e(b_1 + b_2 + b);$$

dividing,

$$\frac{g + g_1 + g_2}{b + b_1 + b_2} = \frac{e_1 g_1 + e_2 g_2 + e_1' b_1 + e_2' b_2}{e_1 b_1 + e_2 b_2 - e_1' g_1 - e_2' g_2};$$

substituting

$$\begin{array}{lll} g = v \cos \alpha & e_1 = a_1 \cos \hat{\omega}_1 & e_2 = a_2 \cos \hat{\omega}_2 \\ b = v \sin \alpha & e_1' = a_1 \sin \hat{\omega}_1 & e_2' = a_2 \sin \hat{\omega}_2 \end{array}$$

gives

$$\frac{g + g_1 + g_2}{b + b_1 + b_2} = \frac{a_1 v_1 \cos(\alpha_1 - \hat{\omega}_1) + a_2 v_2 \cos(\alpha_2 - \hat{\omega}_2)}{a_1 v_1 \sin(\alpha_1 - \hat{\omega}_1) + a_2 v_2 \sin(\alpha_2 - \hat{\omega}_2)}$$

as the equation between the phase displacement angles $\hat{\omega}_1$ and $\hat{\omega}_2$ in parallel operation.

The power supplied to the external circuit is

$$p = e^2 g,$$

of which that supplied by the first machine is,

$$p_1 = e i_1;$$

by the second machine,

$$p_2 = e i_2.$$

The total electrical work done by both machines is,

$$P = P_1 + P_2,$$

of which that done by the first machine is,

$$P_1 = e_1 i_1 - e_1' i_1';$$

by the second machine,

$$P_2 = e_2 i_2 - e_2' i_2'.$$

The difference of output of the two machines is,

$$\Delta p = p_1 - p_2 = e\,(i_1 - i_2);$$

denoting

$$\frac{\hat{\omega}_1 + \hat{\omega}_2}{2} = \epsilon \qquad \frac{\hat{\omega}_1 - \hat{\omega}_2}{2} = \delta.$$

$\Delta p / \Delta\delta$ may be called the synchronizing power of the machines, or the power which is transferred from one machine to the other by a change of the relative phase angle.

196. Special Case. — *Two equal alternators of equal excitation.*

$$a_1 = a_2 = a$$
$$Z_1 = Z_2 = Z_0.$$

Substituting this in the eight initial equations, these assume the form, —

$$\begin{aligned} e_1 &= e + i_1 r_0 + i_1' x_0 \\ e_2 &= e + i_2 r_0 + i_2' x_0 \\ e_1' &= i_1 x_0 - i_1 r_0 \\ e_2' &= i_2 x_0 - i_2' r_0. \\ eg &= i_1 + i_2 \\ eb &= i_1' + i_2' \\ e_1^2 + e_1'^2 &= e_2^2 + e_2'^2 = a^2. \end{aligned}$$

Combining these equations by twos,

$$\begin{aligned} e_1 + e_2 &= 2e + e\,(r_0 g + x_0 b) \\ e_1' + e_2' &= e\,(x_0 g - r_0 b); \end{aligned}$$

substituting

$$\begin{aligned} e_1 &= a \cos \hat{\omega}_1 \\ e_1' &= a \sin \hat{\omega}_1 \\ e_2 &= a \cos \hat{\omega}_2 \\ e_2' &= a \sin \hat{\omega}_2, \end{aligned}$$

we have

$$\begin{aligned} a\,(\cos \hat{\omega}_1 + \cos \hat{\omega}_2) &= e\,(2 + r_0 g + x_0 b) \\ a\,(\sin \hat{\omega}_1 + \sin \hat{\omega}_2) &= e\,(x_0 g - r_0 b); \end{aligned}$$

expanding and substituting —

$$\epsilon = \frac{\hat{\omega}_1 + \hat{\omega}_2}{2}$$

$$\delta = \frac{\hat{\omega}_1 - \hat{\omega}_2}{2},$$

or $$a \cos \epsilon \cos \delta = e\left(1 + \frac{r_0 g + x_0 b}{2}\right)$$

$$a \sin \epsilon \cos \delta = e\frac{x_0 g - r_0 b}{2};$$

hence $$\tan \epsilon = \frac{x_0 g - r_0 b}{2 + r_0 g + x_0 b} = \text{constant.}$$

That is $$\hat{\omega}_1 + \hat{\omega}_2 = \text{constant};$$

and $$\cos \delta = \frac{e}{a}\sqrt{\left(1 + \frac{r_0 g + x_0 b}{2}\right)^2 + \left(\frac{x_0 g - r_0 b}{2}\right)^2};$$

or, $$e = \frac{a \cos \delta}{\sqrt{\left(1 + \frac{r_0 g + x_0 b}{2}\right)^2 + \left(\frac{x_0 g - r_0 b}{2}\right)^2}};$$

at no-phase displacement between the alternators, or,

$$\delta = \frac{\hat{\omega}_1 - \hat{\omega}_2}{2} = 0;$$

we have $$e = \frac{a}{\sqrt{\left(1 + \frac{r_0 g + x_0 b}{2}\right)^2 + \left(\frac{x_0 g - r_0 b}{2}\right)^2}}.$$

From the eight initial equations we get, by combination —

$$e_1 r_0 + e_1' x_0 = e\,(r_0 + x_0) + i_1\,(r_0^2 + x_0^2)$$
$$e_2 r_0 + e_2' x_0 = e\,(r_0 + x_0) + i_2\,(r_0^2 + x_0^2);$$

subtracted and expanded —

$$i_1 - i_2 = \frac{r_0(e_1 - e_2) + x_0(e_1' - e_2')}{Z_0^2};$$

or, since

$$e_1 - e_2 = a\,(\cos \hat{\omega}_1 - \cos \hat{\omega}_2) = -\,2\,a^2 \sin \epsilon \sin \delta$$
$$e_1' - e_2' = a\,(\sin \hat{\omega}_1 - \sin \hat{\omega}_2) = \quad 2\,a\,\cos \epsilon \sin \delta;$$

we have

$$i_1 - i_2 = \frac{2\,a \sin \delta}{z_0^2}\{x_0 \cos \epsilon - r_0 \sin \epsilon\}$$
$$= 2\,a y_0 \sin \delta \cos (\epsilon + \alpha),$$

where

$$\tan \alpha = \frac{x_0}{r_0}.$$

The difference of output of the two alternators is

$$\Delta p = p_1 - p_2 = e\,(i_1 - i_2);$$

hence, substituting,

$$\Delta p = \frac{2\,ae \sin \delta}{z_0^2}\{x_0 \cos \epsilon - r_0 \sin \epsilon\};$$

substituting,

$$e = \frac{a \cos \delta}{\sqrt{\left(1 + \frac{r_0 g + x_0 b}{2}\right)^2 + \left(\frac{x_0 g - r_0 b}{2}\right)^2}};$$

$$\sin \epsilon = \frac{\frac{x_0 g - r_0 b}{2}}{\sqrt{\left(1 + \frac{r_0 g + x_0 b}{2}\right)^2 + \left(\frac{x_0 g - r_0 b}{2}\right)^2}};$$

$$\cos \epsilon = \frac{1 + \frac{r_0 g + x_0 b}{2}}{\sqrt{\left(1 + \frac{r_0 g + x_0 b}{2}\right)^2 + \left(\frac{x_0 g - r_0 b}{2}\right)^2}}$$

we have,

$$\Delta p = \frac{2\,a^2 \sin \delta \cos \delta \left\{ x_0\left(1 + \frac{r_0 g + x_0 b}{2}\right) - r_0\left(\frac{x_0 g - r_0 b}{2}\right)\right\}}{z^2 \left\{\left(1 + \frac{r_0 g + x_0 b}{2}\right)^2 + \left(\frac{x_0 g - r_0 b}{2}\right)^2\right\}};$$

expanding,

$$\Delta p = \frac{a^2 \sin 2\,\delta \left\{ x_0 + \frac{g\,z_0^2}{2}\right\}}{z_0^2\,\{1 + r_0 g + x_0 b + z_0^2 v^2\}};$$

or

$$\Delta p = \frac{a^2 \sin 2\,\delta \left\{ b_0 + \frac{b}{2}\right\}}{y_0^2 + g g_0 + b b_0 + v^2};$$

$$\frac{\Delta p}{\Delta \delta} = \frac{2\,a^2 \cos 2\,\delta \left\{ b_0 + \frac{b}{2}\right\}}{y_0^2 + g g_0 + b b_0 + y^2}$$

Hence, the transfer of power between the alternators, Δp, is a maximum, if $\delta = 45°$; or $\hat{\omega}_1 - \hat{\omega}_2 = 90°$; that is, when the alternators are in quadrature.

The synchronizing power, $\Delta p / \Delta \delta$, is a maximum if $\delta = 0$; that is, the alternators are in phase with each other.

197. As an instance, curves may be plotted for,

$a = 2500,$

$Z_0 = r_1 - j x_0 = 1 - 10 j$; or $Y_0 = g_0 + j b_0 = .01 + .1 j$,

with the angle $\delta = \frac{\hat{\omega}_1 - \hat{\omega}_2}{2}$ as abscissæ, giving

the value of terminal voltage, e;
the value of current in the external circuit, $i = e y$;
the value of interchange of current between the alternators, $i_1 - i_2$;
the value of interchange of power between the alternators, $\Delta p = p_1 - p_2$;
the value of synchronizing power, $\frac{\Delta p}{\Delta \delta}$.

For the condition of external circuit,

$g = 0,$	$b = 0,$	$y = 0,$
.05,	0,	.05,
.08,	0,	.08,
.03,	+ .04,	.05,
.03,	− .04,	.05.

CHAPTER XIX.

SYNCHRONOUS MOTOR.

198. In the chapter on synchronizing alternators we have seen that when an alternator running in synchronism is connected with a system of given E.M.F., the work done by the alternator can be either positive or negative. In the latter case the alternator consumes electrical, and consequently produces mechanical, power; that is, runs as a synchronous motor, so that the investigation of the synchronous motor is already contained essentially in the equations of parallel-running alternators.

Since in the foregoing we have made use mostly of the symbolic method, we may in the following, as an instance of the graphical method, treat the action of the synchronous motor diagrammatically.

Let an alternator of the E.M.F., E_1, be connected as synchronous motor with a supply circuit of E.M.F., E_0, by a circuit of the impedance Z.

If E_0 is the E.M.F. impressed upon the motor terminals, Z is the impedance of the motor of induced E.M.F., E_1. If E_0 is the E.M.F. at the generator terminals, Z is the impedance of motor and line, including transformers and other intermediate apparatus. If E_0 is the induced E.M.F. of the generator, Z is the sum of the impedances of motor, line, and generator, and thus we have the problem, generator of induced E.M.F. E_0, and motor of induced E.M.F. E_1; or, more general, two alternators of induced E.M.Fs., E_0, E_1, connected together into a circuit of total impedance, Z.

Since in this case several E.M.Fs. are acting in circuit

with the same current, it is convenient to use the current, I, as zero line $\overline{OI}$ of the polar diagram. Fig. 138.

If $I = i =$ current, and $Z =$ impedance, $r =$ effective resistance, $x =$ effective reactance, and $z = \sqrt{r^2 + x^2} =$ absolute value of impedance, then the E.M.F. consumed by the resistance is $E_{,,} = ri$, and in phase with the current, hence represented by vector $OE_{,,}$; and the E.M.F. consumed by the reactance is $E_2 = xi$, and 90° ahead of the current, hence the E.M.F. consumed by the impedance is $E = \sqrt{(E_{,,})^2 + (E_2)^2}$, or $= i\sqrt{r^2 + x^2} = iz$, and ahead of the current by the angle δ, where $\tan \delta = x / r$.

We have now acting in circuit the E.M.Fs., E, E_1, E_0; or E_1 and E are components of E_0; that is, E_0 is the diagonal of a parallelogram, with E_1 and E as sides.

Since the E.M.Fs. E_1, E_2, E, are represented in the diagram, Fig. 138, by the vectors $\overline{OE_1}$, $\overline{OE_2}$, $\overline{OE}$, to get the parallelogram of E_0, E_1, E, we draw arcs of circles around 0 with E_0, and around E with E_1. Their point of intersection gives the impressed E.M.F., $\overline{OE_0} = E_0$, and completing the parallelogram $OE\ E_0\ E_1$ we get, $\overline{OE_1} = E_1$, the induced E.M.F. of the motor.

$\measuredangle\ IOE_0$ is the difference of phase between current and impressed E.M.F., or induced E.M.F. of the generator.

$\measuredangle\ IOE_1$ is the difference of phase between current and induced E.M.F. of the motor.

And the power is the current i times the projection of the E.M.F upon the current, or the zero line $\overline{OI}$.

Hence, dropping perpendiculars, $\overline{E_0 E_0^1}$ and $\overline{E_1 E_1^1}$, from E_0 and E_1 upon $\overline{OI}$, it is —

$P_0 = i \times \overline{OE_0^1} =$ power supplied by induced E.M.F. of generator.

$P_1 = i \times \overline{OE_1^1} =$ electric power transformed in mechanical power by the motor.

$P = i \times \overline{OE_1} =$ power consumed in the circuit by effective resistance.

Obviously $P_0 = P_1 + P$.

Since the circles drawn with E_0 and E_1 around O and E respectively intersect twice, two diagrams exist. In general, in one of these diagrams shown in Fig. 138 in drawn

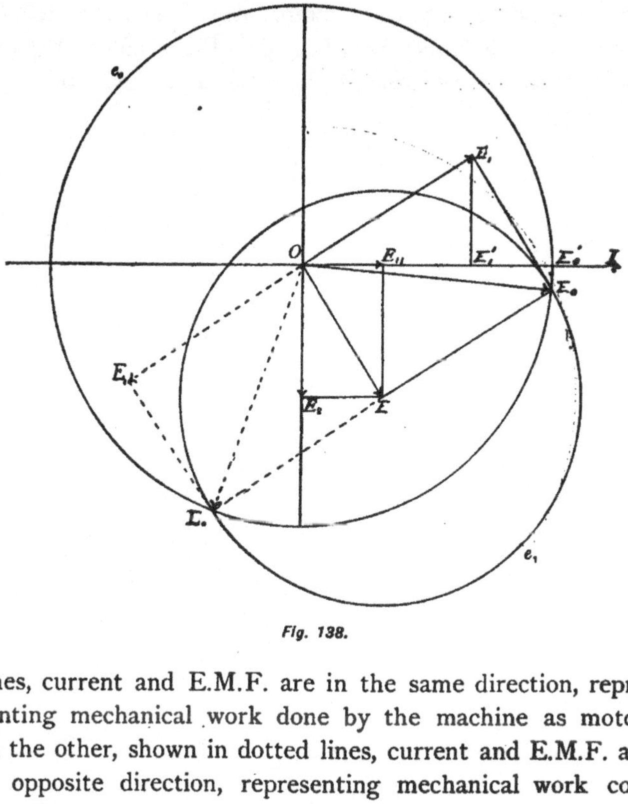

Fig. 138.

lines, current and E.M.F. are in the same direction, representing mechanical work done by the machine as motor. In the other, shown in dotted lines, current and E.M.F. are in opposite direction, representing mechanical work consumed by the machine as generator.

Under certain conditions, however, E_0 is in the same, E_1 in opposite direction, with the current; that is, both machines are generators.

199. It is seen that in these diagrams the E.M.Fs. are considered from the point of view of the motor; that is, work done as synchronous motor is considered as positive, work done as generator is negative. In the chapter on synchronizing generators we took the opposite view, from the generator side.

In a single unit-power transmission, that is, one generator supplying one synchronous motor over a line, the E.M.F. consumed by the impedance, $E = \overline{OE}$, Figs. 139 to 141, consists of three components; the E.M.F. $\overline{OE_2'} = E_2$, consumed

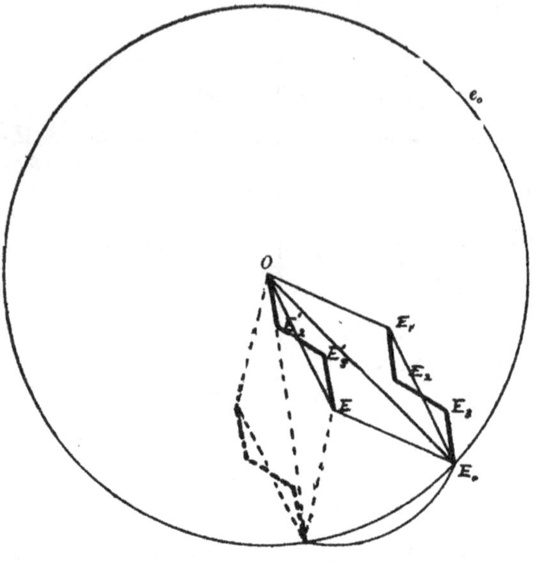

Fig. 139.

by the impedance of the motor, the E.M.F. $\overline{E_2'E_3'} = E_3$ consumed by the impedance of the line, and the E.M.F. $\overline{E_3'E} = E_4$ consumed by the impedance of the generator. Hence, dividing the opposite side of the parallelogram $\overline{E_1E_0}$, in the same way, we have: $\overline{OE_1} = E_1 =$ induced E.M.F. of the motor, $\overline{OE_2} = E_2 =$ E.M.F. at motor terminals or at end of line, $\overline{OE_3} = E_3 =$ E.M.F. at generator terminals, or at beginning of line. $\overline{OE_0} = E_0 =$ induced E.M.F. of generator.

The phase relation of the current with the E.M.Fs. E_1, E_0, depends upon the current strength and the E.M.Fs. E_1 and E_0.

200. Figs. 139 to 141 show several such diagrams for different values of E_1, but the same value of I and E_0. The motor diagram being given in drawn line, the generator diagram in dotted line.

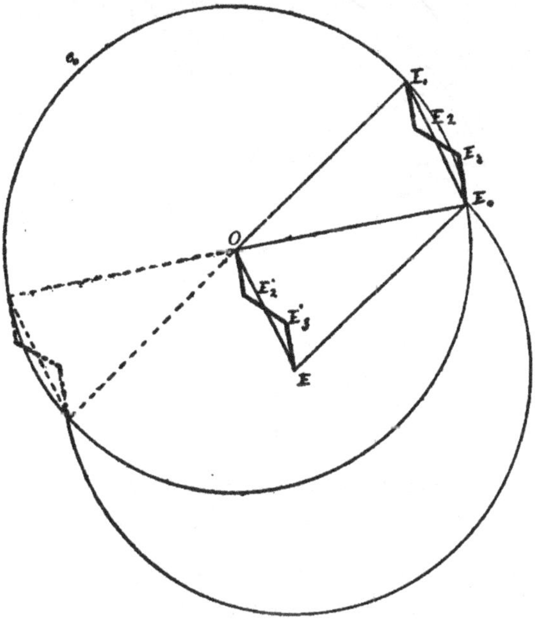

Fig. 140.

As seen, for small values of E_1 the potential drops in the alternator and in the line. For the value of $E_1 = E_0$ the potential rises in the generator, drops in the line, and rises again in the motor. For larger values of E_1, the potential rises in the alternator as well as in the line, so that the highest potential is the induced E.M.F. of the motor, the lowest potential the induced E.M.F. of the generator.

It is of interest now to investigate how the values of these quantities change with a change of the constants.

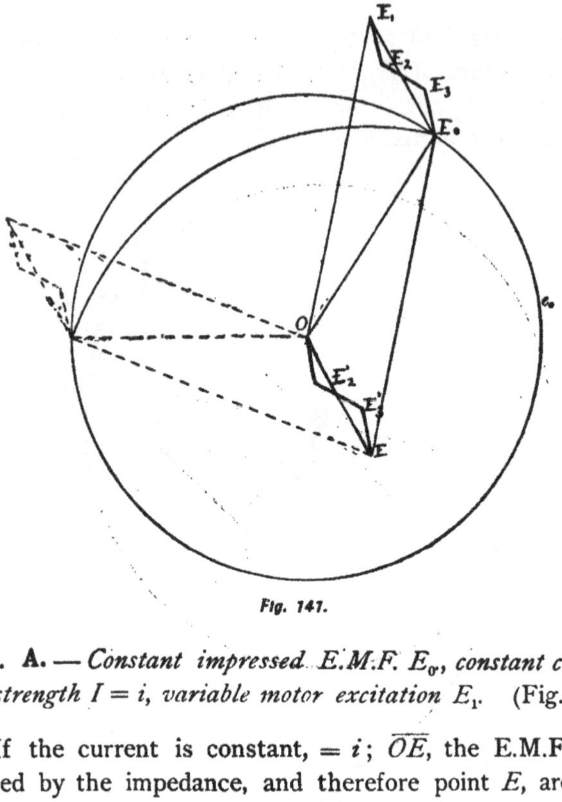

Fig. 147.

201. A. — *Constant impressed E.M.F. E_0, constant current strength $I = i$, variable motor excitation E_1.* (Fig. 142.)

If the current is constant, $= i$; $\overline{OE}$, the E.M.F. consumed by the impedance, and therefore point E, are constant. Since the intensity, but not the phase of E_0 is constant, E_0 lies on a circle e_0 with E_0 as radius. From the parallelogram, $OE\ E_0\ E_1$ follows, since $\overline{E_1 E_0}$ parallel and $= \overline{OE}$, that E_1 lies on a circle e_1 congruent to the circle e_0, but with E_i, the image of E, as center : $\overline{OE_i} = \overline{OE}$.

We can construct now the variation of the diagram with the variation of E_1; in the parallelogram $OE\ E_0\ E_1$, O and E are fixed, and E_0 and E_1 move on the circles e_0 e_1 so that $\overline{E_0\ E_1}$ is parallel to $\overline{OE}$.

The smallest value of E_1 consistent with current strength I is $\overline{01_1} = E_1$, $\overline{01} = E_0$. In this case the power of the motor is $01_1{}^1 \times I$, hence already considerable. Increasing E_1 to $\overline{02_1}$, $\overline{03_1}$, etc., the impressed E.M.Fs. move to $\overline{02}$, $\overline{03}$, etc., the power is $I \times \overline{02_1{}^1}$, $I \times \overline{03_1{}^1}$, etc., increases first,

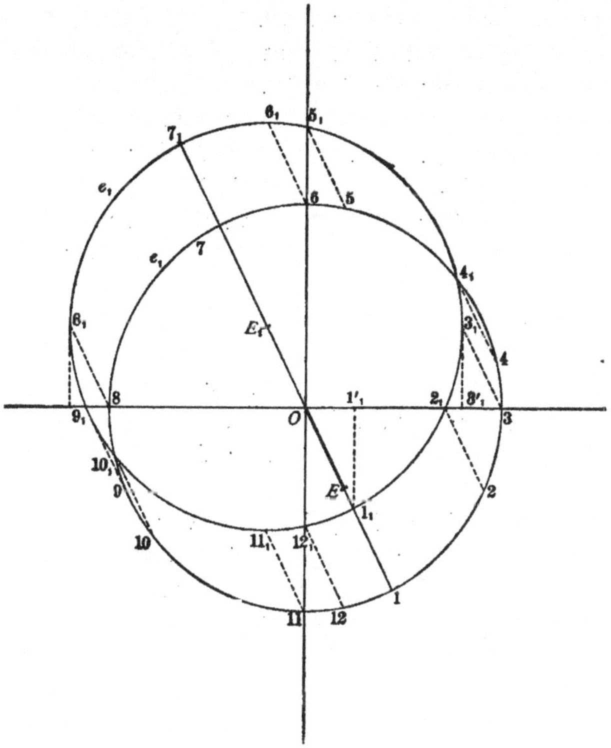

Fig. 142.

reaches the maximum at the point 3_1, 3, the most extreme point at the right, with the impressed E.M.F. in phase with the current, and then decreases again, while the induced E.M.F. of the motor E_1 increases and becomes $= E_0$ at 4_1, 4. At 5_1, 5, the power becomes zero, and further on negative; that is, the motor has changed to a dynamo, and produces electrical energy, while the impressed E.M.F. E_0 still furnishes electrical energy, that is, both machines as generators feed into the line, until at 6_1, 6, the power of the impressed E.M.F. E_0 becomes zero, and further on power begins to flow back; that is, the motor is changed to a generator and the generator to a motor, and we are on the generator side of the diagram. At 7_1, 7, the maximum value of E_1, consistent with the current I, has been reached, and passing still further the E.M.F. E_1 decreases again, while the power still increases up to the maximum at 8_1, 8, and then decreases again, but still E_1 remaining generator, E_0 motor, until at 11_1, 11, the power of E_0 becomes zero; that is, E_0 changes again to a generator, and both machines are generators, up to 12_1, 12, where the power of E_1 is zero, E_1 changes from generator to motor, and we come again to the motor side of the diagram, and while E_1 still decreases, the power of the motor increases until 1_1, 1, is reached.

Hence, there are two regions, for very large E_1 from 5 to 6, and for very small E_1 from 11 to 12, where both machines are generators; otherwise the one is generator, the other motor.

For small values of E_1 the current is lagging, begins, however, at 2 to lead the induced E.M.F. of the motor E_1, at 3 the induced E.M.F. of the generator E_0.

It is of interest to note that at the smallest possible value of E_1, 1_1, the power is already considerable. Hence, the motor can run under these conditions only at a certain load. If this load is thrown off, the motor cannot run with the same current, but the current must increase. We have here the curious condition that loading the motor reduces, unloading increases, the current within the range between 1 and 12.

The condition of maximum output is 3, current in phase with impressed E.M.F. Since at constant current the loss is constant, this is at the same time the condition of maximum efficiency: no displacement of phase of the impressed

E.M.F., or self-induction of the circuit compensated by the effect of the lead of the motor current. This condition of maximum efficiency of a circuit we have found already in the Chapter on Inductance and Capacity.

202. B. *E_0 and E_1 constant, I variable.*

Obviously E_0 lies again on the circle e_0 with E_0 as radius and O as center.

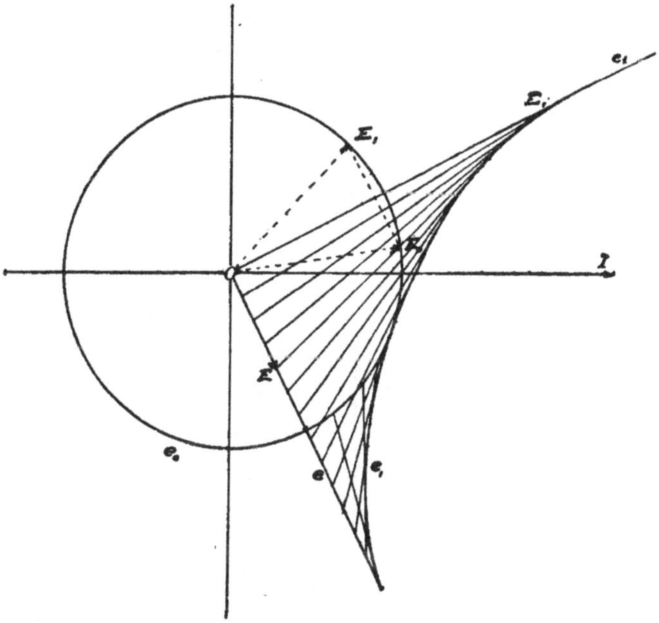

Fig. 148.

E lies on a straight line e, passing through the origin.

Since in the parallelogram $OE\ E_0\ E_1$, $\overline{EE_0} = E_1$, we derive E_0 by laying a line $\overline{EE_0} = E_1$ from any point E in the circle e_0, and complete the parallelogram.

All these lines $\overline{EE_0}$ envelop a certain curve e_1, which can be considered as the characteristic curve of this problem, just as circle e_1 in the former problem.

These curves are drawn in Figs. 143, 144, 145, for the three cases: 1st, $E_1 = E_0$; 2d, $E_1 < E_0$; 3d, $E_1 > E_0$.

In the first case, $E_1 = E_0$ (Fig. 127), we see that at

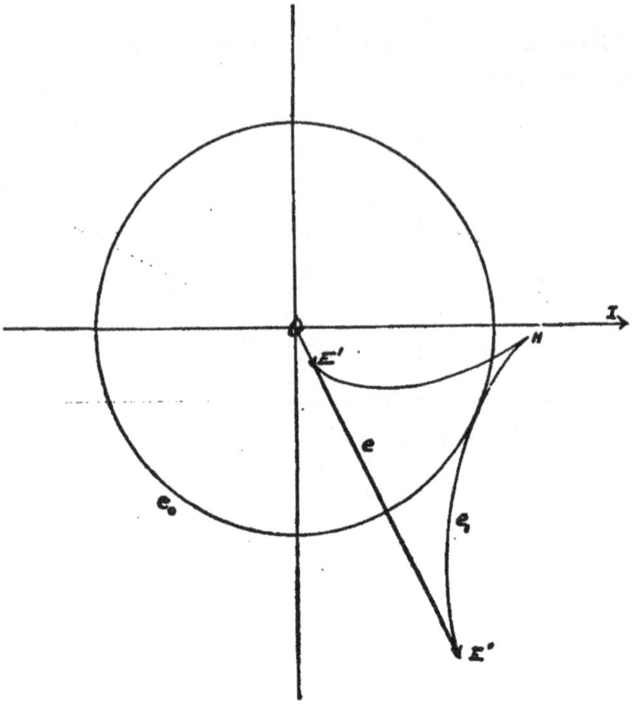

Fig. 144.

very small current, that is very small $\overline{OE}$, the current I leads the impressed E.M.F. E_0 by an angle $E_0OI = \hat{\omega}_0$. This lead decreases with increasing current, becomes zero, and afterwards for larger current, the current lags. Taking now any pair of corresponding points E, E_0, and producing $\overline{EE_0}$ until it intersects e_i, in E_i, we have $\measuredangle E_i\ OE = 90°$, $\overline{E_1 = E_0}$, thus: $\overline{OE_1} = \overline{EE_0} = \overline{OE_0} = \overline{E_0E_i}$; that is, $\overline{EE_i} =$

$2E_0$. That means the characteristic curve e_1 is the envelope of lines $\overline{EE_i}$, of constant lengths $2E_0$, sliding between the legs of the right angle $E_i\ OE$; hence, it is the sextic hypocycloid osculating circle e_0, which has the general equation, with e, e_i as axes of coördinates:

$$\sqrt[3]{x^2} + \sqrt[3]{y^2} = \sqrt[3]{4E_0^2}$$

In the next case, $E_1 < E_0$ (Fig. 144) we see first, that the current can never become zero like in the first case,

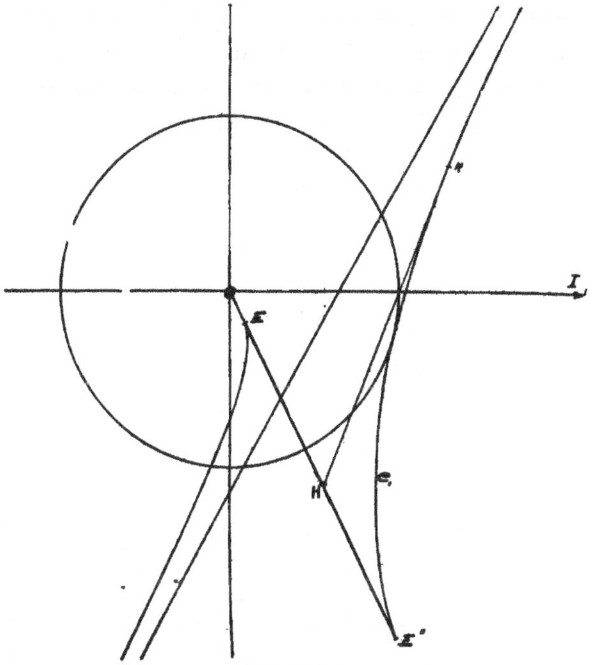

Fig. 145.

$E_1 = E_0$, but has a minimum value corresponding to the minimum value of $\overline{OE_1}$: $I_1' = \dfrac{E_0 - E_1}{z}$, and a maximum value: $I_1'' = \dfrac{E_0 + E_1}{z}$. Furthermore, the current can never lead the impressed E.M.F. E_0, but always lags. The minimum lag is at the point H. The locus e_1, as envelope of the lines $\overline{EE_0}$, is a finite sextic curve, shown in Fig. 144.

If $E_1 < E_0$, at small $E_0 - E_1$, H can be above the zero line, and a range of leading current exist between two ranges of lagging current.

In the case $E_1 > E_0$ (Fig. 145) the current cannot equal zero either, but begins at a finite value C_1', corresponding to the minimum value of $\overline{OE_0}$: $I_1' = \frac{E_1 - E_0}{z}$. At this value however, the alternator E_1 is still generator and changes to a motor, its power passing through zero, at the point corresponding to the vertical tangent, onto e_1, with a very large lead of the impressed E.M.F. against the current. At H the lead changes to lag.

The minimum and maximum value of current in the three conditions are given by:

	Minimum:	*Maximum:*
1st.	$I = 0,$	$I = \frac{2\,E_0}{z}.$
2d.	$I = \frac{E_0 - E_1}{z},$	$I = \frac{E_0 + E_1}{z},$
3d.	$I = \frac{E_1 - E_0}{z},$	$I = \frac{E_0 + E_1}{z},$

Since the current passing over the line at $E_1 = O$, that is, when the motor stands still, is $I_0 = E_0 / z$, we see that in such a synchronous motor-plant, when running at synchronism, the current can rise far beyond the value it has at standstill of the motor, to twice this value at 1, somewhat less at 2, but more at 3.

203. C. $E_0 =$ *constant,* E_1 *varied so that the efficiency is a maximum for all currents.* (Fig. 146.)

Since we have seen that the output at a given current strength, that is, a given loss, is a maximum, and therefore the efficiency a maximum, when the current is in phase with the induced E.M.F. E_0 of the generator, we have as the locus of E_0 the point E_0 (Fig. 146), and when E with increasing current varies on e, E_1 must vary on the straight line e_1 parallel to e.

Hence, at no-load or zero current, $E_1 = E_0$, decreases with increasing load, reaches a minimum at $\overline{OE_1^1}$ perpendicular to e_1, and then increases again, reaches once more

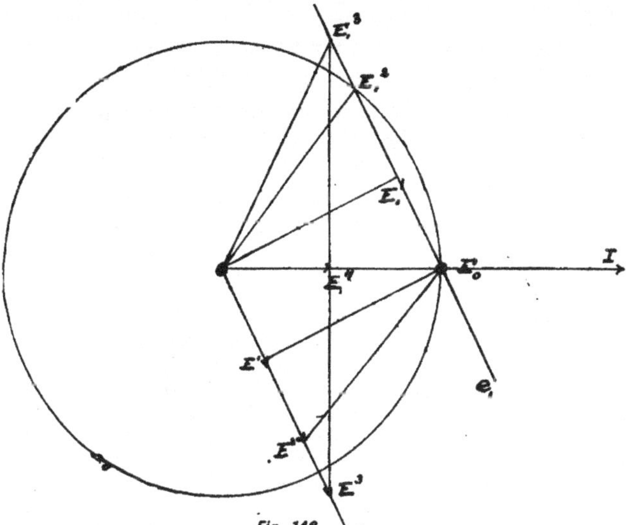

Fig. 146.

$E_1 = E_0$ at E_1^2, and then increases beyond E_0. The current is always ahead of the induced E.M.F. E_1 of the motor, and by its lead compensates for the self-induction of the system, making the total circuit non-inductive.

The power is a maximum at E_1^3, where $\overline{OE_1^4} = \overline{E_1^4 E_0} = 1/2 \times \overline{OE_0}$, and is then $= I \times \overline{E_0}/2$. Hence, since $\overline{OE_1^4} - Ir = E_0/2$, $I = E_0/2\,r$ and $P = E_0^2/4\,r$, hence = the maximum power which, over a non-inductive line of resistance r can be transmitted, at 50 per cent. efficiency, into a non-inductive circuit.

In this case,

$$E_1^s = \frac{z}{r} \times \frac{E_0}{2} = \frac{E_0}{2}\sqrt{1+\left(\frac{x}{r}\right)^2}.$$

In general, it is, taken from the diagram, at the condition of maximum éfficiency:

$$E_1 = \sqrt{(E_0 - Ir)^2 + I^2 x^2}.$$

Comparing these results with those in Chapter IX. on Self-induction and Capacity, we see that the condition of maximum efficiency of the synchronous motor system is the same as in a system containing only inductance and capacity, the lead of the current against the induced E.M.F. E_1 here acting in the same way as the condenser capacity in Chapter IX.

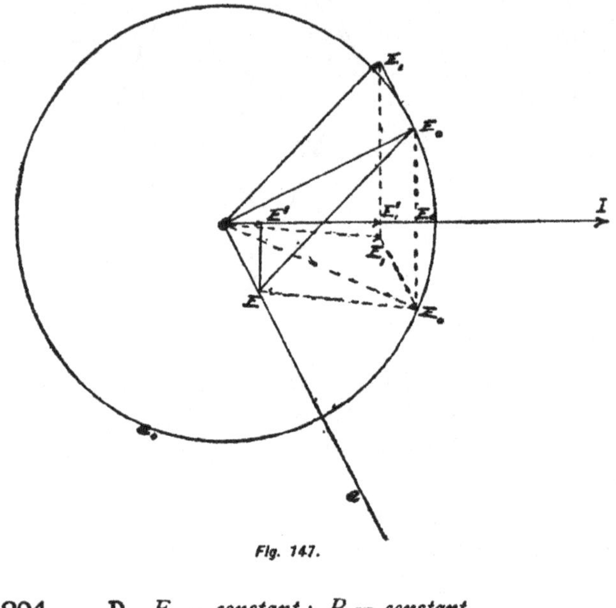

Fig. 147.

204. D. $E_0 = constant$; $P = constant$.

If the power of a synchronous motor remains constant, we have (Fig. 147) $I \times \overline{OE_1}^1 =$ constant, or, since $\overline{OE^1} =$

Ir, $I = \overline{OE^1} / r$, and: $\overline{OE^1} \times \overline{OE_1^{\,1}} = \overline{OE^1} \times \overline{E^1E_0^{\,1}} =$ constant.

Hence we get the diagram for any value of the current I, at constant power P_1, by making $\overline{OE^1} = Ir$, $\overline{E^1E_0^{\,1}} = P^1 / I$ erecting in $E_0^{\,1}$ a perpendicular, which gives two points of intersection with circle e_0, E_0, one leading, the other lagging. Hence, at a given impressed E.M.F. E_0, the same power P_1

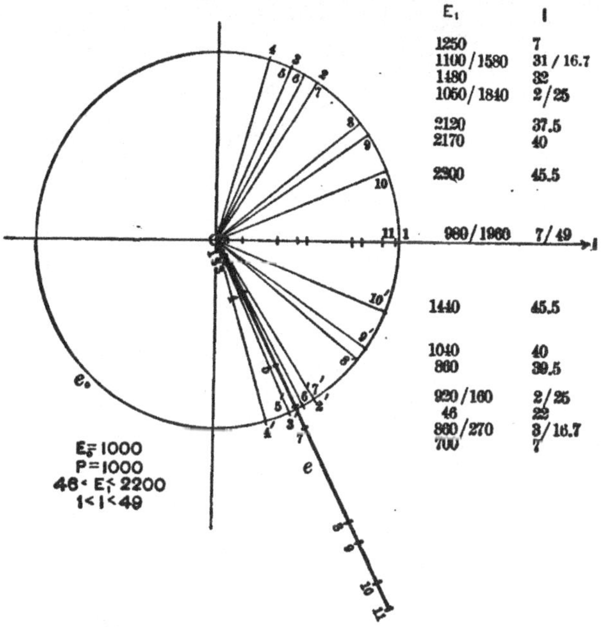

Fig. 148.

can be transmitted by the same current I with two different induced E.M.Fs. E_1 of the motor; one, $\overline{OE_1} = \overline{EE_0}$ small, corresponding to a lagging current; and the other, $\overline{OE_1} = \overline{EE_0}$ large, corresponding to a leading current The former is shown in dotted lines, the latter in drawn lines, in the diagram, Fig. 147.

Hence a synchronous motor can work with a given output, at the same current with two different counter E.M.Fs.

E_1. In one of the cases the current is leading, in the other lagging.

In Figs. 148 to 151 are shown diagrams, giving the points

E_0 = impressed E.M.F., assumed as constant = 1000 volts,
E = E.M.F. consumed by impedance,
E' = E.M.F. consumed by resistance.

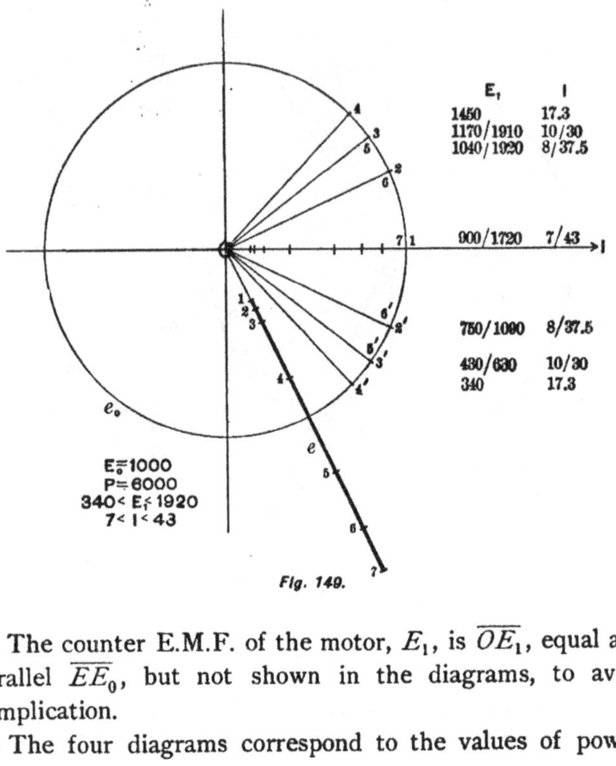

Fig. 149.

The counter E.M.F. of the motor, E_1, is $\overline{OE_1}$, equal and parallel $\overline{EE_0}$, but not shown in the diagrams, to avoid complication.

The four diagrams correspond to the values of power, or motor output,

P = 1,000, 6,000, 9,000, 12,000 watts, and give:

$P = 1,000$	$46 < E_1 < 2,200$,	$1 < I < 49$	Fig. 132.
$P = 6,000$	$340 < E_1 < 1,920$,	$7 < I < 43$	Fig. 133.
$P = 9,000$	$540 < E_1 < 1,750$,	$11.8 < I < 38.2$	Fig. 134.
$P = 12,000$	$920 < E_1 < 1,320$,	$20 < I < 30$	Fig. 153.

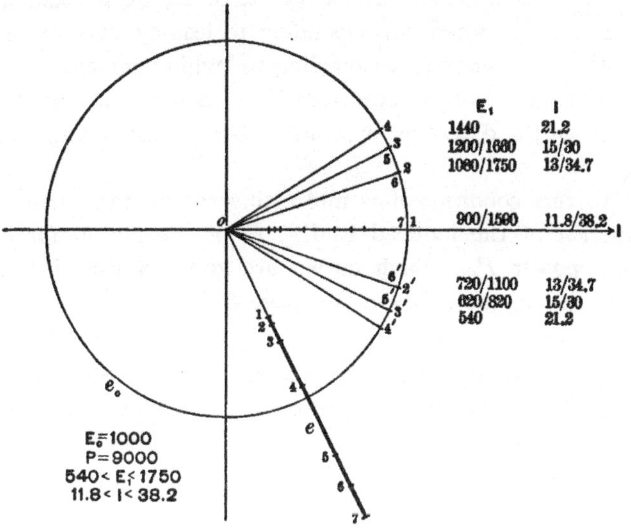

Fig. 150.

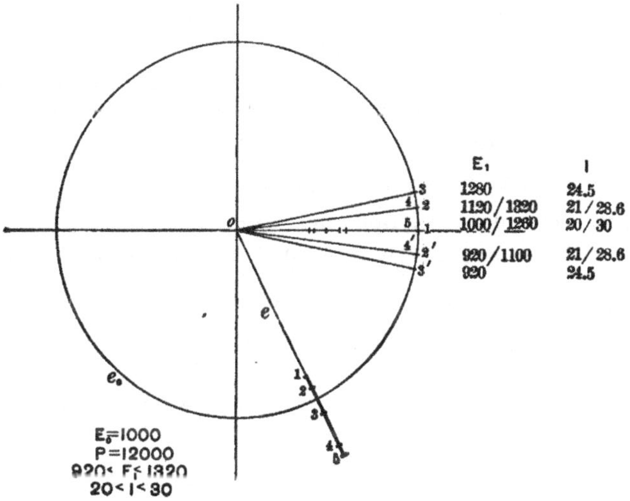

Fig. 151.

As seen, the permissible value of counter E.M.F. E_1 and of current I, becomes narrower with increasing output.

In the diagrams, different points of E_0 are marked with 1, 2, 3 . . ., when corresponding to leading current, with 2^1, 3^1, . . ., when corresponding to lagging current.

The values of counter E.M.F. E_1 and of current I are noted on the diagrams, opposite to the corresponding points E_0.

In this condition it is interesting to plot the current as function of the induced E.M.F. E_1 of the motor, for constant power P_1. Such curves are given in Fig. 155 and explained in the following on page 345.

205. While the graphic method is very convenient to get a clear insight into the interdependence of the different quantities, for numerical calculation it is preferable to express the diagrams analytically.

For this purpose,

Let $z = \sqrt{r^2 + x^2}$ = impedance of the circuit of (equivalent) resistance r and (equivalent) reactance $x = 2 \pi N L$, containing the impressed E.M.F. e_0* and the counter E.M.F. e_1 of the synchronous motor; that is, the E.M.F. induced in the motor armature by its rotation through the (resultant) magnetic field.

Let i = current in the circuit (effective values).

The mechanical power delivered by the synchronous motor (including friction and core loss) is the electric power consumed by the C.E.M.F. e_1; hence —

$$p = i e_1 \cos (i_1, e_1), \qquad (1)$$

thus, —

$$\left.\begin{aligned} \cos(i_1, e_1) &= \frac{p}{i e_1} \\ \sin (i_1 e_1) &= \sqrt{1 - \left(\frac{p}{i e_1}\right)^2}. \end{aligned}\right\} \qquad (2)$$

* If e_0 = E.M.F. at motor terminals, z = internal impedance of the motor; if e_0 = terminal voltage of the generator, z = total impedance of line and motor; if e_0 = E.M.F. of generator, that is, E.M.F. induced in generator armature by its rotation through the magnetic field, z includes the generator impedance also.

The displacement of phase between current i and E.M.F. $e = z\,i$ consumed by the impedance z is:

$$\left.\begin{aligned}\cos(ie) &= \frac{r}{z}\\ \sin(ie) &= \frac{x}{z}\end{aligned}\right\} \qquad (3)$$

Since the three E.M.Fs. acting in the closed circuit:

e_0 = E.M.F. of generator,
e_1 = C.E.M.F. of synchronous motor,
$e = zi$ = E.M.F. consumed by impedance,

form a triangle, that is, e_1 and e are components of e_0, it is (Fig. 152):

$$e_0^2 = e_1^2 + e^2 + 2\,e\,e_1 \cos(e_1 e), \qquad (4)$$

hence, $$\cos(e_1, e) = \frac{e_0^2 - e_1^2 - e^2}{2\,e_1 e} = \frac{e_0^2 - e_1^2 - z^2 i^2}{2\,z\,i\,e_1}, \qquad (5)$$

since, however, by diagram:

$$\begin{aligned}\cos(e_1, e) &= \cos(i, e - i, e_1)\\ &= \cos(i, e)\cos(i, e_1) + \sin(i, e)\sin(i, e_1) \qquad (6)\end{aligned}$$

substitution of (2), (3) and (5) in (6) gives, after some transposition:

$$e_0^2 - e_1^2 - z^2 i^2 - 2\,rp = 2\,x\sqrt{i^2 e_1^2 - p^2}, \qquad (7)$$

the *Fundamental Equation of the Synchronous Motor*, relating impressed E.M.F., e_0; C.E.M.F., e_1; current i; power, p, and resistance, r; reactance, x; impedance z.

This equation shows that, at given impressed E.M.F. e_0, and given impedance $z = \sqrt{r^2 + x^2}$, three variables are left, e_1, i, p, of which two are independent. Hence, at given e_0 and z, the current i is not determined by the load p only, but also by the excitation, and thus the same current i can represent widely different loads p, according to the excitation; and with the same load, the current i can be varied in a wide range, by varying the field excitation e_1.

The meaning of equation (7) is made more perspicuous

by some transformations, which separate e_1 and i, as function of p and of an angular parameter ϕ.

Substituting in (7) the new coördinates:

$$\left.\begin{aligned} \alpha &= \frac{e_1^2 + z^2 i^2}{\sqrt{2}} \\ \beta &= \frac{e_1^2 - z^2 i^2}{\sqrt{2}} \end{aligned}\right\} \text{ or, } \left\{\begin{aligned} e_1^2 &= \frac{\alpha + \beta}{\sqrt{2}} \\ z^2 i^2 &= \frac{\alpha - \beta}{\sqrt{2}} \end{aligned}\right\} \qquad (8)$$

we get

$$e_0^2 - \alpha\sqrt{2} - 2\,r\,p = 2\,\frac{x}{z}\sqrt{\frac{\alpha^2 - \beta^2}{2} - z^2 p^2}\,; \qquad (9)$$

substituting again,

$$\left.\begin{aligned} e_0^2 &= a \\ 2\,z\,p &= b \\ r &= \epsilon z \end{aligned}\right.$$

hence,

$$\left.\begin{aligned} x &= z\sqrt{1 - \epsilon^2} \\ 2\,r\,p &= \epsilon\,b \end{aligned}\right\}$$

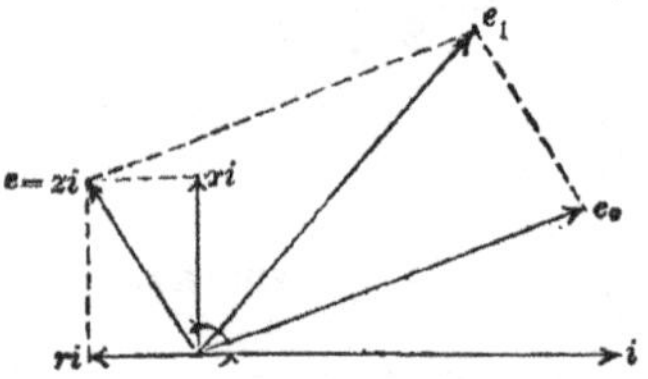

Fig. 152.

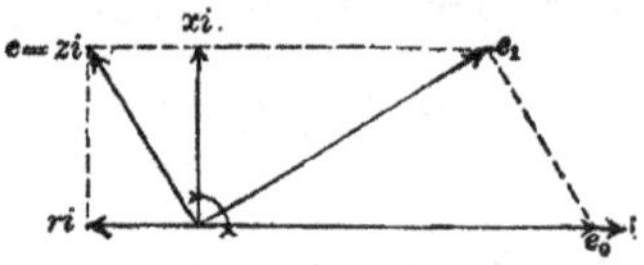

Fig. 153.

we get

$$a - \alpha\sqrt{2} - \epsilon\,b = \sqrt{(1 - \epsilon^2)\,(2\,\alpha^2 - 2\,\beta^2 - b^2)}, \qquad (11)$$

and, squared,

$$\epsilon^2\alpha^2 + (1-\epsilon^2)\,\beta^2 - \alpha\sqrt{2}\,(a - \epsilon b) + \frac{b^2\,(1-\epsilon^2)}{2} + \frac{(a-\epsilon b)^2}{2} = 0, \qquad (12)$$

substituting

$$\left.\begin{aligned} \epsilon\alpha - \frac{(a - \epsilon\,b)\sqrt{2}}{2\,\epsilon} &= v, \\ \beta\sqrt{1 - \epsilon^2} &= w, \end{aligned}\right\} \qquad (13)$$

gives, after some transposition,

$$v^2 + w^2 = \frac{(1 - \epsilon^2)}{2\,\epsilon^2}\,a\,(a - 2\,\epsilon b), \qquad (14)$$

hence, if

$$R = \sqrt{\frac{(1 - \epsilon^2)(a - 2\,\epsilon\,b)\,a}{2\,\epsilon^2}}, \tag{15}$$

it is

$$v^2 + w^2 = R^2 \tag{16}$$

the equation of a circle with radius R.

Substituting now backwards, we get, with some transpositions:

$$\{r^2(e_1^2 + z^2 i^2) - z^2(e_0^2 - 2\,rp)\}^2 + \{r\,x\,(e_1^2 - z^2 i^2)\}^2 = x^2 z^2 e_0^2 (e_0^2 - 4\,rp) \tag{17}$$

the *Fundamental Equation of the Synchronous Motor* in a modified form.

The separation of e_1 and i can be effected by the introduction of a parameter ϕ by the equations:

$$\left.\begin{aligned} r^2(e_1^2 - z^2 i^2) - z^2(e_0^2 - 2\,rp) &= x\,z\,e_0\sqrt{e_0^2 - 4\,rp}\,\cos\phi \\ r\,x\,(e_1^2 - z^2 i^2) &= x\,z\,e_0\sqrt{e_0^2 - 4\,rp}\,\sin\phi \end{aligned}\right\} \tag{18}$$

These equations (18), transposed, give

$$e_1 = \sqrt{\frac{1}{2}\left\{\frac{z^2}{r^2}(e_0^2 - 2\,rp) + \frac{z\,e_0}{r}\left(\frac{x}{r}\cos\phi + \sin\phi\right)\sqrt{e_0^2 - 4\,rp}\right\}}$$

$$= \frac{e_0 z}{r}\sqrt{\frac{1}{2}\left\{\left(1 - \frac{2\,rp}{e_0^2}\right) + \left(\frac{x}{z}\cos\phi + \frac{r}{z}\sin\phi\right)\sqrt{1 - \frac{4\,rp}{e_0^2}}\right\}}. \tag{19}$$

$$i = \sqrt{\frac{1}{2}\left\{\frac{1}{r^2}(e_0^2 - 2\,rp) + \frac{e_0}{r\,z}\left(\frac{x}{r}\cos\phi - \sin\phi\right)\sqrt{e_0^2 - 4\,rp}\right\}}$$

$$= \frac{e_0}{r}\sqrt{\frac{1}{2}\left\{\left(1 - \frac{2\,rp}{e_0^2}\right) + \left(\frac{x}{z}\cos\phi - \frac{r}{z}\sin\phi\right)\sqrt{1 - \frac{4\,rp}{e_0^2}}\right\}}. \tag{20}$$

The parameter ϕ has no direct physical meaning, apparently.

These equations (19) and (20), by giving the values of e_1 and i as functions of p and the parameter ϕ enable us to construct the *Power Characteristics of the Synchronous Motor*, as the curves relating e_1 and i, for a given power p, by attributing to ϕ all different values.

Since the variables v and w in the equation of the circle (16) are quadratic functions of e_1 and i, the *Power Characteristics of the Synchronous Motor are Quartic Curves.*

They represent the action of the synchronous motor under all conditions of load and excitation, as an element of power transmission even including the line, etc.

Before discussing further these Power Characteristics, some special conditions may be considered.

206. A. *Maximum Output.*

Since the expression of e_1 and i [equations (19) and (20)] contain the square root, $\sqrt{e_0^2 - 4rp}$, it is obvious that the maximum value of p corresponds to the moment where this square root disappears by passing from real to imaginary; that is,

$$e_0^2 - 4rp = 0,$$

or,

$$p = \frac{e_0^2}{4r}. \qquad (21)$$

This is the same value which represents the maximum power transmissible by E.M.F., e_0, over a non-inductive line of resistance, r; or, more generally, the maximum power which can be transmitted over a line of impedance,

$$z = \sqrt{r^2 + x^2},$$

into any circuit, shunted by a condenser of suitable capacity.

Substituting (21) in (19) and (20), we get,

$$\left.\begin{aligned} e_1 &= \frac{z}{2r} e_0 \\ i &= \frac{e_0}{2r} \end{aligned}\right\} \qquad (22)$$

and the displacement of phase in the synchronous motor.

$$\cos(e_1, i) = \frac{p}{ie_1} = \frac{r}{z};$$

hence,

$$\tan(e_1, i) = -\frac{x}{r}, \qquad (23)$$

that is, the angle of internal displacement in the synchronous motor is equal, but opposite to, the angle of displacement of line impedance,

$$(e_1, i) = -(e, i),$$
$$= -(z, r), \qquad (24)$$

and consequently,

$$(e_0, i) = 0; \qquad (25)$$

that is, the current, i, is in phase with the impressed E.M.F., e_0.

If $z < 2r$, $e_1 < e_0$; that is, motor E.M.F. < generator E.M.F.
If $z = 2r$, $e_1 = e_0$; that is, motor E.M.F. = generator E.M.F.
If $z > 2r$, $e_1 > e_0$; that is, motor E.M.F. > generator E.M.F.

In either case, the current in the synchronous motor is leading.

207. B. *Running Light,* $p = 0$.

When running light, or for $p = 0$, we get, by substituting in (19) and (20),

$$\left.\begin{aligned} e_1 &= \frac{e_0 z}{r}\sqrt{\frac{1}{2}\left\{1 + \frac{x}{z}\cos\phi + \frac{r}{z}\sin\phi\right\}} \\ i &= \frac{e_0}{r}\sqrt{\frac{1}{2}\left\{1 + \frac{x}{z}\cos\phi - \frac{r}{z}\sin\phi\right\}} \end{aligned}\right\} \qquad (26)$$

Obviously this condition cannot well be fulfilled, since p must at least equal the power consumed by friction, etc.; and thus the true no-load curve merely approaches the curve $p = 0$, being, however, rounded off, where curve (26) gives sharp corners.

Substituting $p = 0$ into equation (7) gives, after squaring and transposing,

$$e_1^4 + e_0^4 + z^4 i^4 - 2e_1^2 e_0^2 - 2z^2 i^2 e_0^2 + 2r^2 i^2 e_1^2 - 2x^2 i^2 e_1^2 = 0. \quad (27)$$

This quartic equation can be resolved into the product of two quadratic equations,

$$\left.\begin{aligned} e_1^2 + z^2 i^2 - e_0^2 + 2xie_1 = 0. \\ e_1^2 + z^2 i^2 - e_0^2 - 2xie_1 = 0. \end{aligned}\right\} \qquad (28)$$

which are the equations of two ellipses, the one the image of the other, both inclined with their axes.

The minimum value of C.E.M.F., e_1, is $e_1 = 0$ at $i = \frac{e_0}{z}$. (29)

The minimum value of current, i, is $i = 0$ at $e_1 = e_0$. (30)

The maximum value of E.M.F., e_1, is given by Equation (28),

$$f = e_1^2 + z^2 i^2 - e_0^2 \pm 2 x i e_1 = 0;$$

by the condition,

$$\frac{de_1}{di} = -\frac{df / di}{df / de_1} = 0, \text{ as } z^2 i \pm x e_1 \neq 0.$$

hence,

$$i = e_0 \frac{x}{rz}, \quad e_1 = \mp e_0 \frac{z}{r} \qquad (31)$$

The maximum value of current, i, is given by equation (28) by

$$\frac{di}{de_1} = 0, \text{ as}$$

$$i = \frac{e_0}{r} \quad e_1 = \mp e_0 \frac{x}{r}. \qquad (32)$$

If, as abscissæ, e_1, and as ordinates, zi, are chosen, the axis of these ellipses pass through the points of maximum power given by equation (22).

It is obvious thus, that in the V-shaped curves of synchronous motors running light, the two sides of the curves are not straight lines, as usually assumed, but arcs of ellipses, the one of concave, the other of convex, curvature.

These two ellipses are shown in Fig. 154, and divide the whole space into six parts — the two parts A and A', whose areas contain the quartic curves (19) (20) of synchronous motor, the two parts B and B', whose areas contain the quartic curves of generator, and the interior space C and exterior space D, whose points do not represent any actual condition of the alternator circuit, but make e_1, i imaginary.

A and A' and the same B and B', are identical conditions of the alternator circuit, differing merely by a simul-

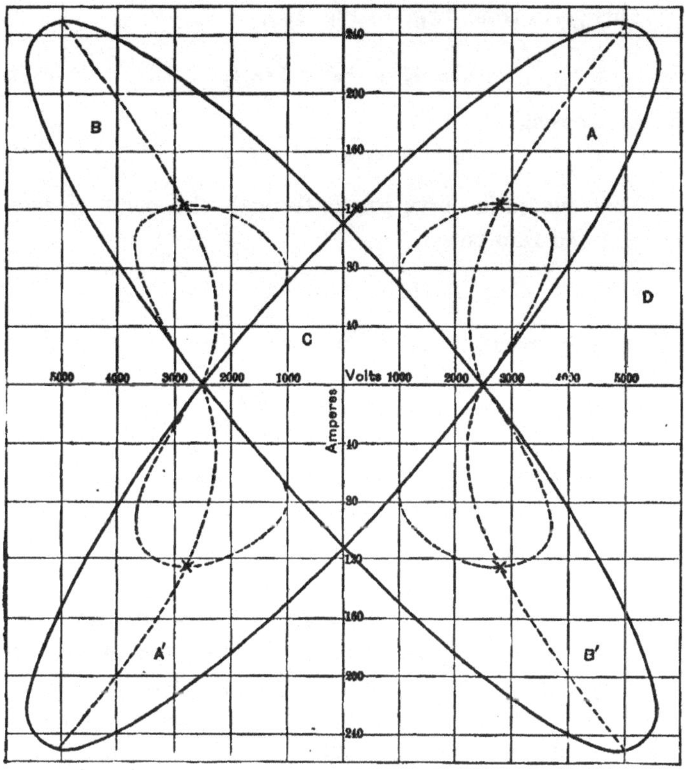

Fig. 154.

taneous reversal of current and E.M.F.; that is, differing by the time of a half period.

Each of the spaces A and B contains one point of equation (22), representing the condition of maximum output of generator, viz., synchronous motor.

208. C. *Minimum Current at Given Power.*

The condition of minimum current, i, at given power, p, is determined by the absence of a phase displacement at the impressed E.M.F. e_0,

$$(e_0, i) = 0.$$

This gives from diagram Fig. 153,

$$e_1^2 = e_0^2 + i^2 z^2 - 2\, i\, e_0 r, \qquad (33)$$

or, transposed,

$$e_1 = \sqrt{(e_0 - i r)^2 + i^2 x^2}. \qquad (34)$$

This quadratic curve passes through the point of zero current and zero power,

$$i = 0, \qquad e_1 = e_0,$$

through the point of maximum power (22),

$$i = \frac{e_0}{2\,r}, \qquad e_1 = \frac{e_0 z}{2\,r},$$

and through the point of maximum current and zero power,

$$i = \frac{e_0}{r}, \qquad e_1 = \frac{e_0 x}{r}, \qquad (35)$$

and divides each of the quartic curves or power characteristics into two sections, one with leading, the other with lagging, current, which sections are separated by the two points of equation 34, the one corresponding to minimum, the other to maximum, current.

It is interesting to note that at the latter point the current can be many times larger than the current which would pass through the motor while at rest, which latter current is,

$$i = \frac{e_0}{z}, \qquad (36)$$

while at no-load, the current can reach the maximum value,

$$i = \frac{e_0}{r}, \qquad (35)$$

the same value as would exist in a non-inductive circuit of the same resistance.

The minimum value at C.E.M.F. e_1, at which coincidence

of phase $(e_0, i) = 0$, can still be reached, is determined from equation (34) by,

$$\frac{de_1}{di} = 0;$$

as

$$i = e_0 \frac{r}{z^2} \quad e_1 = e_0 \frac{x}{z}. \tag{37}$$

The curve of no-displacement, or of minimum current, is shown in Figs. 138 and 139 in dotted lines.*

209. D. *Maximum Displacement of Phase.*

$$(e_0, i) = \text{maximum}.$$

At a given power p the input is,

$$p_0 = p + i^2 r = e_0 i \cos (e_0, i); \tag{38}$$

hence,

$$\cos (e_0, i) = \frac{p + i^2 r}{e_0 i}. \tag{39}$$

At a given power p, this value, as function of the current i, is a maximum when

$$\frac{d}{di}\left(\frac{p + i^2 r}{e_0 i}\right) = 0,$$

this gives,

$$p = i^2 r; \tag{40}$$

or,

$$i = \sqrt{\frac{p}{r}}. \tag{41}$$

That is, the displacement of phase, lead or lag, is a maximum, when the power of the motor equals the power

---

* It is interesting to note that the equation (34) is similar to the value, $e_1 = \sqrt{(e_0 - ir)^2 - i^2 x^2}$, which represents the output transmitted over an inductive line of impedance, $z = \sqrt{r^2 + x^2}$ into a non-inductive circuit.

Equation (34) is identical with the equation giving the maximum voltage, e_1, at current, i, which can be produced by shunting the receiving circuit with a condenser; that is, the condition of "complete resonance" of the line, $z = \sqrt{r^2 + x^2}$, with current, i. Hence, referring to equation (35), $e_1 = e_0 \frac{x}{r}$ is the maximum resonance voltage of the line, reached when closed by a condenser of reactance, $-x$.

consumed by the resistance; that is, at the electrical efficiency of 50 per cent.

Substituting (40) in equation (7) gives, after squaring

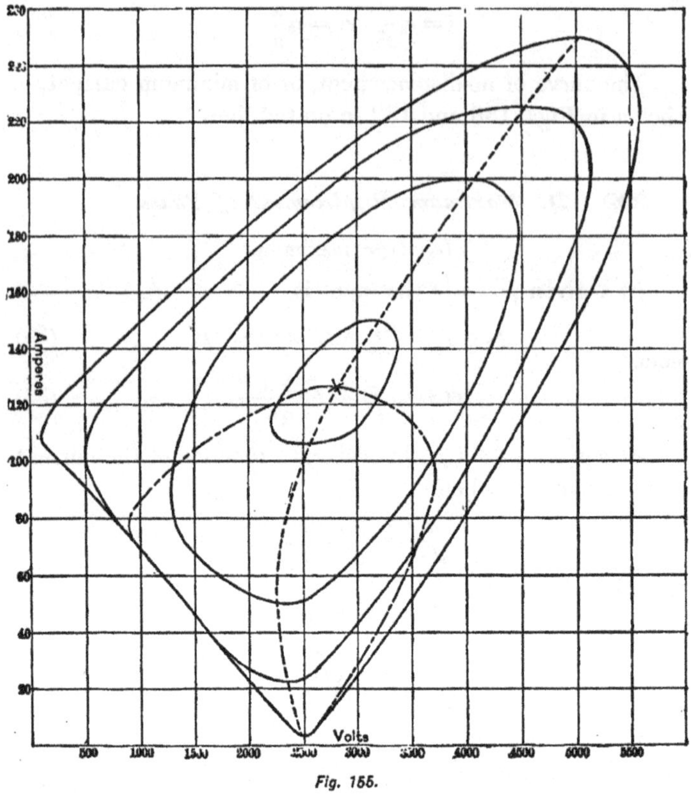

Fig. 155.

and transposing, the Quartic Equation of Maximum Displacement,

$$(e_0^2 - e_1^2)^2 + i^4 z^2 (z^2 + 8\, r^2) + 2\, i^2 e_1^2 (5\, r^2 - z^2) - 2\, i^2 e_0^2 (z^2 + 3\, r^2) = 0. \qquad (42)$$

The curve of maximum displacement is shown in dash-dotted lines in Figs. 154 and 155. It passes through the point of zero current — as singular or nodal point — and through the point of maximum power, where the maximum displacement is zero, and it intersects the curve of zero displacement.

210. E. *Constant Counter E.M.F.*

At constant C.E.M.F., e_1 = constant,

If

$$e_1 < e_0\sqrt{1 - \frac{r^2x^2}{z^4}},$$

the current at no-load is not a minimum, and is lagging. With increasing load, the lag decreases, reaches a minimum, and then increases again, until the motor falls out of step, without ever coming into coincidence of phase.

If

$$e_0\sqrt{1 - \frac{r^2x^2}{z^4}} < e_1 < e_0,$$

the current is lagging at no load; with increasing load the lag decreases, the current comes into coincidence of phase with e_0, then becomes leading, reaches a maximum lead; then the lead decreases again, the current comes again into coincidence of phase, and becomes lagging, until the motor falls out of step.

If $e_0 < e_1$, the current is leading at no load, and the lead first increases, reaches a maximum, then decreases; and whether the current ever comes into coincidence of phase, and then becomes lagging, or whether the motor falls out of step while the current is still leading, depends, whether the C.E.M.F. at the point of maximum output is $> e_0$ or $< e_0$.

211. F. *Numerical Instance.*

Figs. 154 and 155 show the characteristics of a 100-kilowatt motor, supplied from a 2500-volt generator over a distance of 5 miles, the line consisting of two wires, No. 2 B. & S.G., 18 inches apart.

In this case we have,

$$\left.\begin{aligned} e_0 &= 2500 \text{ volts constant at generator terminals;} \\ r &= 10 \text{ ohms, including line and motor;} \\ x &= 20 \text{ ohms, including line and motor;} \\ \text{hence } z &= 22.36 \text{ ohms.} \end{aligned}\right\} \quad (43)$$

Substituting these values, we get,

$$2500^2 - e_1^2 - 500\, i^2 - 20\, p = 40\, \sqrt{i^2 e_1^2 - p^2} \qquad (7)$$

$$\{e_1^2 + 500\, i^2 - 31.25 \times 10^6 + 100\, p\}^2 + \{2\, e_1^2 - 1000\, i^2\}^2 = 7.8125 \times 10^{15} - 5 + 10^9\, p. \qquad (17)$$

$$e_1 = 5590 \sqrt{\tfrac{1}{2}\{(1-3.2\times 10^{-6} p)+(.894 \cos\phi + .447 \sin\phi)\sqrt{1-6.4\times 10^{-6} p}\}}. \qquad (19)$$

$$i = 559 \sqrt{\tfrac{1}{2}\{(1-3.2\times 10^{-6} p)+(.894 \cos\phi - .447 \sin\phi)\sqrt{1-6.4\times 10^{-6} p}\}}. \qquad (20)$$

Maximum output,

$$p = 156.25 \text{ kilowatts} \qquad (21)$$

at

$$\left.\begin{aligned} e_1 &= 2{,}795 \text{ volts} \\ i &= 125 \text{ amperes} \end{aligned}\right\} \qquad (22)$$

Running light,

$$\left.\begin{aligned} e_1^2 + 500\, i^2 - 6.25 \times 10^4 \mp 40\, i\, e_1 &= 0 \\ e_1 &= 20\, i \pm \sqrt{6.25 \times 10^4 - 100\, i^2} \end{aligned}\right\} \qquad (28)$$

At the minimum value of C.E.M.F. $e_1 = 0$ is $i = 112$ (29)
At the minimum value of current, $i = 0$ is $e_1 = 2500$ (30)
At the maximum value of C.E.M.F. $e_1 = 5590$ is $i = 223.5$ (31)
At the maximum value of current $i = 250$ is $e_1 = 5000$ (32)

Curve of zero displacement of phase,

$$e_1 = 10\, \sqrt{(250 - i)^2 + 4\, i^2} \qquad (34)$$
$$= 10\, \sqrt{6.25 \times 10^4 - 500\, i + 5\, i^2}$$

Minimum C.E.M.F. point of this curve,

$$i = 50 \qquad e_1 = 2240 \qquad (35)$$

Curve of maximum displacement of phase,

$$p = 10\, i^2 \qquad (40)$$

$$(6.25 \times 10^6 - e_1^2)^2 + .65 \times 10^6\, i^4 - 10^{10}\, i^2 = 0. \qquad (42)$$

Fig. 154 gives the two ellipses of zero power, in drawn lines, with the curves of zero displacement in dotted, the curves of maximum displacement in dash-dotted lines, and the points of maximum power as crosses.

Fig. 155 gives the motor-power characteristics, for,

$p = 10$ kilowatts.
$p = 50$ kilowatts.
$p = 100$ kilowatts.
$p = 150$ kilowatts.
$p = 156.25$ kilowatts.

together with the curves of zero displacement, and of maximum displacement.

212. G. *Discussion of Results.*

The characteristic curves of the synchronous motor, as shown in Fig. 155, have been observed frequently, with their essential features, the V-shaped curve of no load, with the point rounded off and the two legs slightly curved, the one concave, the other convex; the increased rounding off and contraction of the curves with increasing load; and the gradual shifting of the point of minimum current with increasing load, first towards lower, then towards higher, values of C.E.M.F. e_1.

The upper parts of the curves, however, I have never been able to observe experimentally, and consider it as probable that they correspond to a condition of synchronous motor-running, which is unstable. The experimental observations usually extend about over that part of the curves of Fig. 155 which is reproduced in Fig. 156, and in trying to extend the curves further to either side, the motor is thrown out of synchronism.

It must be understood, however, that these power characteristics of the synchronous motor in Fig. 155 can be considered as approximations only, since a number of assumptions are made which are not, or only partly, fulfilled in practice. The foremost of these are:

1. It is assumed that e_1 can be varied unrestrictedly, while in reality the possible increase of e_1 is limited by magnetic saturation. Thus in Fig. 155, at an impressed E.M.F., $e_0 = 2{,}500$ volts, e_1 rises up to 5,590 volts, which may or may not be beyond that which can be produced by the motor, but certainly is beyond that which can be constantly given by the motor.

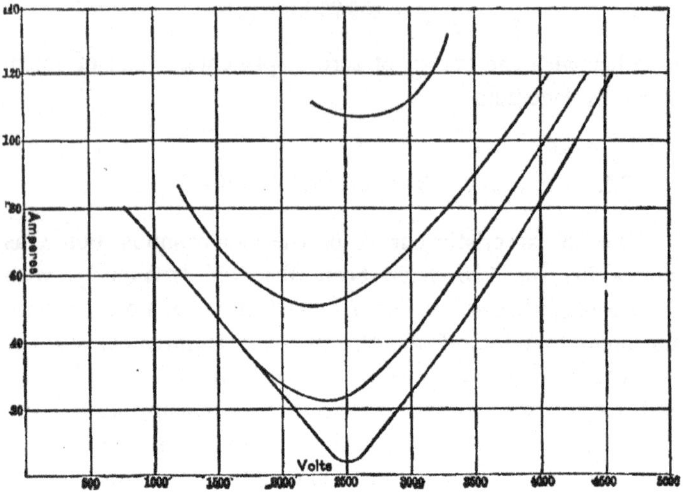

Fig. 156.

2. The reactance, x, is assumed as constant. While the reactance of the line is practically constant, that of the motor is not, but varies more or less with the saturation, decreasing for higher values. This decrease of x increases the current i, corresponding to higher values of e_1, and thereby bends the curves upwards at a lower value of e_1 than represented in Fig. 155.

It must be understood that the motor reactance is not a simple quantity, but represents the combined effect of self-induction, that is, the E.M.F. induced in the armature conductor by the current flowing therein and armature reaction, or the variation of the C.E.M.F. of the motor by the change of the resultant field, due to the superposition of the M.M.F. of the armature current upon the field excitation; that is, it is the "synchronous reactance."

3. These curves in Fig. 155 represent the conditions of constant electric power of the motor, thus including the mechanical and the magnetic friction (core loss). While the mechanical friction can be considered as approximately constant, the magnetic friction is not, but increases with the magnetic induction; that is, with e_1, and the same holds for the power consumed for field excitation.

Hence the useful mechanical output of the motor will on the same curve, p = const., be larger at points of lower C.E.M.F., e_1, than at points of higher e_1; and if the curves are plotted for constant useful mechanical output, the whole system of curves will be shifted somewhat towards lower values of e_1; hence the points of maximum output of the motor correspond to a lower E.M.F. also.

It is obvious that the true mechanical power-characteristics of the synchronous motor can be determined only in the case of the particular conditions of the installation under consideration.

CHAPTER XX

COMMUTATOR MOTORS.

213. Commutator motors — that is, motors in which the current enters or leaves the armature over brushes through a segmental commutator — have been built of various types, but have not found any extensive application, in consequence of the superiority of the induction and synchronous motors, due to the absence of commutators.

The main subdivisions of commutator motors are the repulsion motor, the series motor, and the shunt motor.

REPULSION MOTOR.

214. The repulsion motor is an induction motor or transformer motor; that is, a motor in which the main current enters the primary member or field only, while in the secondary member, or armature, a current is induced, and thus the action is due to the repulsive thrust between induced current and inducing magnetism.

As stated under the heading of induction motors, a multiple circuit armature is required for the purpose of having always secondary circuits in inductive relation to the primary circuit during the rotation. If with a single-coil field, these secondary circuits are constantly closed upon themselves as in the induction motor, the primary circuit will not exert a rotary effect upon the armature while at rest, since in half of the armature coils the current is induced so as to give a rotary effort in the one direction, and in the other half the current is induced to give a rotary effort in the opposite direction, as shown by the arrows in Fig. 157.

In the induction motor a second magnetic field is used to act upon the currents induced by the first, or inducing magnetic field, and thereby cause a rotation. That means the motor consists of a primary electric circuit, inducing

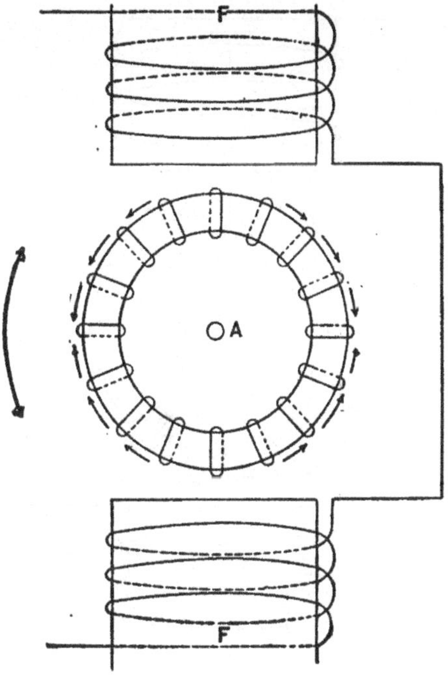

Fig. 157.

in the armature the secondary currents, and a primary magnetizing circuit producing the magnetism to act upon the secondary currents.

In the polyphase induction motor both functions of the primary circuit are usually combined in the same coils; that is, each primary coil induces secondary currents, and produces magnetic flux acting upon secondary currents induced by another primary coil.

215. In the repulsion motor the difficulty due to the equal and opposite rotary efforts, caused by the induced armature currents when acted upon by the inducing magnetic field, is overcome by having the armature coils closed upon themselves, either on short circuit or through resistance, only in that position where the induced currents give

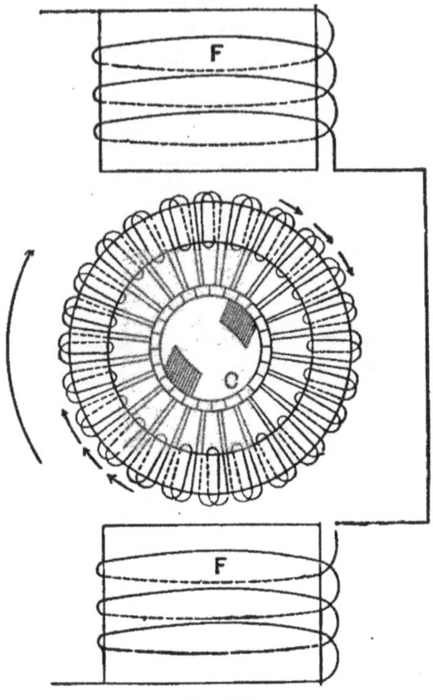

Fig. 158.

a rotary effort in the desired direction, while the armature coils are open-circuited in the position where the rotary effort of the induced currents would be in opposition to the desired rotation. This requires means to open or close the circuit of the armature coils and thereby introduces the commutator.

Thus the general construction of a repulsion motor is as shown in Figs. 158 and 159 diagrammatically as bipolar motor. The field is a single-phase alternating field *F*, the armature shown diagrammatically as ring wound *A* consists of a number of coils connected to a segmental commutator *C*, in general in the same way as in continuous-current machines. Brushes standing under an angle of about 45° with the direction of the magnetic field, short-circuit either a

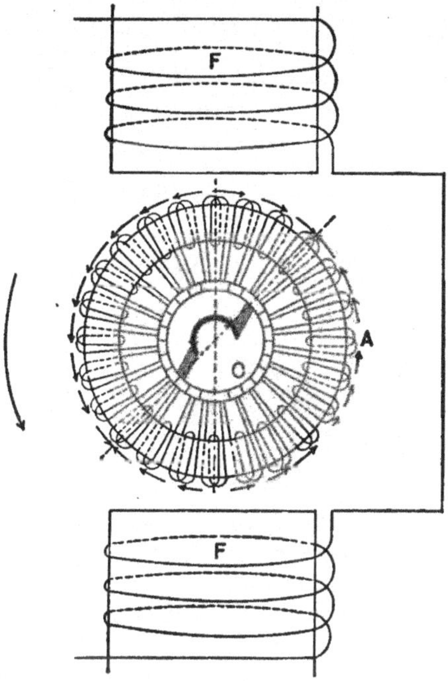

Fig. 159.

part of the armature coils as shown in Fig. 158, or the whole armature by a connection from brush to brush as shown in Fig. 159.

The former arrangement has the disadvantage of using a part of the armature coils only. The second arrangement has the disadvantage that, in the passage of the brush from segment to segment, individual armature coils are short-circuited, and thereby give a torque in opposite direction to the torque developed by the main induced current flowing through the whole armature from brush to brush.

216. Thus the repulsion motor consists of a primary electric circuit, a magnetic circuit interlinked therewith, and a secondary circuit closed upon itself and displaced in

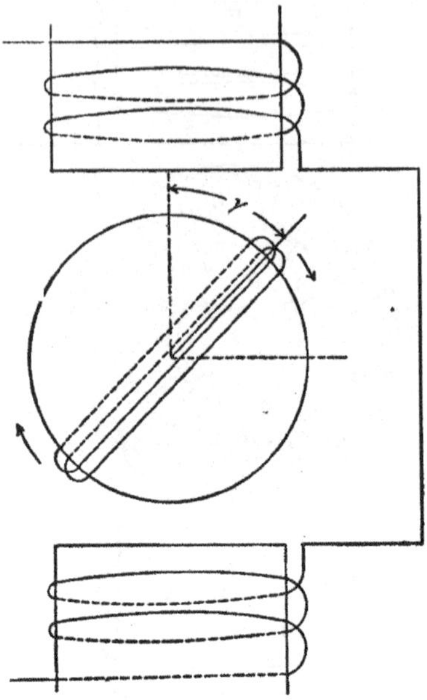

Fig. 160.

space by 45° — in a bipolar motor — from the direction of the magnetic flux, as shown diagrammatically in Fig. 160.

This secondary circuit, while set in motion, still remains in the same position of 45° displacement, with the magnetic flux, or rather, what is theoretically the same, when moving out of this position, is replaced by other secondary circuits entering this position of 45° displacement.

For simplicity, in the following all the secondary quantities, as E.M.F., current, resistance, reactance, etc., are assumed as reduced to the primary circuit by the ratio of turns, in the same way as done in the chapter on Induction Motors.

217. Let

Φ = maximum magnetic flux per field pole;
e = effective E.M.F. induced thereby in the field turns; thus,
$e = \sqrt{2}\,\pi N n \Phi\, 10^{-8}$;
where n = number of turns, N = frequency.

thus, $$\Phi = \frac{e\,10^8}{\sqrt{2}\,\pi n N}.$$

The instantaneous value of magnetism is

$$\phi = \Phi \sin\beta;$$

and the flux interlinked with the armature circuit

$$\phi_1 = \Phi \sin\beta \sin\lambda;$$

when λ is the angle between the plane of the armature coil and the direction of the magnetic flux. (Usually about 45°.)

The E.M.F. induced in the armature circuit, of n turns, (as reduced to primary circuit), is thus,

$$e_1 = -n\frac{d\phi_1}{dt}10^{-8}, \quad = -n\Phi\frac{d}{dt}\sin\beta\sin\lambda\, 10^{-8},$$

$$= -n\Phi\left\{\sin\lambda\cos\beta\frac{d\beta}{dt} + \sin\beta\cos\lambda\frac{d\lambda}{dt}\right\}10^{-8}.$$

If N = frequency in cycles per second, N_1 = frequency of rotation or speed in cycles per second, and $k = N_1 / N$ $= \dfrac{\text{speed}}{\text{frequency}}$ we have

$$\frac{d\beta}{dt} = 2\pi N; \qquad \frac{d\lambda}{dt} = 2\pi N_1 = 2\pi k N;$$

thus, $e_1 = -2\pi n N\Phi\,\{\sin\lambda\cos\beta + k\cos\lambda\sin\beta\}\,10^{-8}$,

or, since $$\Phi = \frac{e\,10^8}{\sqrt{2}\,\pi n N},$$

$$e_1 = e\sqrt{2}\,\{\sin\lambda\cos\beta + k\cos\lambda\sin\beta\}.$$

218. Introducing now complex quantities, and counting the time from the zero value of rising magnetism, the magnetism is represented by $j\Phi$,

the primary induced E.M.F., $\dot{E} = -e$,

the secondary induced E.M.F., $\dot{E}_1 = -e\,\{\sin\lambda + jk\cos\lambda\}$;

hence, if

$Z_1 = r_1 - jx_1 =$ secondary impedance reduced to primary circuit
$Z = r - jx =$ primary impedance,
$Y = g - jb =$ exciting admittance,

we have,

secondary current, $\dot{I}_1 = \frac{\dot{E}_1}{Z_1} = -e\,\frac{\sin\lambda + jk\cos\lambda}{r_1 - jx_1}$,

primary exciting current, $\dot{I}_0 = e\,Y = e\,(g + jb)$,

hence, total primary current,

$$\dot{I} = \dot{I}_0 - \dot{I}_1 = e\left\{g + jb + e\,\frac{\sin\lambda + jk\cos\lambda}{r_1 - jx_1}\right.$$

Primary impressed E.M.F., $\dot{E}_0 = -\dot{E} + \dot{I}Z$;

$$= e\left\{1 + (\sin\lambda + jk\cos\lambda)\frac{r - jx}{r_1 - jx_1} + (g + jb)(r - jx)\right\}$$

Neglecting in E_0 the last term, as of higher order,

$$\dot{E}_0 = e\left\{1 + \sin\lambda + jk\cos\lambda\,\frac{r - jx}{r_1 - jx_1}\right\};$$

or, eliminating imaginary quantities,

$$e_0 = \frac{e\sqrt{(r_1 + r\sin\lambda + kx\cos\lambda)^2 + (x_1 + x\sin\lambda - kr\cos\lambda)^2}}{\sqrt{(r_1^2 + x_1^2)}}$$

The power consumed by the component of primary counter E.M.F., whose flux is interlinked with the secondary $e\sin\lambda$, is,

$$P' = [e\sin\lambda\,\dot{I}]' = \frac{e^2\sin\lambda\,(r_1\sin\lambda - kx_1\cos\lambda)}{r_1^2 + x_1^2},$$

the power consumed by the secondary resistance is,

$$P_1 = I_1^2 r_1 = \frac{e^2 r_1\,(\sin^2\lambda + k^2\cos^2\lambda)}{r_1 + x_1^2},$$

hence the difference, or the mechanical power developed by the motor armature,

$$P = -(P' - P_1) = \frac{e^2 k \cos\lambda}{r_1^2 + x_1^2} (x_1 \sin\lambda + r_1 k \cos\lambda).$$

and substituting for e,

$$P = \frac{e_0^2 k \cos\lambda \, (x_1 \sin\lambda + r_1 k \cos\lambda)}{(r_1 + r \sin\lambda + kx \cos\lambda)^2 + (x_1 + x \sin\lambda - kr \cos\lambda)^2};$$

and the torque in synchronous watts,

$$T = \frac{P}{k} = \frac{e_0^2 \cos\lambda \, (x_1 \sin\lambda + r_1 k \cos\lambda)}{(r_1 + r \sin\lambda + kx \cos\lambda)^2 + (x_1 + x \sin\lambda - kr \cos\lambda)^2}$$

or $\quad T = \sqrt{2}\,\pi N 10^{-8} [j_1 \Phi \sin\lambda \, \dot{I}_1 \cos\lambda]^j = [e_1 \dot{I}_1 \cos\lambda]^j$

$$= \frac{e^2 \cos\lambda \, (x_1 \sin\lambda + r_1 k \cos\lambda)}{r_1^2 + x_1^2} \text{ etc.}$$

The stationary torque is, $k = 0$,

$$T_0 = \frac{e_0^2 x_1 \sin\lambda \cos\lambda}{(r_1 + r \sin\lambda)^2 + (x_1 + x \sin\lambda)^2},$$

and neglecting the primary impedance, $r = 0 = x$,

$$T_0 = \frac{e_0^2 x_1 \sin\lambda \cos\lambda}{r_1^2 + x_1^2} = \frac{e_0^2 x_1 \sin^2\lambda}{2(r_1^2 + x_1^2)},$$

which is a maximum at $\lambda = 45°$.

At speed k, neglecting $r = 0 = x$,

$$T = \frac{e_0^2 \cos\lambda \, (x_1 \sin\lambda + r_1 k \cos\lambda)}{r_1^2 + x_1^2},$$

which is a maximum for $\frac{dT}{d\lambda} = 0$, which gives,

$$\cot 2\lambda = \frac{r_1 k}{x_1}. \quad \text{For } k = 0,\ \lambda = 45°;\ \text{for } k = \infty,\ \lambda = 0.$$

that is, in the repulsion motor, with increasing speed, the angle of secondary closed circuit, λ, has to be reduced to get maximum torque.

219. At $\lambda = 45°$ we have,

$$T = \frac{e_0^2 (x_1 + r_1 k)}{(r_1\sqrt{2} + r + kx)^2 + (x_1\sqrt{2} + x - kr)^2},$$

and the power,

$$P = \frac{e_0^2 k (x_1 + r_1 k)}{(r_1\sqrt{2} + r + kx)^2 + (x_1\sqrt{2} + x - kr)^2},$$

this is a maximum, at constant $\lambda = 45°$, for $\frac{dP}{dk} = 0$, which gives, $k = \frac{x + x_1\sqrt{2}}{x}$.

At $\lambda = 0$ we have,

$$T = \frac{e_0^2 r_1 k}{(r_1 + kx)^2 + (x_1 - kr)^2},$$

that is, $T = 0$ at $k = 0$, or, the motor is not self-starting when $\lambda = 0$.

$$P = \frac{e_0 r_1 k^2}{(r_1 + kx)^2 + (x_1 - kr)^2},$$

which is a maximum at constant $\lambda = 0$ for, $\frac{dP}{dk} = 0$, which gives,

$$k = \frac{r_1^2 + x_1^2}{rx_1 - xr_1}.$$

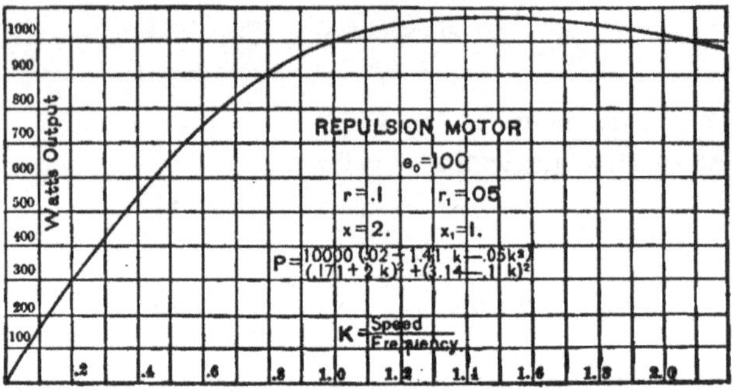

Fig. 161. Repulsion Motor.

As an instance is shown, in Fig. 161, the power output as ordinates, with the speed $k = N_1 / N$ as abscissæ, of a repulsion motor of the constants,

$$\lambda = 45° \qquad e_0 = 100.$$
$$r = .1 \qquad r_1 = .05$$
$$x = 2.0 \qquad x_1 = 1.0$$

giving the power,

$$P = \frac{10{,}000\,\{.02 + 1.41\,k - .05\,k^2\}}{(.171 + 2\,k)^2 + (3.14 - .1\,k)^2}.$$

SERIES MOTOR. SHUNT MOTOR.

220. If, in a continuous-current motor, series motor as well as shunt motor, the current is reversed, the direction of rotation remains the same, since field magnetism and armature current have reversed their sign, and their prod-

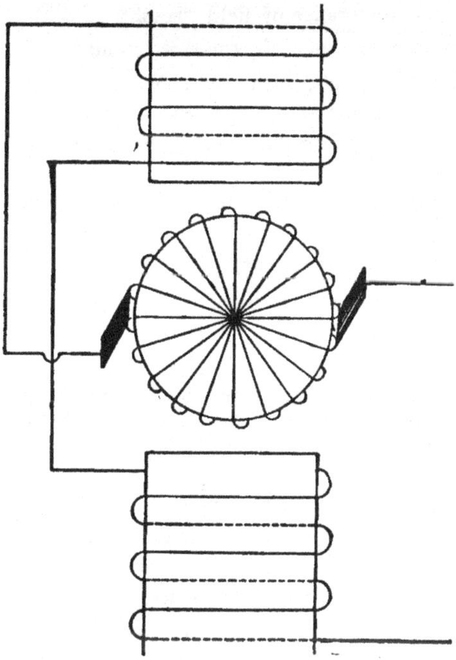

Fig. 162. Series Motor.

uct, the torque, thus maintained the same sign. Therefore such a motor, when supplied by an alternating current, will operate also, provided that the reversals in field and in armature take place simultaneously. In the series motor this is necessarily the case, the same current passing through field and through armature.

With an alternating current in the field, obviously the

magnetic circuit has to be laminated to exclude eddy currents.

Let, in a *series motor*, Fig. 146,

Φ = effective magnetism per pole,
n = number of field turns per pole in series,
n_1 = number of armature turns in series between brushes,
p = number of poles,
$\mathcal{R}$ = magnetic reluctance of field circuit,*
$\mathcal{R}_1$ = magnetic reluctance of armature circuit,†
Φ_1 = effective magnetic flux produced by armature current (cross magnetization) per pole,
r = resistance of field (effective resistance, including hysteresis),
r_1 = resistance of armature (effective resistance, including hysteresis),
N = frequency of alternations,
N_1 = speed in cycles per second.

It is then,

E.M.F. induced in armature conductors by their rotation through the magnetic field (counter E.M.F. of motor).

$$E = 4\, n_1\, N_1\, \Phi\, 10^{-8}$$

E.M.F. of self-induction of field,

$$E' = 2\, \pi\, p\, n\, N \Phi\, 10^{-8},$$

E.M.F. of self-induction of armature,

$$E_1' = 2\, \pi\, n_1\, N \Phi_1\, 10^{-8},$$

E.M.F. consumed by resistance,

$$E_r = (r + r_1)\, I,$$

where

I = current passing through motor, in amperes effective.

Further, it is:

Field magnetism: $\Phi = n\, I 10^8 / \mathcal{R}$

* That is, the main magnetic circuit of the motor.

† That is, the magnetic circuit of the cross magnetization, produced by the armature reaction.

Armature magnetism:

$$\Phi_1 = \frac{n_1 I 10^8}{\mathfrak{R}_1}.$$

Substituting these values,

$$E = \frac{4\, n n_1 N_1 I}{\mathfrak{R}};$$

$$E' = \frac{2\,\pi p n^2 N I}{\mathfrak{R}};$$

$$E_1' = \frac{2\,\pi n_1^2 N I}{\mathfrak{R}_1};$$

$$E_r = (r + r_1)\, I.$$

Thus the impressed E.M.F.,

$$E_0 = \sqrt{(E + E_r)^2 + (E' + E_1')^2}$$

$$= I\sqrt{\left(\frac{4\, n n_1 N_1}{\mathfrak{R}} + r + r_1\right)^2 + 4\,\pi^2 N^2\left(\frac{p n^2}{\mathfrak{R}} + \frac{n_1^2}{\mathfrak{R}_1}\right)^2};$$

or, since

$$x = 2\,\pi N \frac{p n^2}{\mathfrak{R}} = \text{reactance of field};$$

$$x_1 = 2\,\pi N \frac{n_1^2}{\mathfrak{R}_1} = \text{reactance of armature};$$

$$E_0 = I\sqrt{\left(\frac{4\, n n_1 N_1}{\mathfrak{R}} + r + r_1\right)^2 + (x + x_1)^2}$$

$$= I\sqrt{\left(\frac{2}{\pi}\frac{n_1}{p n}\frac{N_1}{N} x + r + r_1\right)^2 + (x + x_1)^2};$$

and

$$I = \frac{E_0}{\sqrt{\left(\frac{4\, n n_1 N_1}{\mathfrak{R}} + r + r_1\right)^2 + 4\,\pi^2 N^2\left(\frac{p n^2}{\mathfrak{R}} + \frac{n_1^2}{\mathfrak{R}_1}\right)^2}}$$

$$= \frac{E_0}{\sqrt{\left(\frac{4\, n n_1 N_1}{\mathfrak{R}} + r + r_1\right)^2 + (x + x_1)^2}}$$

$$= \frac{E_0}{\sqrt{\left(\frac{2}{\pi}\frac{n_1}{p n}\frac{N_1}{N} x + r + r_1\right)^2 + (x + x_1)^2}}.$$

221. The power output at armature shaft is,

$$P = E\,I$$

$$= \frac{\dfrac{4\,n n_1 N_1}{\mathcal{R}}\,E_0^2}{\left(\dfrac{4\,n n_1 N_1}{\mathcal{R}} + r + r_1\right)^2 + 4\,\pi^2 N^2\left(\dfrac{p n^2}{\mathcal{R}} + \dfrac{n_1^2}{\mathcal{R}_1}\right)^2}$$

$$= \frac{\dfrac{4\,n n_1 N_1}{\mathcal{R}}\,E_0^2}{\left(\dfrac{4\,n n_1 N_1}{\mathcal{R}} + r + r_1\right)^2 + (x + x_1)^2}$$

$$= \frac{\dfrac{2}{\pi}\,\dfrac{n_1}{p n}\,\dfrac{N_1}{N}\,x\,E_0^2}{\left(\dfrac{2}{\pi}\,\dfrac{n_1}{p n}\,\dfrac{N_1}{N}\,x + r + r_1\right)^2 + (x + x_1)^2.}$$

The displacement of phase between current and E.M.F is

$$\tan \hat{\omega} = \frac{E' + E_1'}{E + E_r}$$

$$= \frac{2\,\pi\,N\left(\dfrac{p n^2}{\mathcal{R}} + \dfrac{n_1^2}{\mathcal{R}_1}\right)}{\dfrac{4\,n n_1 N_1}{\mathcal{R}} + r + r_1}$$

$$= \frac{x + x_1}{\dfrac{2}{\pi}\,\dfrac{n_1}{p n}\,\dfrac{N_1}{N}\,x + r + r_1.}$$

Neglecting, as approximation, the resistances $r + r_1$, it is,

$$\tan \hat{\omega} = \frac{1 + \dfrac{x_1}{x}}{\dfrac{2}{\pi}\,\dfrac{n_1}{p n}\,\dfrac{N_1}{N}}$$

$$P = \frac{E_0^2}{\dfrac{2}{\pi}\,\dfrac{n_1}{p n}\,\dfrac{N_1}{N} + \dfrac{\left(1 + \dfrac{x_1}{x}\right)^2}{\dfrac{2}{\pi}\,\dfrac{n_1}{p n}\,\dfrac{N_1}{N}}}$$

hence a maximum for,

$$\frac{2}{\pi}\frac{n_1}{pn}\frac{N_1}{N} = \frac{\left(1+\frac{x_1}{x}\right)^2}{\frac{2}{\pi}\frac{n_1}{pn}\frac{N_1}{N}},$$

or,

$$\frac{N_1}{N} = \frac{1+\frac{x_1}{x}}{\frac{2}{\pi}\frac{n_1}{pn}},$$

substituting this in tan $\hat{\omega}$, it is:

$$\tan\hat{\omega} = 1, \qquad \text{or,} \qquad \hat{\omega} = 45°.$$

222. Instance of such an alternating-current motor,

$$E_0 = 100 \qquad N = 60 \qquad p = 2.$$

$$r = .03 \qquad r_1 = .12$$
$$x = .9 \qquad x_1 = .5$$
$$n = 10 \qquad n_1 = 48$$

Special provisions were made to keep the armature reactance a minimum, and overcome the distortion of the field by the armature M.M.F., by means of a coil closely surrounding the armature and excited by a current of equal phase but opposite direction with the armature current (Eickemeyer). Thereby it was possible to operate a two-circuit, 96-turn armature in a bipolar field of 20 turns, at a ratio of

$$\frac{\text{armature ampere-turns}}{\text{field ampere-turns}} = 2.4.$$

It is in this case,

$$I = \frac{100}{\sqrt{(.023\,N_1 + .15)^2 + 1.96}}$$

$$P = \frac{230\,N_1}{(.023\,N_1 + .15)^2 + 1.96}$$

$$\tan\hat{\omega} = \frac{1.4}{.023\,N_1 + .15}, \quad \text{or,} \quad \cos\hat{\omega} = \frac{.023\,N_1 + .15}{\sqrt{(.023\,N_1 + .15)^2 + 1.96}}.$$

In Fig. 163 are given, with the speed N_1 as abscissæ, the values of current I, power P, and power factor cos $\hat{\omega}$ of this motor.

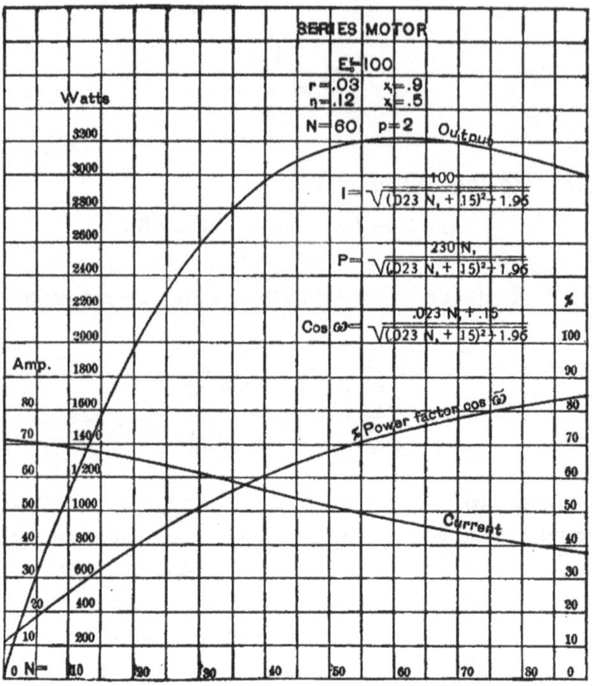

Fig. 163. Series Motor.

223. The *shunt motor* with laminated field will not operate satisfactorily in an alternating-current circuit. It will start with good torque, since in starting the current in armature, as well as in field, are greatly lagging, and thus approximately in phase with each other. With increasing speed, however, the armature current should come more into phase with the impressed E.M.F., to represent power. Since, however, the field current, and thus the field magnetism, lag nearly 90°, the induced E.M.F. of the armature rotation will lag nearly 90°, and thus not represent power.

Hence, to make a shunt motor work on alternating-current circuits, the magnetism of the field should be approximately in phase with the impressed E.M.F., that is, the field reactance negligible. Since the self-induction of the field is far in excess to its resistance, this requires the insertion of negative reactance, or capacity, in the field.

If the self-induction of the field circuit is balanced by capacity, the motor will operate, provided that the armature reactance is low, and that in starting sufficient resistance is inserted in the armature circuit to keep the armature current approximately in phase with the E.M.F. Under these conditions the equations of the motor will be similar to those of the series motor.

However, such motors have not been introduced, due to the difficulty of maintaining the balance between capacity and self-induction in the field circuit, which depends upon the square of the frequency, and thus is disturbed by the least change of frequency.

The main objection to both series and shunt motors is the destructive sparking at the commutator due to the induction of secondary currents in those armature coils which pass under the brushes. As seen in Fig. 162, with the normal position of brushes midway between the field poles, the armature coil which passes under the brush incloses the total magnetic flux. Thus, in this moment no E.M.F. is induced in the armature coil due to its rotation, but the E.M.F. induced by the alternation of the magnetic flux has a maximum at this moment, and the coil, when short-circuited by the brush, acts as a short-circuited secondary to the field coils as primary; that is, an excessive current flows through this armature coil, which either destroys it, or at least causes vicious sparking when interrupted by the motion of the armature.

To overcome this difficulty various arrangements have been proposed, but have not found an application.

224. Compared with the synchronous motor which has practically no lagging currents, and the induction motor which reaches very high power factors, the power factor of the series motor is low, as seen from Fig. 163, which represents about the best possible design of such motors.

In the alternating-series motor, as well as in the shunt motor, no position of an armature coil exists wherein the coil is dead; but in every position E.M.F. is induced in the armature coil: in the position parallel with the field flux an E.M.F. in phase with the current, in the position at right angles with the field flux an E.M.F. in quadrature with the current, intermediate E.M.Fs. in intermediate positions. At the speed $\pi N / 2$ the two induced E.M.Fs. in phase and in quadrature with the current are equal, and the armature coils are the seat of a complete system of symmetrical and balanced polyphase E.M.Fs. Thus, by means of stationary brushes, from such a commutator polyphase currents could be derived.

CHAPTER XXI.

REACTION MACHINES.

225. In the chapters on Alternating-Current Generators and on Induction Motors, the assumption has been made that the reactance x of the machine is a constant. While this is more or less approximately the case in many alternators, in others, especially in machines of large armature reaction, the reactance x is variable, and is different in the different positions of the armature coils in the magnetic circuit. This variation of the reactance causes phenomena which do not find their explanation by the theoretical calculations made under the assumption of constant reactance.

It is known that synchronous motors of large and variable reactance keep in synchronism, and are able to do a considerable amount of work, and even carry under circumstances full load, if the field-exciting circuit is broken, and thereby the counter E.M.F. E_1 reduced to zero, and sometimes even if the field circuit is reversed and the counter E.M.F. E_1 made negative.

Inversely, under certain conditions of load, the current and the E.M.F. of a generator do not disappear if the generator field is broken, or even reversed to a small negative value, in which latter case the current flows against the E.M.F. E_0 of the generator.

Furthermore, a shuttle armature without any winding will in an alternating magnetic field revolve when once brought up to synchronism, and do considerable work as a motor.

These phenomena are not due to remanent magnetism nor to the magnetizing effect of Foucault currents, because they exist also in machines with laminated fields, and exist if the alternator is brought up to synchronism by external means and the remanent magnetism of the field poles destroyed beforehand by application of an alternating current.

226. These phenomena cannot be explained under the assumption of a constant synchronous reactance; because in this case, at no-field excitation, the E.M.F. or counter E.M.F. of the machine is zero, and the only E.M.F. existing in the alternator is the E.M.F. of self-induction; that is, the E.M.F. induced by the alternating current upon itself. If, however, the synchronous reactance is constant, the counter E.M.F. of self-induction is in quadrature with the current and wattless; that is, can neither produce nor consume energy.

In the synchronous motor running without field excitation, always a large lag of the current behind the impressed E.M.F. exists; and an alternating generator will yield an E.M.F. without field excitation, only when closed by an external circuit of large negative reactance; that is, a circuit in which the current leads the E.M.F., as a condenser, or an over-excited synchronous motor, etc.

Self-excitation of the alternator by armature reaction can be explained by the fact that the counter E.M.F. of self-induction is not wattless or in quadrature with the current, but contains an energy component; that is, that the reactance is of the form $X = h - jx$, where x is the wattless component of reactance and h the energy component of reactance, and h is positive if the reactance consumes power, — in which case the counter E.M.F. of self-induction lags more than 90° behind the current, — while h is negative if the reactance produces power, — in which case the counter E.M.F. of self-induction lags less than 90° behind the current.

227. A case of this nature has been discussed already in the chapter on Hysteresis, from a different point of view.

There the effect of magnetic hysteresis was found to distort the current wave in such a way that the equivalent sine wave, that is, the sine wave of equal effective strength and equal power with the distorted wave, is in advance of the wave of magnetism by what is called the angle of hysteretic advance of phase α. Since the E.M.F. induced by the magnetism, or counter E.M.F. of self-induction, lags 90° behind the magnetism, it lags $90 + \alpha$ behind the current; that is, the self-induction in a circuit containing iron is not in quadrature with the current and thereby wattless, but lags more than 90° and thereby consumes power, so that the reactance has to be represented by $X = h - jx$, where h is what has been called the "effective hysteretic resistance."

A similar phenomenon takes place in alternators of variable reactance, or what is the same, variable magnetic reluctance.

228. Obviously, if the reactance or reluctance is variable, it will perform a complete cycle during the time the armature coil moves from one field pole to the next field pole, that is, during one-half wave of the main current. That is, in other words, the reluctance and reactance vary with twice the frequency of the alternating main current. Such a case is shown in Figs. 164 and 165. The impressed E.M.F., and thus at negligible resistance, the counter E.M.F., is represented by the sine wave E, thus the magnetism produced thereby is a sine wave Φ, 90° ahead of E. The reactance is represented by the sine wave x, varying with the double frequency of E, and shown in Fig. 164 to reach the maximum value during the rise of magnetism, in Fig. 165 during the decrease of magnetism. The current I required to produce the magnetism Φ is found from Φ and x in combination with the cycle of molecular magnetic friction of the material, and the power P is the product IE As seen in Fig. 164, the positive part of P is larger than the

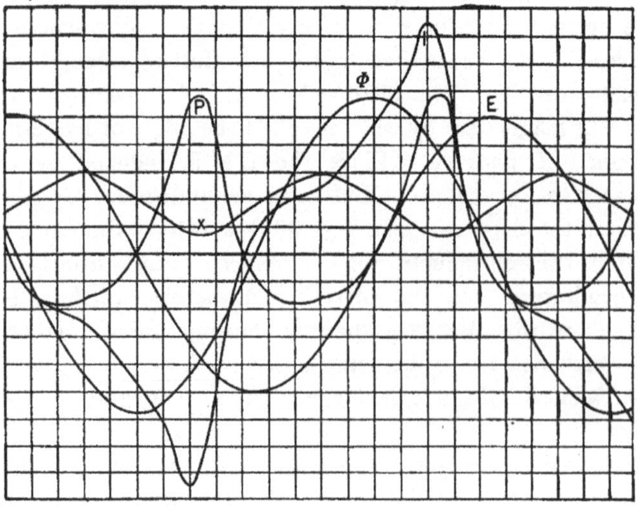

Fig. 164. Variable Reactance, Reaction Machine.

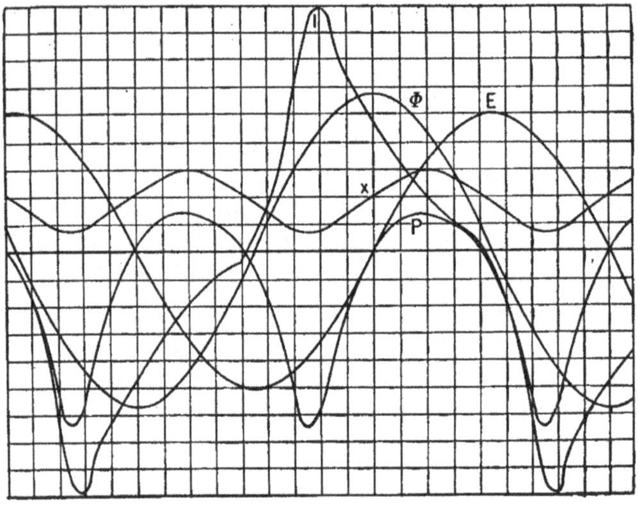

Fig. 165. Variable Reactance, Reaction Machine.

negative part; that is, the machine produces electrical energy as generator. In Fig. 165 the negative part of P is larger than the positive; that is, the machine consumes electrical energy and produces mechanical energy as synchronous motor. In Figs. 166 and 167 are given the two hysteretic cycles or looped curves Φ, I under the two conditions. They show that, due to the variation of reactance x, in the first case the hysteretic cycle has been overturned so as to represent not consumption, but production of electrical

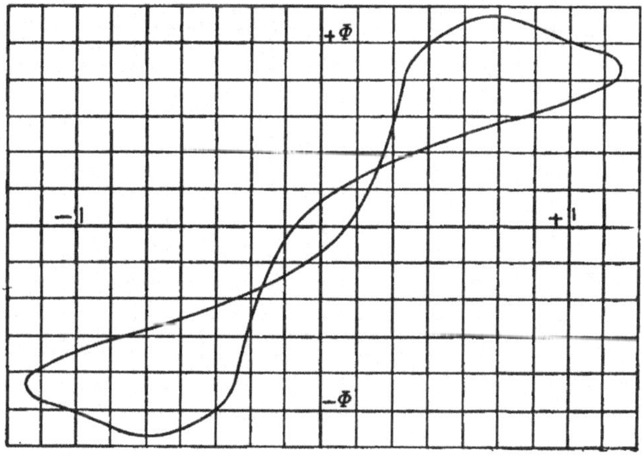

Fig. 166. Hysteretic Loop of Reaction Machine.

energy, while in the second case the hysteretic cycle has been widened, representing not only the electrical energy consumed by molecular magnetic friction, but also the mechanical output.

229. It is evident that the variation of reluctance must be symmetrical with regard to the field poles; that is, that the two extreme values of reluctance, maximum and minimum, will take place at the moment where the armature coil stands in front of the field pole, and at the moment where it stands midway between the field poles.

The effect of this periodic variation of reluctance is a distortion of the wave of E.M.F., or of the wave of current, or of both. Here again, as before, the distorted wave can be replaced by the equivalent sine wave, or sine wave of equal effective intensity and equal power.

The instantaneous value of magnetism produced by the

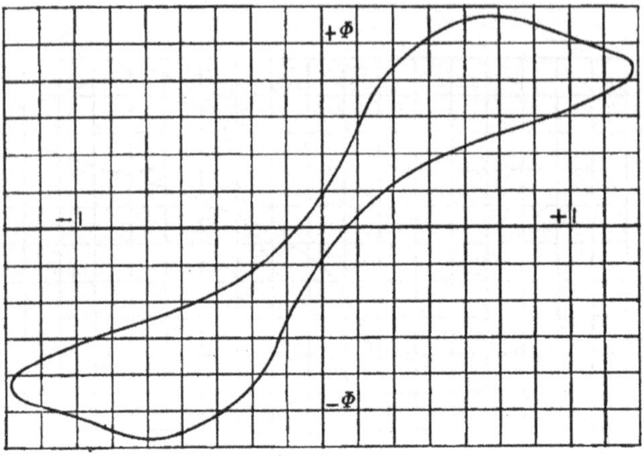

Fig. 167. Hysteretic Loop of Reaction Machine.

armature current — which magnetism induces in the armature conductor the E.M.F. of self-induction — is proportional to the instantaneous value of the current, divided by the instantaneous value of the reluctance. Since the extreme values of the reluctance coincide with the symmetrical positions of the armature with regard to the field poles, — that is, with zero and maximum value of the induced E.M.F., E_0, of the machine, — it follows that, if the current is in phase or in quadrature with the E.M.F. E_0, the reluctance wave is symmetrical to the current wave, and the wave of magnetism therefore symmetrical to the current wave also. Hence the equivalent sine wave of magnetism is of equal phase with the current wave; that is, the E.M.F. of self-induction lags 90° behind the current, or is wattless.

Thus at no-phase displacement, and at 90° phase displacement, a reaction machine can neither produce electrical power nor mechanical power.

230. If, however, the current wave differs in phase from the wave of E.M.F. by less than 90°, but more than zero degrees, it is unsymmetrical with regard to the reluctance wave, and the reluctance will be higher for rising current than for decreasing current, or it will be higher for decreasing than for rising current, according to the phase relation of current with regard to induced E.M.F., E_0.

In the first case, if the reluctance is higher for rising, lower for decreasing, current, the magnetism, which is proportional to current divided by reluctance, is higher for decreasing than for rising current; that is, its equivalent sine wave lags behind the sine wave of current, and the E.M.F. or self-induction will lag more than 90° behind the current; that is, it will consume electrical power, and thereby deliver mechanical power, and do work as synchronous motor.

In the second case, if the reluctance is lower for rising, and higher for decreasing, current, the magnetism is higher for rising than for decreasing current, or the equivalent sine wave of magnetism leads the sine wave of the current, and the counter E.M.F. at self-induction lags less than 90° behind the current; that is, yields electric power as generator, and thereby consumes mechanical power.

In the first case the reactance will be represented by $X = h - jx$, similar as in the case of hysteresis; while in the second case the reactance will be represented by $X = -h - jx$.

231. The influence of the periodical variation of reactance will obviously depend upon the nature of the variation, that is, upon the shape of the reactance curve. Since, however, no matter what shape the wave has, it can always be dissolved in a series of sine waves of double frequency, and its higher harmonics, in first approximation the assumption can be made that the reactance or the reluctance vary with double frequency of the main current; that is, are represented in the form,

$$x = a + b \cos 2\beta.$$

Let the inductance, or the coefficient of self-induction, be represented by —

$$L = l + \phi \cos 2\beta$$
$$= l(1 + \gamma \cos 2\beta)$$

where γ = amplitude of variation of inductance.

Let

$\hat{\omega}$ = angle of lag of zero value of current behind maximum value of inductance L.

It is then, assuming the current as sine wave, or replacing it by the equivalent sine wave of effective intensity I,
Current,

$$i = I\sqrt{2} \sin(\beta - \hat{\omega}).$$

The magnetism produced by this current is,

$$\Phi = \frac{L i}{n},$$

where n = number of turns.
Hence, substituted,

$$\Phi = \frac{l I \sqrt{2}}{n} \sin(\beta - \hat{\omega})(1 + \gamma \cos 2\beta),$$

or, expanded,

$$\Phi = \frac{l I \sqrt{2}}{n} \left\{ \left(1 - \frac{\gamma}{2}\right) \cos\hat{\omega} \sin\beta - \left(1 + \frac{\gamma}{2}\right) \sin\hat{\omega} \cos\beta \right\}$$

when neglecting the term of triple frequency, as wattless.

Thus the E.M.F. induced by this magnetism is,

$$e = -n\frac{d\Phi}{dt}$$

$$= -2\pi N n\frac{d\Phi}{d\beta}$$

hence, expanded —

$$e = -2\pi N l I\sqrt{2}\left\{\left(1-\frac{\gamma}{2}\right)\cos\hat{\omega}\cos\beta + \left(1+\frac{\gamma}{2}\right)\sin\hat{\omega}\sin\beta\right\}$$

and the effective value of E.M.F.,

$$E = 2\pi N l I\sqrt{\left(1-\frac{\gamma}{2}\right)^2\cos^2\hat{\omega} + \left(1+\frac{\gamma}{2}\right)^2\sin^2\hat{\omega}}$$

$$= 2\pi N l I\sqrt{1+\frac{\gamma^2}{4} - \gamma\cos 2\hat{\omega}}.$$

Hence, the apparent power, or the voltamperes -

$$P_0 = IE = 2\pi N l I^2\sqrt{1+\frac{\gamma^2}{4} - \gamma\cos 2\hat{\omega}}$$

$$= \frac{E^2}{2\pi N l\sqrt{1+\frac{\gamma^2}{4} - \gamma\cos 2\hat{\omega}}}.$$

The instantaneous value of power is

$$p = ei$$

$$= -4\pi N l I^2\sin(\beta-\omega)\left\{\left(1-\frac{\gamma}{2}\right)\cos\omega\cos\beta + \left(1+\frac{\gamma}{2}\right)\sin\omega\sin\beta\right\}.$$

and, expanded —

$$p = -2\pi N l I^2\left\{\left(1+\frac{\gamma}{2}\right)\sin 2\omega\sin^2\beta - \left(1-\frac{\gamma}{2}\right)\sin 2\omega\cos^2\beta + \sin 2\beta\left(\cos 2\omega - \frac{\gamma}{2}\right)\right\}$$

Integrated, the effective value of power is

$$P = -\pi N l I^2\gamma\sin 2\hat{\omega};$$

hence, negative, that is, the machine consumes electrical, and produces mechanical, power, as synchronous motor, if $\hat{\omega} > 0$; that is, with lagging current; positive, that is, the machine produces electrical, and consumes mechanical, power, as generator, if $\hat{\omega} > 0$; that is, with leading current.

The power factor is

$$f = \frac{P}{P_0} = \frac{\gamma \sin 2\hat{\omega}}{2\sqrt{1 + \frac{\gamma^2}{4} - \gamma \cos 2\hat{\omega}}};$$

hence, a maximum, if,

$$\frac{df}{d\hat{\omega}} = 0;$$

or, expanded,

$$\cos 2\hat{\omega} = \frac{1}{\gamma} + \frac{\gamma}{4} \pm \frac{1}{4}\sqrt{8 + \gamma^2}.$$

The power, P, is a maximum at given current, I, if

$$\sin 2\hat{\omega} = 1;$$

that is,

$$\hat{\omega} = 45^\circ$$

at given E.M.F., E, the power is

$$P = -\frac{E^2 \gamma \sin 2\hat{\omega}}{4\pi N l\left(1 + \frac{\gamma^2}{4} - \gamma \cos 2\hat{\omega}\right)};$$

hence, a maximum at

$$\frac{\delta P}{\delta \omega} = 0;$$

or, expanded,

$$\cos 2\hat{\omega} = \frac{\pm \gamma}{1 + \frac{\gamma^2}{4}}.$$

232. We have thus, at impressed E.M.F., E, and negligible resistance, if we denote the mean value of reactance,

$$x = 2\pi N l.$$

Current

$$I = \frac{E}{x\sqrt{1 + \frac{\gamma^2}{4} - \gamma \cos 2\hat{\omega}}}.$$

Voltamperes,

$$P_0 = \frac{E^2}{x\sqrt{1 + \frac{\gamma^2}{4} - \gamma \cos 2\hat{\omega}}}.$$

Power,

$$P = -\frac{E^2 \gamma \sin 2\hat{\omega}}{2x\left(1 + \frac{\gamma^2}{4} - \gamma \cos 2\hat{\omega}\right)}.$$

Power factor,

$$f = \cos(E, I) = \frac{\gamma \sin 2\hat{\omega}}{2\sqrt{1 + \frac{\gamma^2}{4} - \gamma \cos 2\hat{\omega}}}$$

Maximum power at

$$\cos 2\hat{\omega} = \frac{\gamma}{1 + \frac{\gamma^2}{4}}.$$

Maximum power factor at

$$\cos 2\hat{\omega} = \frac{1}{\gamma} + \frac{\gamma}{4} \pm \frac{1}{4}\sqrt{8 + \gamma^2}$$

$\hat{\omega} > 0$: synchronous motor, with lagging current,
$\hat{\omega} < 0$: generator, with leading current.

As an instance is shown in Fig. 168, with angle $\hat{\omega}$ as abscissæ, the values of current, power, and power factor, for the constants, —

$$E = 110$$
$$x = 3$$
$$\gamma = .8$$

hence

$$I = \frac{41}{\sqrt{1.45 - \cos 2\hat{\omega}}}$$

$$P = \frac{-2017 \sin 2\hat{\omega}}{1.45 - \cos 2\hat{\omega}}$$

$$f = \cos(E, I) = \frac{.447 \sin 2\hat{\omega}}{\sqrt{1.45 - \cos 2\hat{\omega}}}$$

As seen from Fig. 152, the power factor f of such a machine is very low — does not exceed 40 per cent in this instance.

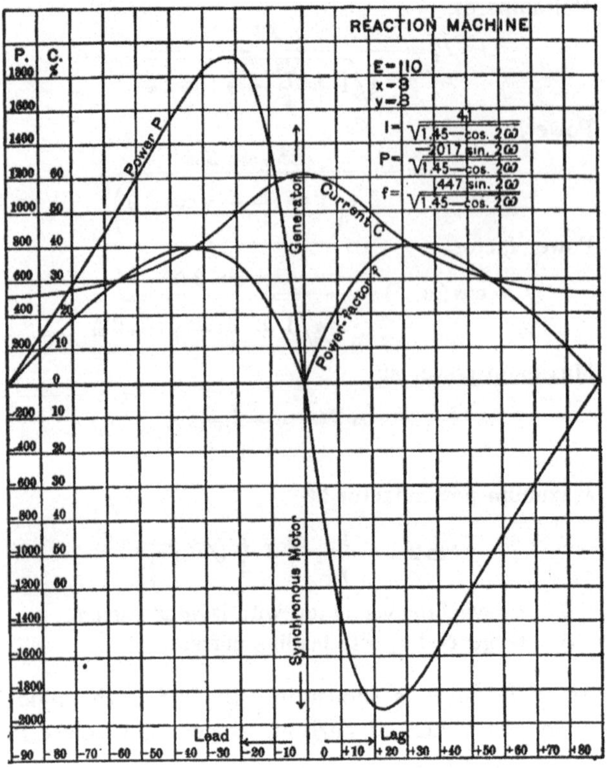

Fig. 168. Reaction Machine.

CHAPTER XXII.

DISTORTION OF WAVE-SHAPE AND ITS CAUSES.

233. In the preceding chapters we have considered the alternating currents and alternating E.M.Fs. as sine waves or as replaced by their equivalent sine waves.

While this is sufficiently exact in most cases, under certain circumstances the deviation of the wave from sine shape becomes of importance, and with certain distortions it may not be possible to replace the distorted wave by an equivalent sine wave, since the angle of phase displacement of the equivalent sine wave becomes indefinite. Thus it becomes desirable to investigate the distortion of the wave, its causes and its effects.

Since, as stated before, any alternating wave can be represented by a series of sine functions of odd orders, the investigation of distortion of wave-shape resolves itself in the investigation of the higher harmonics of the alternating wave.

In general we have to distinguish between higher harmonics of E.M.F. and higher harmonics of current. Both depend upon each other in so far as with a sine wave of impressed E.M.F. a distorting effect will cause distortion of the current wave, while with a sine wave of current passing through the circuit, a distorting effect will cause higher harmonics of E.M.F.

234. In a conductor revolving with uniform velocity through a uniform and constant magnetic field, a sine wave of E.M.F. is induced. In a circuit with constant resistance and constant reactance, this sine wave of E.M.F. produces a sine wave of current. Thus distortion of the wave-shape or higher harmonics may be due to: lack of uniformity of the velocity of the revolving conductor; lack of uniformity or pulsation of the magnetic field; pulsation of the resistance; or pulsation of the reactance.

The first two cases, lack of uniformity of the rotation or of the magnetic field, cause higher harmonics of E.M.F. at open circuit. The last, pulsation of resistance and reactance, causes higher harmonics only with a current flowing in the circuit, that is, under load.

Lack of uniformity of the rotation is of no practical interest as cause of distortion, since in alternators, due to mechanical momentum, the speed is always very nearly uniform during the period.

Thus as causes of higher harmonics remain:

1st. Lack of uniformity and pulsation of the magnetic field, causing a distortion of the induced E.M.F. at open circuit as well as under load.

2d. Pulsation of the reactance, causing higher harmonics under load.

3d. Pulsation of the resistance, causing higher harmonics under load also.

Taking up the different causes of higher harmonics we have: —

Lack of Uniformity and Pulsation of the Magnetic Field.

235. Since most of the alternating-current generators contain definite and sharply defined field poles covering in different types different proportions of the pitch, in general the magnetic flux interlinked with the armature coil will not vary as simply sine wave, of the form:

$$\Phi \cos \beta,$$

but as a complex harmonic function, depending on the shape and the pitch of the field poles, and the arrangement of the armature conductors. In this case, the magnetic flux issuing from the field pole of the alternator can be represented by the general equation,

$$\Phi = A_0 + A_1 \cos \beta + A_2 \cos 2\beta + A_3 \cos 3\beta + \ldots$$
$$+ B_1 \sin \beta + B_2 \sin 2\beta + B_3 \sin 3\beta + \ldots$$

If the reluctance of the armature is uniform in all directions, so that the distribution of the magnetic flux at the field-pole face does not change by the rotation of the armature, the rate of cutting magnetic flux by an armature conductor is Φ, and the E.M.F. induced in the conductor thus equal thereto in wave shape. As a rule $A_0, A_2, A_4 \ldots B_2, B_4$ equal zero; that is, successive field poles are equal in strength and distribution of magnetism, but of opposite polarity. In some types of machines, however, especially induction alternators, this is not the case.

The E.M.F. induced in a full-pitch armature turn — that is, armature conductor and return conductor distant from former by the pitch of the armature pole (corresponding to the distance from field pole center to pole center) is,

$$\delta = \Phi_0 - \Phi_{180}$$
$$= 2 \{A_1 \cos \beta + A_3 \cos 3\beta + A_5 \cos 5\beta + \ldots$$
$$+ B_1 \sin \beta + B_3 \sin 3\beta + B_5 \sin 5\beta + \ldots \}$$

Even with an unsymmetrical distribution of the magnetic flux in the air-gap, the E.M.F. wave induced in a full-pitch armature coil is symmetrical; the positive and negative half waves equal, and correspond to the mean flux distribution of adjacent poles. With fractional pitch windings — that is, windings whose turns cover less than the armature pole pitch — the induced E.M.F. can be unsymmetrical with unsymmetrical magnetic field, but as a rule is symmetrical also. In unitooth alternators the total induced E.M.F. has the same shape as that induced in a single turn.

With the conductors more or less distributed over the surface of the armature, the total induced E.M.F. is the resultant of several E.M.Fs. of different phases, and is thus more uniformly varying; that is, more sinusoidal, approaching sine shape, to within 3% or less, as for instance the curves Fig. 169 and Fig. 170 show, which represent the no-load and full-load wave of E.M.F. of a three-phase multitooth alternator. The principal term of these harmonics is the third harmonic, which consequently appears more or less in all alternator waves. As a rule these harmonics can be considered together with the harmonics due to the varying reluctance of the magnetic circuit. In ironclad alternators with few slots and teeth per pole, the passage of slots across the field poles causes a pulsation of the magnetic reluctance, or its reciprocal, the magnetic inductance of the circuit. In consequence thereof the magnetism per field pole, or at least that part of the magnetism passing through the armature, will pulsate with a frequency 2γ if γ = number of slots per pole.

Thus, in a machine with one slot per pole, the instantaneous magnetic flux interlinked with the armature conductors can be expressed by the equation:

$$\phi = \Phi \cos \beta \{1 + \epsilon \cos [2\beta - \omega]\}$$

where, Φ = average magnetic flux,

ϵ = amplitude of pulsation,

and ω = phase of pulsation.

In a machine with γ slots per pole, the instantaneous flux interlinked with the armature conductors will be:

$$\phi = \Phi \cos \beta \{1 + \epsilon \cos [2\gamma\beta - \omega]\},$$

if the assumption is made that the pulsation of the magnetic flux follows a simple sine law, as first approximation.

In general the instantaneous magnetic flux interlinked with the armature conductors will be:

$$\phi = \Phi \cos \beta \{1 + \epsilon_1 \cos (2\beta - \hat{\omega}_1) + \epsilon_2 \cos (4\beta - \hat{\omega}_2) + \dots\},$$

where the term ϵ_γ is predominating if γ = number of armature slots per pole. This general equation includes also the effect of lack of uniformity of the magnetic flux.

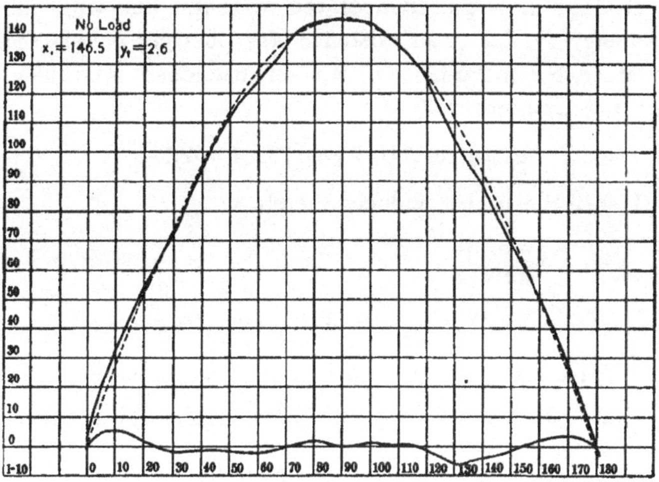

Fig. 169. No-load Wave of E.M.F. of Multitooth Three-phaser.

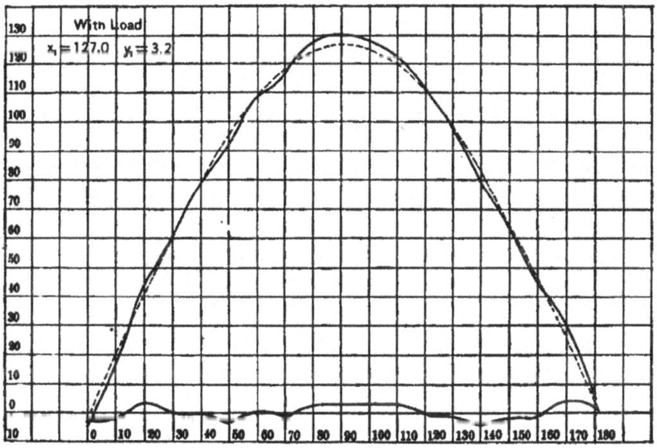

Fig. 170. Full-Load Wave of E.M.F. of Multitooth Three-phaser.

In case of a pulsation of the magnetic flux with the frequency 2γ, due to an existence of γ slots per pole in the armature, the instantaneous value of magnetism interlinked with the armature coil is:

$$\phi = \Phi \cos \beta \{1 + \epsilon \cos [2 \gamma \beta - \hat{\omega}]\}.$$

Hence the E.M.F. induced thereby:

$$e = -n \frac{d\phi}{dt}$$

$$= -\sqrt{2} \pi N n \Phi \frac{d}{d\beta} \{\cos \beta (1 + \epsilon \cos [2 \gamma \beta - \hat{\omega}])\}$$

And, expanded:

$$e = \sqrt{2} \pi N n \Phi \{\sin \beta + \epsilon \frac{2\gamma - 1}{2} \sin [(2\gamma - 1) \beta - \hat{\omega}] + \epsilon \frac{2\gamma + 1}{2} \sin [(2\gamma + 1) \beta - \hat{\omega}]\}.$$

Hence, the pulsation of the magnetic flux with the frequency 2γ, as due to the existence of γ slots per pole, introduces two harmonics, of the orders $(2\gamma - 1)$ and $(2\gamma + 1)$.

236. If $\gamma = 1$ it is:

$$e = \sqrt{2} \pi N n \Phi \{\sin \beta + \frac{\epsilon}{2} \sin (\beta - \hat{\omega}) + \frac{3\epsilon}{2} \sin (3\beta - \hat{\omega})\};$$

that is: In a unitooth single-phaser a pronounced triple harmonic may be expected, but no pronounced higher harmonics.

Fig. 171 shows the wave of E.M.F. of the main coil of a monocyclic alternator at no load, represented by:

$$e = E\{\sin \beta - .242 \sin (3\beta - 6.3) - .046 \sin (5\beta - 2.6) + .068 \sin (7\beta - 3.3) - .027 \sin (9\beta - 10.0) - .018 \sin (11\beta - 6.6) + .029 \sin (13\beta - 8.2)\};$$

hence giving a pronounced triple harmonic only, as expected.

If $\gamma = 2$, it is:

$$e = \sqrt{2} \pi N n \Phi \left\{ \sin \beta + \frac{3\epsilon}{2} \sin (3\beta - \hat{\omega}) + \frac{5\epsilon}{2} \sin (5\beta - \hat{\omega}) \right\}$$

the no-load wave of a unitooth quarter-phase machine, having pronounced triple and quintuple harmonics.

If $\gamma = 3$, it is:

$$e = \sqrt{2}\,\pi N n \Phi \left\{ \sin\beta + \frac{5\epsilon}{2}\sin(5\beta - \hat{\omega}) + \frac{7\epsilon}{2}\sin(7\beta - \hat{\omega}) \right\}.$$

That is: In a unitooth three-phaser, a pronounced quintuple and septuple harmonic may be expected, but no pronounced triple harmonic.

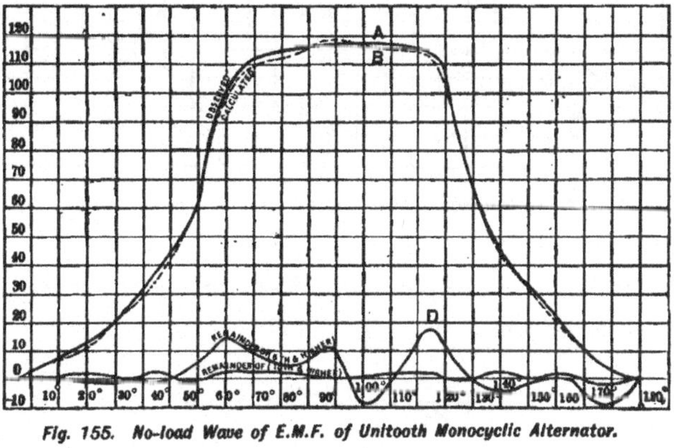

Fig. 155. No-load Wave of E.M.F. of Unitooth Monocyclic Alternator.

Fig. 156 shows the wave of E.M.F. of a unitooth three-phaser at no load, represented by:

$$e = E\,\{\sin\beta - .12\sin(3\beta - 2.3) - .23\sin(5\beta - 1.5) + .134\sin(7\beta - 6.2) - .002\sin(9\beta + 27.7) - .046\sin(11\beta - 5.5) + .031\sin(13\beta - 61.5)\}.$$

Thus giving a pronounced quintuple and septuple and a lesser triple harmonic, probably due to the deviation of the field from uniformity, as explained above, and deviation of the pulsation of reluctance from sine shape. In some especially favorable cases, harmonics as high as the 23d and 25th have been observed, caused by pulsation of the reluctance.

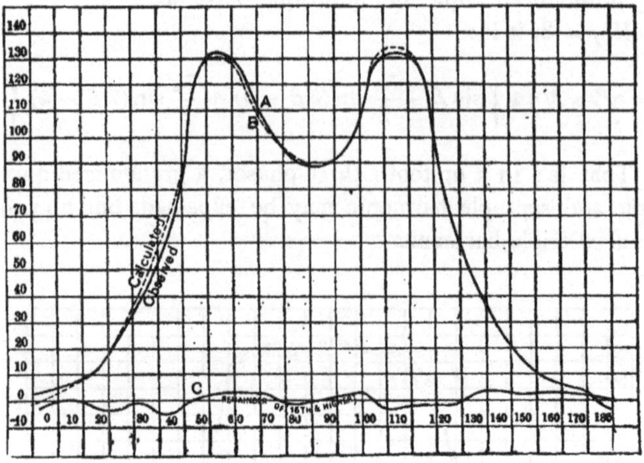

Fig. 172. No-load Wave of E.M.F. of Unitooth Three-phase Alternator.

In general, if the pulsation of the magnetic inductance is denoted by the general expression:

$$1 + \sum_1^\infty {}_\gamma\, \epsilon_\gamma \cos(2\gamma\beta - \hat{\omega}_\gamma),$$

the instantaneous magnetic flux is:

$$\phi = \Phi \cos\beta \left\{ 1 + \sum_1^\infty {}_\gamma\, \epsilon_\gamma \cos(2\gamma\beta - \hat{\omega}_\gamma) \right\}.$$

$$= \Phi \left\{ \cos\beta + \frac{\epsilon_1}{2}\cos(\beta - \hat{\omega}_1) + \sum_1^\infty {}_\gamma \left[\frac{\epsilon_\gamma}{2}\cos((2\gamma + 1)\beta - \hat{\omega}_\gamma) + \frac{\epsilon_{\gamma+1}}{2}\cos((2\gamma+1)\beta - \hat{\omega}_{\gamma+1}) \right] \right\}$$

hence, the E.M.F.

$$e = \sqrt{2}\,\pi N n \Phi \left\{ \sin\beta + \frac{\epsilon_1}{2}\sin(\beta - \hat{\omega}_1) + \sum_1^\infty {}_\gamma \frac{2\gamma+1}{2} \left[\epsilon_\gamma \sin((2\gamma+1)\beta - \hat{\omega}_\gamma) + \epsilon_{\gamma+1}\sin((2\gamma+1)\beta - \hat{\omega}_{\gamma+1}) \right] \right\}$$

Pulsation of Reactance.

237. The main causes of a pulsation of reactance are: magnetic saturation and hysteresis, and synchronous motion. Since in an ironclad magnetic circuit the magnetism is not proportional to the M.M.F., the wave of magnetism and thus the wave of E.M.F. will differ from the wave of current. As far as this distortion is due to the variation of permeability, the distortion is symmetrical and the wave of induced E.M.F. represents no power. The distortion caused by hysteresis, or the lag of the magnetism behind the M.M.F., causes an unsymmetrical distortion of the wave which makes the wave of induced E.M.F. differ by more than 90° from the current wave and thereby represents power, — the power consumed by hysteresis.

In practice both effects are always superimposed; that is, in a ferric inductance, a distortion of wave-shape takes place due to the lack of proportionality between magnetism and M.M.F. as expressed by the variation in the hysteretic cycle.

This pulsation of reactance gives rise to a distortion consisting mainly of a triple harmonic. Such current waves distorted by hysteresis, with a sine wave of impressed E.M.F., are shown in Figs. 66 to 69, Chapter X., on Hysteresis. Inversely, if the current is a sine wave, the magnetism and the E.M.F. will differ from sine shape.

For further discussion of this distortion of wave-shape by hysteresis, Chapter X. may be consulted.

238. Distortion of wave-shape takes place also by the pulsation of reactance due to synchronous rotation, as discussed in chapter on Reaction Machines.

In Figs. 148 and 149, at a sine wave of impressed E.M.F., the distorted current waves have been constructed.

Inversely, if a sine wave of current,

$$i = I \cos \beta,$$

passes through a circuit of synchronously varying reactance; as for instance, the armature of a unitooth alternator or synchronous motor — or, more general, an alternator whose armature reluctance is different in different positions with regard to the field poles — and the reactance is expressed by

$$X = x\{1 + \epsilon \cos(2\beta - \hat{\omega})\};$$

or, more general,

$$X = x\left\{1 + \sum_{1}^{\infty}{}_{\gamma}\, \epsilon_{\gamma} \cos(2\gamma\beta - \hat{\omega}_{\gamma})\right\};$$

the wave of magnetism is

$$\phi = \frac{X}{2\pi Nn}\cos\beta = \frac{x}{2\pi Nn}\left\{\cos\beta + \sum_{1}^{\infty}{}_{\gamma}\, \epsilon_{\gamma}\cos\beta\cos(2\gamma\beta - \hat{\omega}_{\gamma})\right\}$$

$$= \frac{x}{2\pi Nn}\left\{\cos\beta + \frac{\epsilon_1}{2}\cos(\beta - \hat{\omega}_1) + \sum_{1}^{\infty}{}_{\gamma}\left[\frac{\epsilon_{\gamma}}{2}\cos((2\gamma + 1)\right.\right.$$

$$\left.\left.\beta - \hat{\omega}_{\gamma}) + \frac{\epsilon_{\gamma} + 1}{2}\cos((2\gamma + 1)\beta - \hat{\omega}_{\gamma} + 1)\right]\right\};$$

hence the wave of induced E.M.F.

$$e = -n\frac{d\phi}{dt} = -2\pi Nn\frac{d\phi}{d\beta}$$

$$= x\left\{\sin\beta + \frac{\epsilon_1}{2}\sin(\beta - \hat{\omega}_1) + \sum_{1}^{\infty}{}_{\gamma}\frac{2\gamma + 1}{2}\left[\epsilon_{\gamma}\sin((2\gamma + 1)\right.\right.$$

$$\left.\left.\beta - \hat{\omega}_{\gamma}) + \epsilon_{\gamma} + 1\sin((2\gamma + 1)\beta - \hat{\omega}_{\gamma} + 1)\right]\right\};$$

that is, the pulsation of reactance of frequency, 2γ, introduces two higher harmonics of the order $(2\gamma - 1)$, and $(2\gamma + 1)$.

If $X = x\{1 + \epsilon\cos(2\beta - \omega)\}$,

it is
$$\phi = \frac{x}{2\pi Nn}\left\{\cos\beta + \frac{\epsilon}{2}\cos(\beta - \hat{\omega}) + \frac{\epsilon}{2}\cos(3\beta - \hat{\omega})\right\};$$

$$e = x\left\{\sin\beta + \frac{\epsilon}{2}\sin(\beta - \hat{\omega}) + \frac{3\epsilon}{2}\sin(3\beta - \omega)\right\}.$$

Since the pulsation of reactance due to magnetic saturation and hysteresis is essentially of the frequency, $2N$,

— that is, describes a complete cycle for each half-wave of current, — this shows why the distortion of wave-shape by hysteresis consists essentially of a triple harmonic.

The phase displacement between e and i, and thus the power consumed or produced in the electric circuit, depend upon the angle, $\hat{\omega}$, as discussed before.

239. In case of a distortion of the wave-shape by reactance, the distorted waves can be replaced by their equivalent sine waves, and the investigation with sufficient exactness for most cases be carried out under the assumption of sine waves, as done in the preceding chapters.

Similar phenomena take place in circuits containing polarization cells, leaky condensers, or other apparatus representing a synchronously varying negative reactance. Possibly dielectric hysteresis in condensers causes a distortion similar to that due to magnetic hysteresis.

Pulsation of Resistance.

240. To a certain extent the investigation of the effect of synchronous pulsation of the resistance coincides with that of reactance; since a pulsation of reactance, when unsymmetrical with regard to the current wave, introduces an energy component which can be represented by an "effective resistance."

Inversely, an unsymmetrical pulsation of the ohmic resistance introduces a wattless component, to be denoted by "effective reactance."

A typical case of a synchronously pulsating resistance is represented in the alternating arc.

The apparent resistance of an arc depends upon the current passing through the arc; that is, the apparent resistance of the arc $= \frac{\text{potential difference between electrodes}}{\text{current}}$ is high for small currents, low for large currents. Thus in an alternating arc the apparent resistance will vary during every half-wave of current between a maximum value at zero current and a minimum value at maximum current, thereby describing a complete cycle per half-wave of current.

Let the effective value of current passing through the arc be represented by I.

Then the instantaneous value of current, assuming the current wave as sine wave, is represented by

$$i = I\sqrt{2}\sin\beta;$$

and the apparent resistance of the arc, in first approximation, by

$$R = r(1 + \epsilon\cos 2\beta);$$

thus the potential difference at the arc is

$$\begin{aligned} e = iR &= I\sqrt{2}\, r\sin\beta\,(1 + \epsilon\cos 2\beta) \\ &= rI\sqrt{2}\left\{\left(1 - \frac{\epsilon}{2}\right)\sin\beta + \frac{\epsilon}{2}\sin 3\beta\right\}. \end{aligned}$$

Hence the effective value of potential difference,

$$\begin{aligned} E &= rI\sqrt{\left(1 - \frac{\epsilon}{2}\right)^2 + \frac{\epsilon^2}{4}} \\ &= rI\sqrt{1 - \epsilon + \frac{\epsilon^2}{2}}. \end{aligned}$$

and the apparent resistance of the arc

$$r_0 = \frac{E}{I} = r\sqrt{1 - \epsilon + \frac{\epsilon^2}{2}}.$$

The instantaneous power consumed in the arc is,

$$p = ie = 2rI^2\left\{\left(1 - \frac{\epsilon}{2}\right)\sin^2\beta + \frac{\epsilon}{2}\sin\beta\sin 3\beta\right\}.$$

Hence the effective power,

$$P = rI^2\left(1 - \frac{\epsilon}{2}\right).$$

The apparent power, or volt amperes consumed by the arc, is,

$$IE = rI^2\sqrt{1 - \epsilon + \frac{\epsilon^2}{2}}$$

thus the power factor of the arc,

$$f = \frac{P}{IE} = \frac{1 - \frac{\epsilon}{2}}{\sqrt{1 - \epsilon + \frac{\epsilon^2}{2}}}.$$

that is, less than unity.

241. We find here a case of a circuit in which the power factor — that is, the ratio of watts to volt amperes — differs from unity without any displacement of phase; that is, while current and E.M.F. are in phase with each other, but are distorted, the alternating wave cannot be replaced by an equivalent sine wave; since the assumption of equivalent sine wave would introduce a phase displacement,

$$\cos \hat{\omega} = f$$

of an angle, $\hat{\omega}$, whose sign is indefinite.

As an instance are shown, in Fig. 173 for the constants,

$$I = 12$$
$$r = 3$$
$$\epsilon = .9$$

the resistance,

$$R = 3\,\{1 + .9 \cos 2\,\beta\};$$

the current,

$$i = 17 \sin \beta;$$

the potential difference,

$$e = 28\,(\sin \beta + .82 \sin 3\,\beta).$$

In this case the effective E.M.F. is

$$E = 25.5;$$

the apparent resistance,

$$r_0 = 2.13;$$

the power,

$$P = 244;$$

the apparent power,

$$E I = 307;$$

the power factor,

$$f = .796.$$

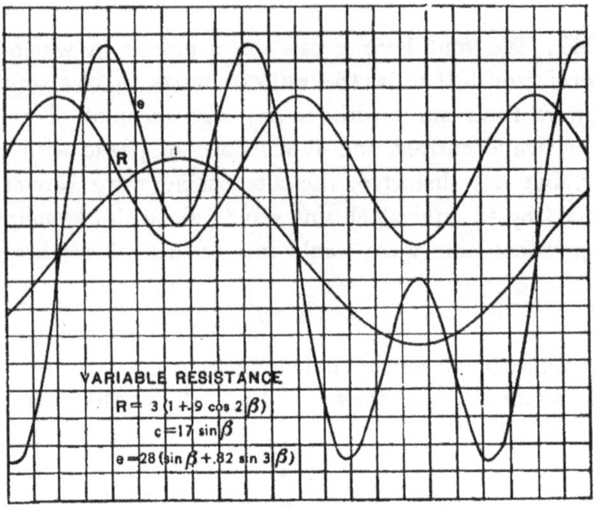

Fig. 173. Periodically Varying Resistance.

As seen, with a sine wave of current the E.M.F. wave in an alternating arc will become double-peaked, and rise very abruptly near the zero values of current. Inversely, with a sine wave of E.M.F. the current wave in an alternating arc will become peaked, and very flat near the zero values of E.M.F.

242. In reality the distortion is of more complex nature; since the pulsation of resistance in the arc does not follow a simple sine law of double frequency, but varies much more abruptly near the zero value of current, making thereby the variation of E.M.F. near the zero value of current much more abruptly, or, inversely, the variation of current more flat.

A typical wave of potential difference, with a sine wave of current passing through the arc, is given in Fig. 174.*

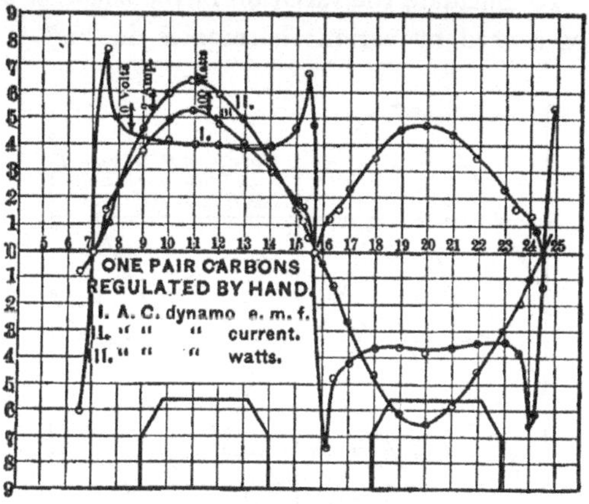

Fig. 174. Electric Arc.

243. The value of ϵ, the amplitude of the resistance pulsation, largely depends upon the nature of the electrodes and the steadiness of the arc, and with soft carbons and a steady arc is small, and the power factor f of the arc near unity. With hard carbons and an unsteady arc, ϵ rises greatly, higher harmonics appear in the pulsation of resistance, and the power factor f falls, being in extreme cases even as low as .6.

The conclusion to be drawn herefrom is, that photometric tests of alternating arcs are of little value, if, besides current and voltage, the power is not determined also by means of electro-dynamometers.

* From American Institute of Electrical Engineers, Transactions, 1890, p. 376. Tobey and Walbridge, on the Stanley Alternate Arc Dynamo.

CHAPTER XXIII.

EFFECTS OF HIGHER HARMONICS.

244. To elucidate the variation in the shape of alternating waves caused by various harmonics, in Figs. 175 and

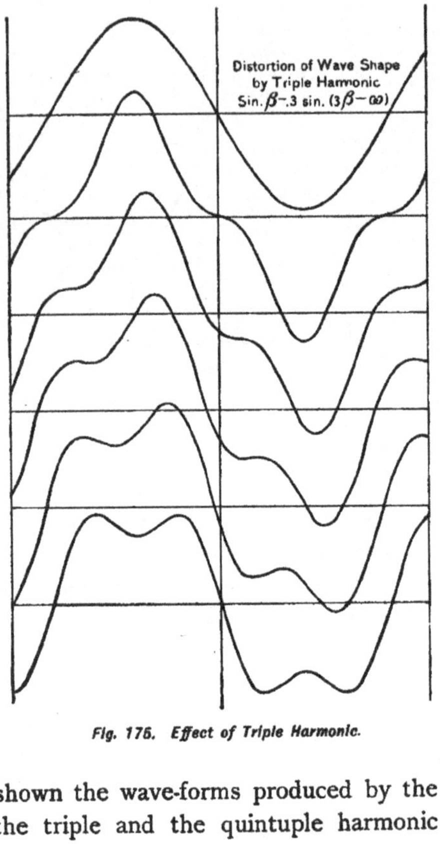

Fig. 176. Effect of Triple Harmonic.

176 are shown the wave-forms produced by the superposition of the triple and the quintuple harmonic upon the fundamental sine wave.

In Fig. 175 is shown the fundamental sine wave and the complex waves produced by the superposition of a triple harmonic of 30 per cent the amplitude of the fundamental, under the relative phase displacements of 0°, 45°, 90°, 135°, and 180°, represented by the equations:

$$\begin{aligned}
&\sin \beta \\
&\sin \beta - .3 \sin 3 \beta \\
&\sin \beta - .3 \sin (3 \beta - 45°) \\
&\sin \beta - .3 \sin (3 \beta - 90°) \\
&\sin \beta - .3 \sin (3 \beta - 135°) \\
&\sin \beta - .3 \sin (3 \beta - 180°).
\end{aligned}$$

As seen, the effect of the triple harmonic is in the first figure to flatten the zero values and point the maximum values of the wave, giving what is called a peaked wave. With increasing phase displacement of the triple harmonic, the flat zero rises and gradually changes to a second peak, giving ultimately a flat-top or even double-peaked wave with sharp zero. The intermediate positions represent what is called a saw-tooth wave.

In Fig. 176 are shown the fundamental sine wave and the complex waves produced by superposition of a quintuple harmonic of 20 per cent the amplitude of the fundamental, under the relative phase displacement of 0°, 45°, 90°, 135°, 180°, represented by the equations:

$$\begin{aligned}
&\sin \beta \\
&\sin \beta - .2 \sin 5 \beta \\
&\sin \beta - .2 \sin (5 \beta - 45°) \\
&\sin \beta - .2 \sin (5 \beta - 90°) \\
&\sin \beta - .2 \sin (5 \beta - 135°) \\
&\sin \beta - .2 \sin (5 \beta - 180°).
\end{aligned}$$

The quintuple harmonic causes a flat-topped or even double-peaked wave with flat zero. With increasing phase displacement, the wave becomes of the type called saw-tooth wave also. The flat zero rises and becomes a third peak, while of the two former peaks, one rises, the other decreases, and the wave gradually changes to a triple-peaked wave with one main peak, and a sharp zero.

As seen, with the triple harmonic, flat-top or double-peak coincides with sharp zero, while the quintuple harmonic flat-top or double-peak coincides with flat zero.

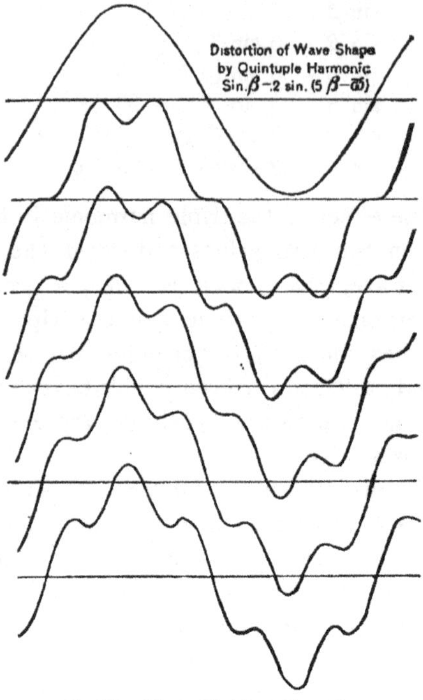

Fig. 176. Effect of Quintuple Harmonic.

Sharp peak coincides with flat zero in the triple, with sharp zero in the quintuple harmonic. With the triple harmonic, the saw-tooth shape appearing in case of a phase difference between fundamental and harmonic is single, while with the quintuple harmonic it is double.

Thus in general, from simple inspection of the wave shape, the existence of these first harmonics can be discovered. Some characteristic shapes are shown in Fig. 177.

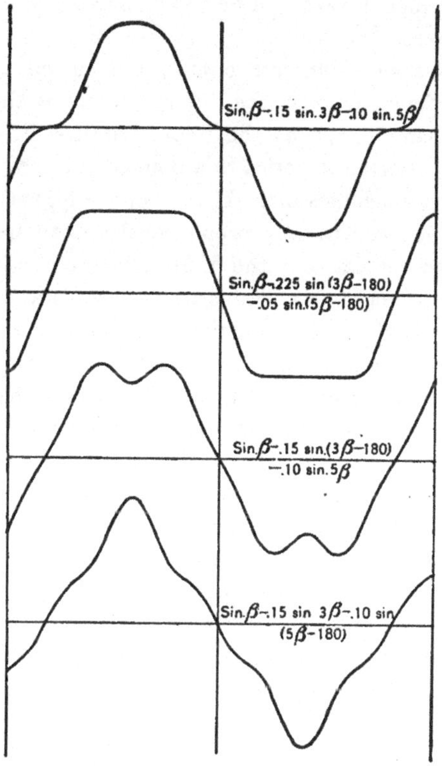

Fig. 177. Some Characteristic Wave Shapes.

Flat top with flat zero:

$$\sin \beta - .15 \sin 3 \beta - .10 \sin 5 \beta.$$

Flat top with sharp zero:

$$\sin \beta - .225 \sin (3 \beta - 180°) - .05 \sin (5 \beta - 180°).$$

Double peak, with sharp zero:

$$\sin \beta - .15 \sin (3 \beta - 180°) - .10 \sin 5 \beta.$$

Sharp peak with sharp zero:

$$\sin \beta - .15 \sin 3 \beta - .10 \sin (5 \beta - 180°).$$

245. Since the distortion of the wave-shape consists in the superposition of higher harmonics, that is, waves of higher frequency, the phenomena taking place in a circuit supplied by such a wave will be the combined effect of the different waves.

Thus in a non-inductive circuit, the current and the potential difference across the different parts of the circuit are of the same shape as the impressed E.M.F. If self-induction is inserted in series to a non-inductive circuit, the self-induction consumes more E.M.F. of the higher harmonics, since the reactance is proportional to the frequency, and thus the current and the E.M.F. in the non-inductive part of the circuit shows the higher harmonics in a reduced amplitude. That is, self-induction in series to a non-inductive circuit reduces the higher harmonics or smooths out the wave to a closer resemblance with sine shape. Inversely, capacity in series to a non-inductive circuit consumes less E.M.F. at higher than at lower frequency, and thus makes the higher harmonics of current and of potential difference in the non-inductive part of the circuit more pronounced — intensifies the harmonics.

Self-induction and capacity in series may cause an increase of voltage due to complete or partial resonance with higher harmonics, and a discrepancy between volt-amperes and watts, without corresponding phase displacement, as will be shown hereafter.

246. In long-distance transmission over lines of noticeable inductance and capacity, rise of voltage due to resonance may occur with higher harmonics, as waves of higher frequency, while the fundamental wave is usually of too low a frequency to cause resonance.

An approximate estimate of the possible rise by resonance with various harmonics can be obtained by the investigation of a numerical instance. Let in a long-distance line, fed by step-up transformers at 60 cycles,

The resistance drop in the transformers at full load = 1%.
The inductance voltage in the transformers at full load = 5% with the fundamental wave.
The resistance drop in the line at full load = 10%.

The inductance voltage in the line at full load = 20% with the fundamental wave.

The capacity or charging current of the line = 20% of the full-load current I at the frequency of the fundamental.

The line capacity may approximately be represented by a condenser shunted across the middle of the line. The E.M.F. at the generator terminals E is assumed as maintained constant.

The E.M.F. consumed by the resistance of the circuit from generator terminals to condenser is

$$Ir = .06\,E,$$

or,

$$r = .06\,\frac{E}{I}.$$

The reactance E.M.F. between generator terminals and condenser is, for the fundamental frequency,

$$Ix = .15\,E,$$

or,

$$x = .15\,\frac{E}{I},$$

thus the reactance corresponding to the frequency $(2k-1)\,N$ of the higher harmonic is:

$$x\,(2\,k-1) = .15\,(2\,k-1)\,\frac{E}{I}.$$

The capacity current at fundamental frequency is:

$$i = .2\,I,$$

hence, at the frequency: $(2\,k-1)\,N$:

$$i = .2\,(2\,k-1)\,e'\,\frac{I}{E},$$

if:

e' = E.M.F. of the $(2\,k-1)^{\text{th}}$ harmonic at the condenser,
e = E.M.F. of the $(2\,k-1)^{\text{th}}$ harmonic at the generator terminals.

The E.M.F. at the condenser is:—

$$e' = \sqrt{e^2 - i^2 r^2} + ix\,(2\,k-1)\;;$$

hence, substituted:

$$a = \frac{e'}{e} = \frac{1}{\sqrt{1 - .059856\,(2\,k - 1)^2 + .0009\,(2\,k - 1)^4}},$$

the rise of voltage by inductance and capacity.

Substituting:

	$k =$	1	2	3	4	5	6
or,	$2\,k - 1 =$	1	3	5	7	9	11
it is,	$a =$	1.03	1.36	3.76	2.18	.70	.38

That is, the fundamental will be increased at open circuit by 3 per cent, the triple harmonic by 36 per cent, the quintuple harmonic by 276 per cent, the septuple harmonic by 118 per cent, while the still higher harmonics are reduced.

The maximum possible rise will take place for:

$$\frac{d\,a}{d\,(2\,k - 1)} = 0, \text{ or, } 2\,k - 1 = 5.77$$

That is, at a frequency: $N = 346$, and $a = 14.4$.

That is, complete resonance will appear at a frequency between quintuple and septuple harmonic, and would raise the voltage at this particular frequency 14.4 fold.

If the voltage shall not exceed the impressed voltage by more than 100 per cent, even at coincidence of the maximum of the harmonic with the maximum of the fundamental,

the triple harmonic must be less than 70 per cent of the fundamental,

the quintuple harmonic must be less than 26.5 per cent of the fundamental,

the septuple harmonic must be less than 46 per cent of the fundamental.

The voltage will not exceed twice the normal, even at a frequency of complete resonance with the higher harmonic, if none of the higher harmonics amounts to more than 7 per cent. of the fundamental. Herefrom it follows that the danger of resonance in high potential lines is in general greatly over-estimated, since the conditions assumed in this instance are rather more severe than found in practice, the capacity current of the line very seldom reaching 20% of the main current.

247. The power developed by a complex harmonic wave in a non-inductive circuit is the sum of the powers of the individual harmonics. Thus if upon a sine wave of alternating E.M.F. higher harmonic waves are superposed, the effective E.M.F., and the power produced by this wave in a given circuit or with a given effective current, are increased. In consequence hereof alternators and synchronous motors of ironclad unitooth construction — that is, machines giving waves with pronounced higher harmonics — give with the same number of turns on the armature, and the same magnetic flux per field pole at the same frequency, a higher output than machines built to produce sine waves.

248. This explains an apparent paradox:

If in the three-phase star-connected generator with the magnetic field constructed as shown diagrammatically in Fig. 162, the magnetic flux per pole $= \Phi$, the number of turns in series per circuit $= n$, the frequency $= N$, the E.M.F. between any two collector rings is:

$$E = \sqrt{2}\,\pi N\, 2\, n\, \Phi\, 10^{-8}.$$

since $2n$ armature turns simultaneously interlink with the magnetic flux Φ.

The E.M.F. per armature circuit is:

$$e = \sqrt{2}\,\pi N n \Phi\, 10^{-8};$$

hence the E.M.F. between collector rings, as resultant of two E.M.Fs. e displaced by 60° from each other, is:

$$E = e\sqrt{3} = \sqrt{2}\,\pi N \sqrt{3}\, n\, \Phi\, 10^{-8};$$

while the same E.M.F. was found by direct calculation from number of turns, magnetic flux, and frequency to be equal to $2e$; that is the two values found for the same E.M.F. have the proportion $\sqrt{3} : 2 = 1 : 1.154$.

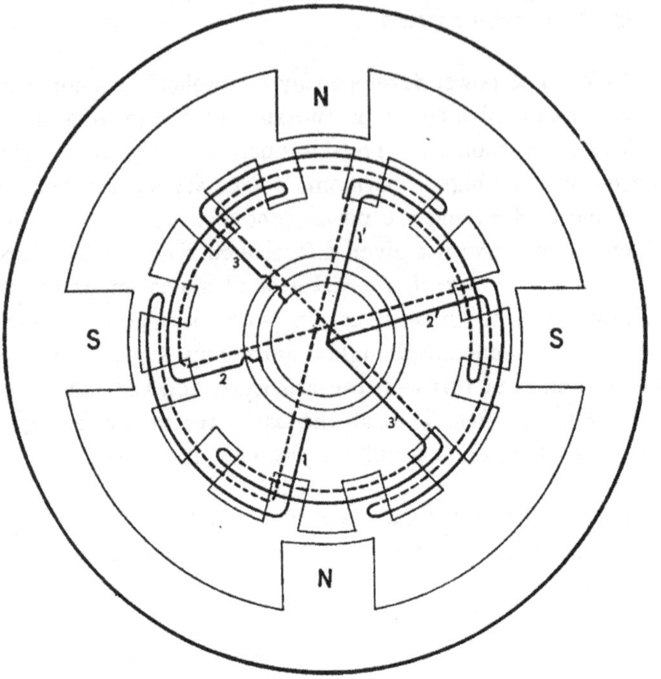

Fig. 178. Three-phase Star-connected Alternator.

This discrepancy is due to the existence of more pronounced higher harmonics in the wave e than in the wave $E = e \times \sqrt{3}$, which have been neglected in the formula:

$$e = \sqrt{2}\, \pi N n \Phi 10^{-8}.$$

Hence it follows that, while the E.M.F. between two collector rings in the machine shown diagrammatically in Fig. 178 is only $e \times \sqrt{3}$, by massing the same number of turns in one slot instead of in two slots, we get the E.M.F. $2e$ or 15.4 per cent higher E.M.F., that is, larger output.

It follows herefrom that the distorted E.M.F. wave of a unitooth alternator is produced by lesser magnetic flux per pole — that is, in general, at a lesser hysteretic loss in the armature or at higher efficiency — than the same effective E.M.F. would be produced with the same number of armature turns if the magnetic disposition were such as to produce a sine wave.

249. Inversely, if such a distorted wave of E.M.F. is impressed upon a magnetic circuit, as, for instance, a transformer, the wave of magnetism in the primary will repeat in shape the wave of magnetism interlinked with the armature coils of the alternator, and consequently, with a lesser maximum magnetic flux, the same effective counter E.M.F. will be produced, that is, the same power converted in the transformer. Since the hysteretic loss in the transformer depends upon the maximum value of magnetism, it follows that the hysteretic loss in a transformer is less with a distorted wave of a unitooth alternator than with a sine wave.

Thus with the distorted waves of unitooth machines, generators, transformers, and synchronous motors — and induction motors in so far as they are transformers — operate more efficiently.

250. From another side the same problem can be approached.

If upon a transformer a sine wave of E.M.F. is impressed, the wave of magnetism will be a sine wave also. If now upon the sine wave of E.M.F. higher harmonics, as sine waves of triple, quintuple, etc., frequency are superposed in such a way that the corresponding higher harmonic sine waves of magnetism do not increase the maximum value of magnetism, or even lower it by a coincidence of their negative maxima with the positive maximum of the fundamental, — in this case all the power represented by these higher harmonics of E.M.F. will be transformed without an increase of the hysteretic loss, or even with a decreased hysteretic loss.

Obviously, if the maximum of the higher harmonic wave of magnetism coincides with the maximum of the fundamental, and thereby makes the wave of magnetism more pointed, the hysteretic loss will be increased more than in proportion to the increased power transformed, i.e., the efficiency of the transformer will be lowered.

That is: Some distorted waves of E.M.F. are transformed at a lesser, some at a larger, hysteretic loss than the sine wave, if the same effective E.M.F. is impressed upon the transformer.

The unitooth alternator wave and the first wave in Fig. 175 belong to the former class; the waves derived from continuous-current machines, tapped at two equi-distant points of the armature, in general, to the latter class.

251. Regarding the loss of energy by Foucault or eddy currents, this loss is not affected by distortion of wave shape, since the E.M.F. of eddy currents, as induced E.M.F., is proportional to the secondary E.M.F.; and thus at constant impressed primary E.M.F., the energy consumed by eddy currents bears a constant relation to the output of the secondary circuit, as obvious, since the division of power between the two secondary circuits — the eddy current circuit, and the useful or consumer circuit — is unaffected by wave-shape or intensity of magnetism.

252. In high potential lines, distorted waves whose maxima are very high above the effective values, as peaked waves, may be objectionable by increasing the strain on the insulation. It is, however, not settled yet beyond doubt whether the striking-distance of a rapidly alternating potential depends upon the maximum value or upon some value between effective and maximum. Since disruptive phenomena do not always take place immediately after application of the potential, but the time element plays an important part, it is possible that insulation-strain and striking-distance is, in a certain range, dependent upon the effective potential, and thus independent of the wave-shape.

In this respect it is quite likely that different insulating materials show a different behavior, and homogeneous solid substances, as paraffin, depend in their disruptive strength upon the maximum value of the potential difference, while heterogeneous materials, as mica, laminated organic substances, air, etc., that is substances in which the disruptive strength decreases with the time application of the potential difference, are less affected by very high peaks of E.M.F. of very short duration.

In general, as conclusions may be derived that the importance of a proper wave-shape is generally greatly overrated, but that in certain cases sine waves are desirable, in other cases certain distorted waves are preferable.

CHAPTER XXIV.

SYMBOLIC REPRESENTATION OF GENERAL ALTERNATING WAVES.

253. The vector representation,

$$\dot{A} = a^1 + ja^{11} = a\,(\cos a + j \sin a)$$

of the alternating wave,

$$A = a_0 \cos (\phi - a)$$

applies to the sine wave only.

The general alternating wave, however, contains an infinite series of terms, of odd frequencies,

$$A = A_1 \cos (\phi - a_1) + A_3 \cos (3\,\phi - a_3) + A_5 \cos (5\,\phi - a_5) +$$

thus cannot be directly represented by one complex vector quantity.

The replacement of the general wave by its equivalent sine wave, as before discussed, that is a sine wave of equal effective intensity and equal power, while sufficiently accurate in many cases, completely fails in other cases, especially in circuits containing capacity, or in circuits containing periodically (and in synchronism with the wave) varying resistance or reactance (as alternating arcs, reaction machines, synchronous induction motors, oversaturated magnetic circuits, etc.).

Since, however, the individual harmonics of the general alternating wave are independent of each other, that is, all products of different harmonics vanish, each term can be represented by a complex symbol, and the equations of the general wave then are the resultants of those of the individual harmonics.

This can be represented symbolically by combining in one formula symbolic representations of different frequencies, thus,

$$\dot{A} = \sum_{1}^{\infty}{}_{2n-1} (a_n^{1} + j_n a_n^{11})$$

where,

$$j_n = \sqrt{-1}$$

and the index of the j_n merely denotes that the j's of differentindices n, while algebraically identical, physically represent different frequencies, and thus cannot be combined.

The general wave of E.M.F. is thus represented by,

$$\dot{E} = \sum_{1}^{\infty}{}_{2n-1} (e_n^{1} + j_n e_n^{11})$$

the general wave of current by,

$$\dot{I} = \sum_{1}^{\infty}{}_{2n-1} (i_n^{1} + j_n i_n^{11})$$

If,

$$Z_1 = r - j (x_m + x_0 + x_c)$$

is the impedance of the fundamental harmonic, where

x_m is that part of the reactance which is proportional to the frequency (inductance, etc.).

x_o is that part of the reactance which is independent of the frequency (mutual induction, synchronous motion, etc.).

x_c is that part of the reactance which is inversely proportional to the frequency (capacity, etc.).

The impedance for the nth harmonic is,

$$Z = r - j_n\left(n\,x_m + x_0 + \frac{x_c}{n}\right)$$

This term can be considered as the general symbolic expression of the impedance of a circuit of general wave shape.

Ohm's law, in symbolic expression, assumes for the general alternating wave the form,

$$\dot{I} = \frac{\dot{E}}{\dot{Z}} \text{ or,}$$

$$\sum_1^{\infty}{}^{2n-1}(i_n^1 + j_n i_n^{11}) = \sum_1^{\infty}{}^{2n-1} \frac{e_n^1 + j_n e_n^{11}}{r - j_n\left(n x_m + x_0 + \frac{x_c}{n}\right)}$$

$$\dot{E} = \dot{I}\,\dot{Z} \text{ or,}$$

$$\sum_1^{\infty}{}^{2n-1}(e_n^1 + j_n e_n^{11}) = \sum_1^{\infty}{}^{2n-1}\left[r - j_n\left(n x_m + x_0 + \frac{x_c}{n}\right)\right]$$
$$(i_n^1 + j_n i_n^{11})$$

$$\dot{Z} = \frac{\dot{E}}{\dot{I}} \text{ or,}$$

$$Z = r - j_n\left(n x_m + x_0 + \frac{x_c}{n}\right) = \frac{e_n + j_n e_n^{11}}{i_n^1 + j_n i_n^{11}}$$

The symbols of multiplication and division of the terms $\dot{E}$, $\dot{I}$, Z, thus represent not algebraic operation, but multiplication and division of corresponding terms of E, I, $\dot{Z}$, that is, terms of the same index *n*, or, in algebraic multiplication and division of the series $\dot{E}$, $\dot{I}$, all compound terms, that is terms containing two different *n*'s, vanish.

254. The effective value of the general wave:

$$a = A_1 \cos(\phi - a_1) + A_3 \cos(3\phi - a_3) + A_5 \cos(5\phi - a_5) + \ldots$$

is the square root of the sum of mean squares of individual harmonics,

$$A = \sqrt{\tfrac{1}{2}\left\{A_1^2 + A_3^2 + A_5^2 + \ldots\right\}}$$

Since, as discussed above, the compound terms, of two different indices *n*, vanish, the absolute value of the general alternating wave,

$$\dot{A} = \sum_1^{\infty}{}^{2n-1} \frac{a_n^1 + j_n a_n^{11}}{b_n^1 + j_n b_n^{11}}$$

is thus,

$$A = \sqrt{\sum_1^{\infty}{}_{2n-1} \frac{a_n^{1\,2} + a_n^{11\,2}}{b_n^{1\,2} + b_n^{11\,2}}}$$

which offers an easy means of reduction from symbolic to absolute values

Thus, the absolute value of the E.M.F.

$$\dot{E} = \sum_1^{\infty}{}_{2n-1} (e_n^1 + j_n e_n^{11})$$

is,

$$E = \sqrt{\sum_1^{\infty}{}_{2n-1} (e_n^{1\,2} + e_n^{11\,2})}$$

the absolute value of the current,

$$\dot{I} = \sum_1^{\infty}{}_{2n-1} (i_n^1 + j_n i_n^{11})$$

is,

$$I = \sqrt{\sum_1^{\infty}{}_{2n-1} (i_n^{1\,2} + i_n^{11\,2})}$$

255. The double frequency power (torque, etc.) equation of the general alternating wave has the same symbolic expression as with the sine wave:

$$\begin{aligned}
\dot{P} &= [\dot{E}\dot{I}] \\
&= P^1 + jP^j \\
&= [\dot{E}\dot{I}]^1 + j[\dot{E}\dot{I}]^j \\
&= \sum_1^{\infty}{}_{2n-1} (e_n^1 i_n^1 + e_n^{11} i_n^{11}) + \sum_1^{\infty}{}_{2n-1} j_n (e_n^{11} i_n^1 - e_n i_n^{11})
\end{aligned}$$

where,

$$P^1 = [\dot{E}\dot{I}]^1 = \sum_1^{\infty}{}_{2n-1} (e_n^1 i_n^1 + e_n^{11} i_n^{11})$$

$$P^j = [\dot{E}\dot{I}]^j = \sum_1^{\infty}{}_{2n-1} \frac{j_n}{j} (e_n^{11} i_n^1 - e_n^1 i_n^{11})$$

The j_n enters under the summation sign of the "wattless power" P^j, so that the wattless powers of the different harmonics cannot be algebraically added.

Thus,

The total "true power" of a general alternating current circuit is the algebraic sum of the powers of the individual harmonics.

The total "wattless power" of a general alternating current circuit is not the algebraic, but the absolute sum of the wattless powers of the individual harmonics.

Thus, regarding the wattless power as a whole, in the general alternating circuit no distinction can be made between lead and lag, since some harmonics may be leading, others lagging.

The apparent power, or total volt-amperes, of the circuit is,

$$\therefore \quad Q = EI = \sqrt{\sum_1^{\infty}{}_{2n-1} \left(e_n^{1\,2} + e_n^{11\,2}\right) \sum_1^{\infty}{}_{2n-1} \left(i_n^{1\,2} + i_n^{11\,2}\right)}$$

The power factor of the circuit is,

$$p = \frac{P^1}{Q} = \frac{\sum_1^{\infty}{}_{2n-1} \left(e_n^1 i_n^1 + e_n^{11} i_n^{11}\right)}{\sqrt{\sum_1^{\infty}{}_{2n-1} \left(e_n^{1\,2} + e_n^{11\,2}\right) \sum_1^{\infty}{}_{2n-1} \left(i_n^{1\,2} + i_n^{11\,2}\right)}}$$

The term "inductance factor," however, has no meaning any more, since the wattless powers of the different harmonics are not directly comparable.

The quantity,

$$q_0 = \sqrt{1 - p^2}$$

has no physical significance, and is not $\dfrac{\text{wattless power}}{\text{total apparent power}}$.

The term,

$$\frac{P^j}{EI}$$

$$= \sum_1^\infty {}_{2n-1} \frac{j_n}{j} \frac{e_n^{11} i_n^1 - e_n^1 i_n^{11}}{EI}$$

$$= \sum_1^\infty {}_{2n-1} \frac{j_n}{j} q_n$$

where,

$$q_n = \frac{e_n^{11} i_n^1 - e_n^1 i_n^{11}}{EI}$$

consists of a series of inductance factors q_n of the individual harmonics.

As a rule, if

$$q^2 = \sum_1^\infty {}_{2n-1} q_n^2,$$

$$p^2 + q^2 < 1$$

for the general alternating wave, that is q differs from

$$q_0 = \sqrt{1 - p^2}$$

The complex quantity,

$$\dot{U} = \frac{\dot{P}}{Q} = \frac{[\dot{E}\dot{I}]}{EI} = \frac{[\dot{E}\dot{I}]^1 + j[\dot{E}\dot{I}]^j}{EI}$$

$$= \frac{\sum_1^\infty {}_{2n-1} (e_n^1 i_n^1 + e_n^{11} i_n^{11}) + \sum_1^\infty {}_{2n-1} j_n \, (e_n^{11} i_n^1 - e_n^1 i_n^{11})}{\sqrt{\sum_1^\infty {}_{2n-1} ({e_n^1}^2 + {e_n^{11}}^2) \sum_1^\infty {}_{2n-1} ({i_n^1}^2 + {i_n^{11}}^2)}}$$

$$= p + \sum_1^\infty {}_{2n-1} j_n q_n$$

takes in the circuit of the general alternating wave the same position as power factor and inductance factor with the sine wave.

$\dot{U} = \frac{\dot{P}}{Q}$ may be called the "*circuit factor.*"

It consists of a real term p, the power factor, and a series of imaginary terms $j_n \, q_n$, the inductance factors of the individual harmonics.

The absolute value of the circuit factor:

$$u = \sqrt{p^2 + \sum_1^{\infty}{}_{2n-1}\, q_n^2}$$

as a rule, is < 1.

256. Some applications of this symbolism will explain its mechanism and its usefulness more fully.

1st *Instance:* Let the E.M.F.,

$$E = \sum_1^{5}{}_{2n-1} \left(e_n^{1} + j_n e_n^{11}\right)$$

be impressed upon a circuit of the impedance,

$$Z = r - j_n \left(n x_m - \frac{x_c}{n}\right)$$

that is, containing resistance r, inductive reactance x_m and capacity reactance x_c in series.

Let

$$e_1^{1} = 720 \qquad e_1^{11} = 540$$
$$e_3^{1} = 283 \qquad e_3^{11} = -283$$
$$e_5^{1} = -104 \qquad e_5^{11} = 138$$

or,

$$e_1 = 900 \qquad \tan \omega_1 = .75$$
$$e_3 = 400 \qquad \tan \omega_3 = -1$$
$$e_5 = 173 \qquad \tan \omega_5 = -1.33$$

It is thus in symbolic expression,

$$Z_1 = 10 + 80 j_1 \qquad z_1 = 80.6$$
$$Z_3 = 10 \qquad z_3 = 10$$
$$Z_5 = 10 - 32 j_5 \qquad z_5 = 33.5$$

and, E.M.F.,

$$E = (720 + 540 j_1) + (283 - 283 j_3) + (-104 + 138 j_5)$$

or absolute,

$$E = 1000$$

and current,

$$\dot{I} = \frac{\dot{E}}{Z} = \frac{720 + 540 j_1}{10 + 80 j_1} + \frac{283 - 283 j_3}{10} + \frac{-104 + 138 j_5}{10 - 32 j_5}$$

$$= (7.76 - 8.04 j_1) + (28.3 - 28.3 j_3) + (-4.86 - 1.73 j_5)$$

or, absolute,

$$I = 41.85$$

of which is of fundamental frequency,		$I_1 = 11.15$
" " " " triple	"	$I_3 = 40$
" " " " quintuple	"	$I_5 = 5.17$

The total apparent power of the circuit is,

$$Q = E\,I = 41{,}850$$

The true power of the circuit is:

$$P^1 = [E\,I]^1 = 1240 + 16{,}000 + 270$$
$$= 17{,}510$$

the wattless power,

$$j\,P^j = j\,[E\,I]^j = 10{,}000\,j_1 - 850\,j_5$$

thus, the total power,

$$\dot{P} = 17{,}510 + 10{,}000 j_1 - 850\,j_5$$

That is, the wattless power of the first harmonic is leading, that of the third harmonic zero, and that of the fifth harmonic lagging.

$$17{,}510 = I^2\,r, \text{ as obvious.}$$

The circuit factor is,

$$\dot{U} = \frac{\dot{P}}{Q} = \frac{[\dot{E}\,\dot{I}]}{E\,I}$$
$$= .418 + .239\,j_1 - .0203\,j_5$$

or, absolute,

$$u = \sqrt{.418^2 + .239^2 + .0203^2}$$
$$= .482$$

The power factor is,

$$p = .418$$

The inductance factor of the first harmonic is : $q_1 = .239$, that of the third harmonic $q_3 = 0$, and of the fifth harmonic $q_5 = -.0203$.

Considering the waves as replaced by their equivalent sine waves, from the sine wave formula,

$$p^2 + q_0^2 = 1$$

the inductance factor would be,

$$q_0 = .914$$

and the phase angle,

$$\tan \omega = \frac{q_0}{p} = \frac{.914}{.418} = 2.8 \qquad \omega = 65.4°$$

giving apparently a very great phase displacement, while in reality, of the 41.85 amperes total current, 40 amperes (the current of the third harmonic) are in phase with their E.M.F.

We thus have here a case of a circuit with complex harmonic waves which cannot be represented by their equivalent sine waves. The relative magnitudes of the different harmonics in the wave of current and of E.M.F. differ essentially, and the circuit has simultaneously a very low power factor and a very low inductance factor; that is, a low power factor exists without corresponding phase displacement, the circuit factor being less than one-half.

Such circuits, for instance, are those including alternating arcs, reaction machines, synchronous induction motors, reactances with over-saturated magnetic circuit, high potential lines in which the maximum difference of potential exceeds the voltage at which brush discharges begin, polarization cells, and in general electrolytic conductors above the dissociation voltage of the electrolyte, etc. Such circuits cannot correctly, and in many cases not even approximately, be treated by the theory of the equivalent sine waves, but require the symbolism of the complex harmonic wave.

257. *2d instance:* A condenser of capacity $C_0 = 20$ m.f. is connected into the circuit of a 60-cycle alternator giving a wave of the form,

$$e = E\,(\cos\phi - .10\cos 3\phi - .08\cos 5\phi + .06\cos 7\phi)$$

or, in symbolic expression,

$$\dot{E} = e\,(1_1 - .10_3 - .08_5 + .06_7)$$

The synchronous impedance of the alternator is,

$$Z_0 = r_0 - j_n n x_0 = .3 - 5\,n j_n$$

What is the apparent capacity C of the condenser (as calculated from its terminal volts and amperes) when connected directly with the alternator terminals, and when connected thereto through various amounts of resistance and inductive reactance.

The capacity reactance of the condenser is,

$$x_c = \frac{10^6}{2\,\pi N C_0} = 132 \text{ ohms},$$

or, in symbolic expression,

$$+ j_n \frac{x_c}{n} = \frac{132}{n} j_n$$

Let

$Z_1 = r - j_n\, n x$ = impedance inserted in series with the condenser.

The total impedance of the circuit is then,

$$Z = Z_0 + Z_1 + j_n \frac{x_c}{n}$$

$$= (.3 + r) - j_n \left([5 + x]\,n - \frac{132}{n}\right)$$

The current in the circuit is,

$$\dot{I} = \frac{\dot{E}}{Z}$$

$$= e\left[\frac{1}{(.3 + r) - j(x - 132)} - \frac{.1}{(.3 + r) - j_3(3x - 29)} - \frac{.08}{(.3 + r) - j_5(5x - 1.4)} + \frac{.06}{(.3 + r) - j_7(7x + 16.1)}\right]$$

and the E.M.F. at the condenser terminals,

$$\dot{E}_1 = j_n \frac{x_c \dot{I}}{n}$$

$$= e \left[\frac{132 j_1}{(.3 + r) - j_1 (x - 132)} - \frac{4.4 j_3}{(.3 + r) - j_3 (3x - 29)} - \frac{2.11 j_5}{(.3 + r) - j_5 (5x - 1.4)} + \frac{1.13 j_7}{(.3 + r) - j_7 (7x + 16.1)} \right]$$

thus the apparent capacity reactance of the condenser is,

$$x_1 = \frac{E_1}{I}$$

and the apparent capacity,

$$C = \frac{10^6}{2 \pi N x_1}$$

(*a.*) $x = 0$: Resistance r in series with the condenser Reduced to absolute values, it is,

$$\frac{1}{x_1^2} = \frac{\frac{1}{(.3+r)+17424} + \frac{.01}{(.3+r)^2+841} + \frac{.0064}{(.3+r)^2+1.96} + \frac{.0036}{(.3+r)^2+259}}{\frac{17424}{(.3+r)^2+17424} + \frac{19.4}{(.3+r)^2+841} + \frac{4.45}{(.3+r)^2+1.96} + \frac{1.28}{(.3+r)^2+259}}$$

(*b.*) $r = 0$: Inductive reactance x in series with the condenser. Reduced to absolute values, it is,

$$\frac{1}{x_1^2} = \frac{\frac{1}{.09+(x-132)^2} + \frac{.01}{.09+(3x-29)^2} + \frac{.0064}{.09+(5x-1.4)^2} + \frac{.0036}{.09+(7x+16.1)^2}}{\frac{17424}{.09+(x-132)^2} + \frac{19.4}{.09+(3x-29)^2} + \frac{4.45}{.09+(5x-1.4)^2} + \frac{1.28}{.09+(7x+16.1)^2}}$$

From $\frac{1}{x_1^2}$ are derived the values of apparent capacity,

$$C = \frac{10^6}{2 \pi N x_1}$$

and plotted in Fig. 179 for values of r and x respectively varying from 0 to 22 ohms.

As seen, with neither additional resistance nor reactance in series to the condenser, the apparent capacity with this generator wave is 84 m.f., or 4.2 times the true capacity, and gradually decreases with increasing series resistance, to $C = 27.5$ m.f. $= 1.375$ times the true capacity at $r = 13.2$ ohms, or $\frac{1}{10}$ the true capacity reactance, with $r = 132$ ohms, or with an additional resistance equal to the capacity reactance, $C = 20.5$ m.f. or only 2.5% in excess of the true capacity C_0, and at $r = \infty$, $C = 20,3$ m.f. or 1.5% in excess of the true capacity.

With reactance x, but no additional resistance r in series, the apparent capacity C rises from 4.2 times the true capacity at $x = 0$, to a maximum of 5,03 times the true capacity, or $C = 100.6$ m.f. at $x = .28$, the condition of resonance of the fifth harmonic, then decreases to a minimum of 27 m.f., or 35 % in excess of the true capacity, rises again to 60.2 m.f., or 3.01 times the true capacity at $x = 9.67$, the condition of resonance with the third harmonic, and finally decreases, reaching 20 m.f., or the true capacity at $x = 132$, or an inductive reactance equal to the capacity reactance, then increases again to 20.2 m.f. at $x = \infty$.

This rise and fall of the apparent capacity is within certain limits independent of the magnitude of the higher harmonics of the generator wave of E.M.F., but merely depends upon their presence. That is, with such a reactance connected in series as to cause resonance with one of the higher harmonics, the increase of apparent capacity is approximately the same, whatever the value of the harmonic, whether it equals 25% of the fundamental or less than 5%, provided the resistance in the circuit is negligible. The only effect of the amplitude of the higher harmonic is that when it is small, a lower resistance makes itself felt by reducing the increase of apparent capacity below the value it would have were the amplitude greater.

It thus follows that the true capacity of a condenser cannot even approximately be determined by measuring volts and amperes if there are any higher harmonics present in the generator wave, except by inserting a very large resistance or reactance in series to the condenser.

258. *3d instance:* An alternating current generator of the wave,

$$E_0 = 2000\,[1_1 + .12_3 - .23_5 - .13_7]$$

and of synchronous impedance,

$$Z_0 = .3 - 5\,nj_n$$

feeds over a line of impedance,

$$Z_1 = 2 - 4\,nj_n$$

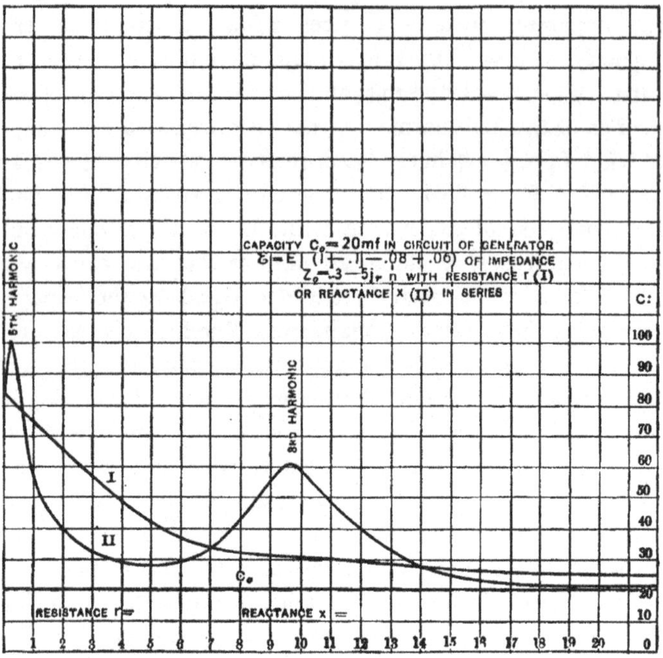

Fig. 179.

a synchronous motor of the wave,

$$E_1 = 2250\,[(\cos\omega + j_1 \sin\omega) + .24\,(\cos 3\,\omega + j_3 \sin 3\,\omega)]$$

and of synchronous impedance,

$$Z_2 = .3 - 6\,nj_n$$

The total impedance of the system is then,

$$Z = Z_0 + Z_1 + Z_2$$
$$= 2.6 - 15\,nj_n$$

thus the current,

$$\dot{I} = \frac{\dot{E}_0 - \dot{E}}{Z}$$

$$= \frac{2000 - 2250 \cos\omega - 2250 j_1 \sin\omega}{2.6 - 15 j_1} + \frac{240 - 540 \cos 3\omega - 540 j_3 \sin 3\omega}{2.6 - 45 j_3}$$

$$- \frac{460}{2.6 - 75 j_5} - \frac{260}{2.6 - 105 j_7}$$

$$= (a_1^{1} + j_1 a_1^{11}) + (a_3^{1} + j_3 a_3^{11}) + (a_5^{1} + j_5 a_5^{11}) + (a_7^{1} + j_7 a_7^{11})$$

where,

$$a_1^{1} = 22.5 - 25.2 \cos\omega + 146 \sin\omega$$
$$a_3^{1} = .306 - .69 \cos 3\omega + 11.9 \sin 3\omega$$
$$a_5^{1} = -.213$$
$$a_7^{1} = -.061$$
$$a_1^{11} = 130 - 146 \cos\omega - 25.2 \sin\omega$$
$$a_3^{11} = 5.3 - 11.9 \cos 3\omega - .69 \sin 3\omega$$
$$a_5^{11} = -6.12$$
$$a_7^{11} = -2.48$$

or, absolute,

1st harmonic,

$$a_1 = \sqrt{{a_1^{1}}^2 + {a_1^{11}}^2}$$

3d harmonic,

$$a_3 = \sqrt{{a_3^{1}}^2 + {a_3^{11}}^2}$$

5th harmonic,

$$a_5 = 6.12$$

7th harmonic,

$$a_7 = 2.48$$

$$I = \sqrt{a_1^2 + a_3^2 + a_5^2 + a_7^2}$$

while the total current of higher harmonics is,

$$I_0 = \sqrt{a_3^2 + a_5^2 + a_7^2}$$

The true input of the synchronous motor is,

$$P^1 = [E_1 I]^1$$
$$= (2250\, a_1{}^1 \cos\omega + 2250\, a_1{}^{11} \sin\omega) + (540\, a_1{}^1 \cos 3\omega + 540\, a_3{}^{11} \sin 3\omega)$$
$$= P_1^1 + P_3^1$$
$$P_1^1 = 2250\, (a_1{}^1 \cos\omega + a_1{}^{11} \sin\omega)$$

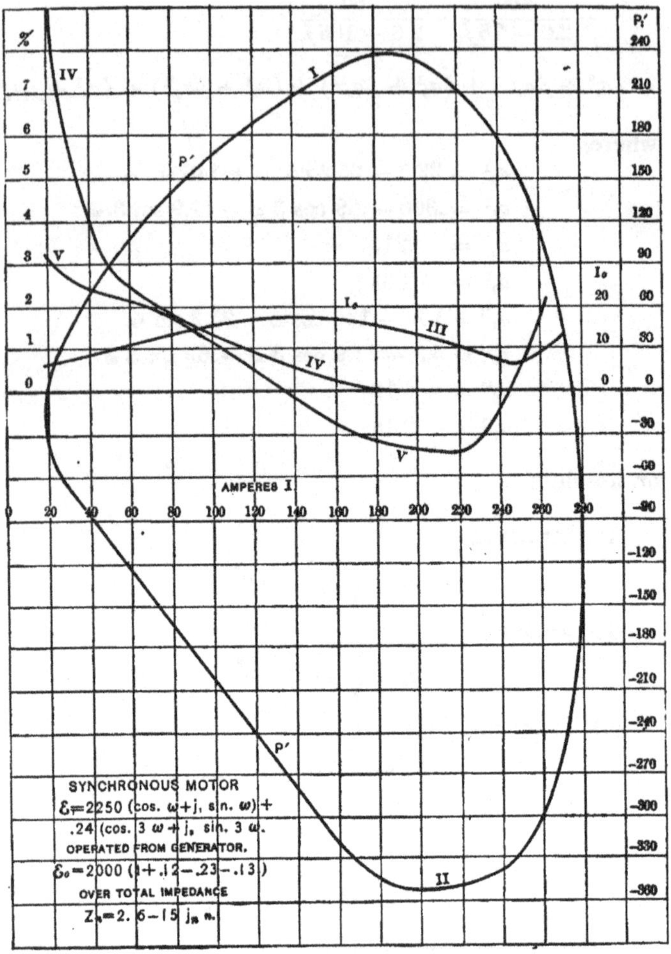

Fig. 180. Synchronous Motor.

is the power of the fundamental wave,

$$P_3^1 = 540\ (a_3^{1} \cos 3\ \omega + a_3^{11} \sin 3\ \omega)$$

the power of the third harmonic.

The 5th and 7th harmonics do not give any power, since they are not contained in the synchronous motor wave. Substituting now different numerical values for ω the phase angle between generator E.M.F. and synchronous motor counter E.M.F., corresponding values of the currents $I\ I_0$, and the powers P^1, P_1^1, P_3^1 are derived. These are plotted in Fig. 180 with the total current I as abcissae. To each value of the total current I correspond two values of the total power P^1, a positive value plotted as Curve I.—synchronous motor — and a negative value plotted as Curve II. — alternating current generator.—. Curve III. gives the total current of higher frequency I_0, Curve IV., the difference between the total current and the current of fundamental frequency, $I - a_1$, in percentage of the total current I, and V the power of the third harmonic, P_3^1, in percentage of the total power P^1.

Curves III., IV. and V. correspond to the positive or synchronous motor part of the power curve P^1. As seen, the increase of current due to the higher harmonics is small, and entirely disappears at about 180 amperes. The power of the third harmonic is positive, that is, adds to the work of the synchronous motor up to about 140 amperes, or near the maximum output of the motor, and then becomes negative.

It follows herefrom that higher harmonics in the E.M.F. waves of generators and synchronous motors do not represent a mere waste of current, but may contribute more or less to the output of the motor. Thus at 75 amperes total current, the percentage of increase of power due to the higher harmonic is equal to the increase of current, or in other words the higher harmonics of current do work with the same efficiency as the fundamental wave.

259. *4th Instance:* In a small three-phase induction motor, the constants per delta circuit are

Primary admittance $Y = .002 + .03j$
Self-inductive impedance $Z_0 = Z_1 = .6 - 2.4j$

and a sine wave of E.M.F. $e_0 = 110$ volts is impressed upon the motor.

The power output P, current input I_s, and power factor p, as function of the slip s are given in the first columns of the following table, calculated in the manner as described in the chapter on Induction Motors.

To improve the power factor of the motor and bring it to unity at an output of 500 watts, a condenser capacity is required giving 4.28 amperes leading current at 110 volts, that is, neglecting the energy loss in the condenser, capacity susceptance

$$\frac{4.28}{110} = .039$$

In this case, let $\dot{I}_s$ = current input into the motor per delta circuit at slip s, as given in the following table.

The total current supplied by the circuit with a sine wave of impressed E.M.F., is

$$\dot{I}^1 = \dot{I}_s - 4.28j$$

and herefrom the power factor $= \frac{\text{energy current}}{\text{total current}}$, given in the second columns of the table.

If the impressed E.M.F. is not a sine wave but a wave of the shape

$$\dot{E}_0 = e_0 (1_1 + .12_3 - .23_5 - .134_7)$$

to give the same output, the fundamental wave must be the same: $e_0 = 110$ volts, when assuming the higher harmonics in the motor as wattless, that is

$$\begin{aligned} \dot{E}_0 &= 110_1 + 13.2_3 - 25.3_5 - 14.7_7 \\ &= e_0 + \dot{E}_0^1 \end{aligned}$$

where $\dot{E}_0^1 = 13.2_3 - 25.3_5 - 14.7_7$
= component of impressed E.M.F. of higher frequency.

The effective value is:

$$E_0 = 114.5 \text{ volts.}$$

The condenser admittance for the general alternating wave is

$$Y_c = -.039\, n j_n$$

Since the frequency of rotation of the motor is very small compared with the frequency of the higher harmonics, as total impedance of the motor for these higher harmonics can be assumed the stationary impedance, and by neglecting the resistance it is

$$\begin{aligned} Z^1 &= -n j_n (x_0 + x_1) \\ &= -4.8\, n j_n \end{aligned}$$

The exciting admittance of the motor, for these higher harmonics, is, by neglecting the conductance,

$$\begin{aligned} Y^1 &= \frac{b j_n}{n} \\ &= \frac{.05\, j_n}{n} \end{aligned}$$

and the higher harmonics of counter E.M.F.

$$\dot{E}^1 = \frac{E_0^1}{2}$$

Thus we have,

Current input in the condenser,

$$\begin{aligned} \dot{I}_c &= E_0 Y_c \\ &= -4.28 j_1 - 1.54 j_3 + 4.93 j_5 + 4.02 j_7 \end{aligned}$$

High frequency component of motor impedance current,

$$\frac{\dot{E}_0^1}{Z^1} = .92 j_3 - 1.06 j_5 - .44 j_7$$

High frequency component of motor exciting current,

$$\begin{aligned} \dot{E}^1 Y^1 &= \frac{\dot{E}_0^1 Y^1}{2} \\ &= .07 j_3 - .08 j_5 - .03 j_7 \end{aligned}$$

thus, total high frequency component of motor current,

$$\dot{I}_0^1 = \frac{\dot{E}_0^1}{\dot{Z}^1} + \dot{E}^1 Y^1$$
$$= .99 j_3 - 1.14 j_5 - .47 j_7$$

and total current,

without condenser,

$$\dot{I}_0 = \dot{I}_s + \dot{I}_0^1$$
$$= \dot{I}_s + .99 j_3 - 1.14 j_5 - .47 j_7$$

with condenser,

$$\dot{I} = \dot{I}_s + \dot{I}_0^1 - \dot{I}_c$$
$$= \dot{I}_s - 4.28 j_1 - .55 j_3 + 3.79 j_5 + 3.55 j_7$$

and herefrom the power factor.

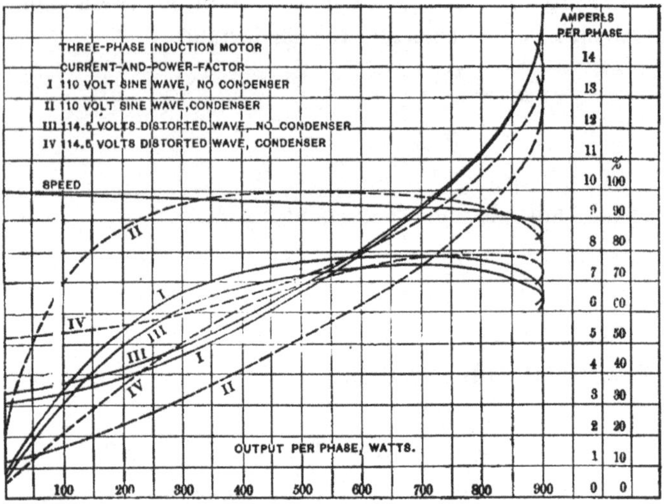

Fig. 181.

In the following table and in Fig. 181 are given the values of current and power factor:—

I. With sine wave of E.M.F., of 110 volts, and no condenser.
II. With sine wave of E.M.F, of 110 volts, and with condenser.
III. With distorted wave of E.M.F., of 114.5 volts, and no condenser.
IV. With distorted wave of E.M.F., of 114.5 volts, and with condenser.

TABLE.

		I.			II.		III.		IV.	
s	P	$\dot{I}_0$	I_0	p	I^1	p	I_0	p	I	p
0	0	.24 + 3.10j	3.1	7.8	1.2	20	3.5	6.6	5.2	4.4
.01	160	1.73 + 3.16j	3.6	48	2.1	84	3.9	43	5.5	31
.02	320	3.32 + 3.47j	4.8	69	3.4	97.2	5.1	64	6.1	54
.035	500	5.16 + 4.28j	6.7	77	5.2	100	6.9	72.5	7.2	68
.05	660	6.95 + 5.4j	8.8	79	7.0	98.7	8.9	76	8.6	77
.07	810	8.77 + 7.3j	11.4	77	9.3	94.5	11.5	73.5	10.6	80
.10	885	10.1 + 9.85j	14.1	71.5	11.5	87	14.2	68	12.6	77
.13	900	10.45 + 11.45j	15.5	67.5	12.7	82	15.6	64.5	13.7	73
.15	890	10.75 + 12.9j	16.8	64	13.8	78	16.9	61	14.7	70

The curves II. and IV. with condenser are plotted in dotted lines in Fig. 181. As seen, even with such a distorted wave the current input and power factor of the motor are not much changed if no condenser is used. When using a condenser in shunt to the motor, however, with such a wave of impressed E.M.F. the increase of the total current, due to higher frequency currents in the condenser, is greater than the decrease, due to the compensation of lagging currents, and the power factor is actually lowered by the condenser, over the total range of load up to overloads, and especially at light loads.

Where a compensator or transformer is used for feeding the condenser, due to the internal self-induction of the compensator, the higher harmonics of current are still more accentuated, that is the power factor still more lowered.

In the preceding the energy loss in the condenser and compensator and that due to the higher harmonics of current in the motor has been neglected. The effect of this energy loss is a slight decrease of efficiency and corresponding increase of power factor. The power produced by the higher harmonics has also been neglected; it may be positive or negative, according to the index of the harmonic, and the winding of the motor primary. Thus for instance, the effect of the triple harmonic is negative in the quarter-phase motor, zero in the three-phase motor, etc., altogether, however, the effect of these harmonics is very small.

CHAPTER XXV.

GENERAL POLYPHASE SYSTEMS.

260. A polyphase system is an alternating-current system in which several E.M.Fs. of the same frequency, but displaced in phase from each other, produce several currents of equal frequency, but displaced phases.

Thus any polyphase system can be considered as consisting of a number of single circuits, or branches of the polyphase system, which may be more or less interlinked with each other.

In general the investigation of a polyphase system is carried out by treating the single-phase branch circuits independently.

Thus all the discussions on generators, synchronous motors, induction motors, etc., in the preceding chapters, apply to single-phase systems as well as polyphase systems, in the latter case the total power being the sum of the powers of the individual or branch circuits.

If the polyphase system consists of n equal E.M.Fs. displaced from each other by $1 / n$ of a period, the system is called a *symmetrical system*, otherwise an *unsymmetrical system*.

Thus the three-phase system, consisting of three equal E.M.Fs. displaced by one-third of a period, is a symmetrical system. The quarter-phase system, consisting of two equal E.M.Fs. displaced by 90°, or one-quarter of a period, is an unsymmetrical system.

261. The flow of power in a single-phase system is pulsating; that is, the watt curve of the circuit is a sine wave of double frequency, alternating between a maximum value and zero, or a negative maximum value. In a polyphase system the watt curves of the different branches of the system are pulsating also. Their sum, however, or the total flow of power of the system, may be either constant or pulsating. In the first case, the system is called a *balanced system*, in the latter case an *unbalanced system.*

The three-phase system and the quarter-phase system, with equal load on the different branches, are balanced systems; with unequal distribution of load between the individual branches both systems become unbalanced systems.

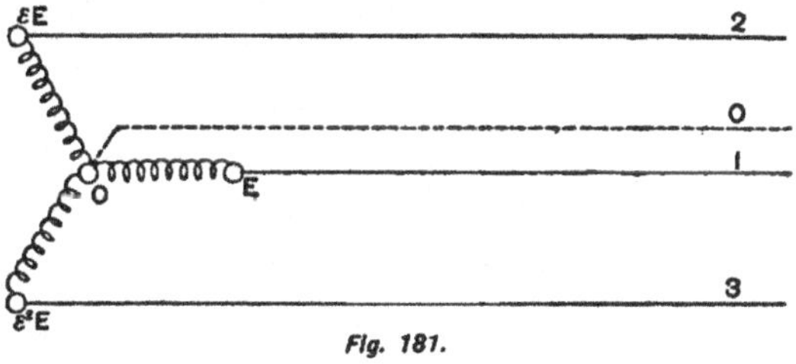

Fig. 181.

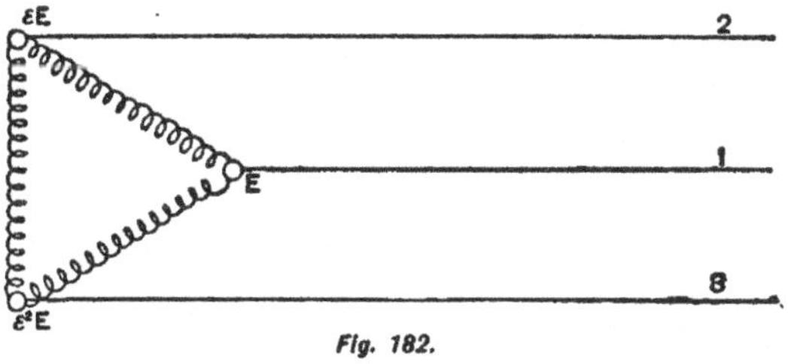

Fig. 182.

The different branches of a polyphase system may be either independent from each other, that is, without any electrical interconnection, or they may be interlinked with each other. In the first case, the polyphase system is called an *independent system*, in the latter case an *interlinked system.*

The three-phase system with star-connected or ring-connected generator, as shown diagrammatically in Figs. 181 and 182, is an interlinked system.

The four-phase system as derived by connecting four equidistant points of a continuous-current armature with four collector rings, as shown diagrammatically in Fig. 183,

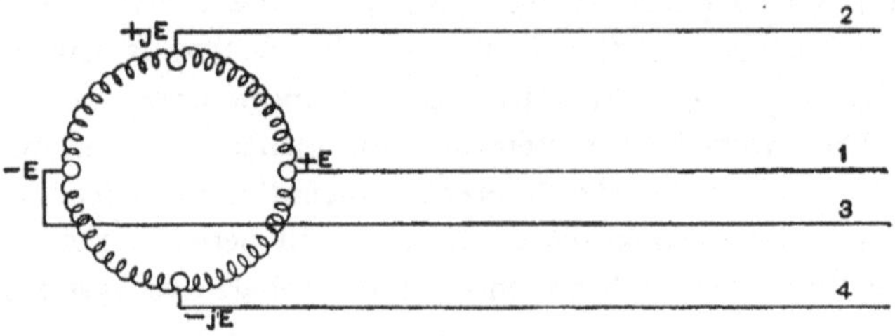

Fig. 183.

is an interlinked system also. The four-wire quarter-phase system produced by a generator with two independent armature coils, or by two single-phase generators rigidly connected with each other in quadrature, is an independent system. As interlinked system, it is shown in Fig. 184, as star-connected four-phase system.

Fig. 184.

262. Thus, polyphase systems can be subdivided into:

Symmetrical systems and unsymmetrical systems.

Balanced systems and unbalanced systems.

Interlinked systems and independent systems.

The only polyphase systems which have found practical application are:

The three-phase system, consisting of three E.M.Fs. displaced by one-third of a period, used exclusively as interlinked system.

The quarter-phase system, consisting of two E.M.Fs. in quadrature, and used with four wires, or with three wires, which may be either an interlinked system or an independent system.

The six-phase system, consisting of two three-phase systems in opposition to each other, and derived by transformation from a three-phase system, in the alternating supply circuit of large synchronous converters.

The inverted three-phase system, consisting of two E.M.F.'s displaced from each other by 60°, and derived from two phases of a three-phase system by transformation with two transformers, of which the secondary of one is reversed with regard to its primary (thus changing the phase difference from 120° to 180° − 120° = 60°), finds a limited application in low tension distribution.

CHAPTER XXVI.

SYMMETRICAL POLYPHASE SYSTEMS.

263. If all the E.M.Fs. of a polyphase system are equal in intensity, and differ from each other by the same angle of difference of phase, the system is called a symmetrical polyphase system.

Hence, a symmetrical n-phase system is a system of n E.M.Fs. of equal intensity, differing from each other in phase by $1/n$ of a period:

$$\begin{aligned} e_1 &= E \sin \beta; \\ e_2 &= E \sin \left(\beta - \frac{2\pi}{n}\right); \\ e_3 &= E \sin \left(\beta - \frac{4\pi}{n}\right); \\ &\vdots \\ e_n &= E \sin \left(\beta - \frac{2(n-1)\pi}{n}\right). \end{aligned}$$

The next E.M.F. is again:

$$e_1 = E \sin (\beta - 2\pi) = E \sin \beta.$$

In the polar diagram the n E.M.Fs. of the symmetrical n-phase system are represented by n equal vectors, following each other under equal angles.

Since in symbolic writing, rotation by $1/n$ of a period, or angle $2\pi/n$, is represented by multiplication with:

$$\cos \frac{2\pi}{n} + j \sin \frac{2\pi}{n} = \epsilon,$$

the E.M.Fs. of the symmetrical polyphase system are:

$$\dot{E};$$

$$\dot{E}\left(\cos\frac{2\pi}{n}+j\sin\frac{2\pi}{n}\right)=\dot{E}\,\epsilon;$$

$$\dot{E}\left(\cos\frac{4\pi}{n}+j\sin\frac{4\pi}{n}\right)=\dot{E}\,\epsilon^2;$$

$$\dot{E}\left(\cos\frac{2(n-1)\pi}{n}+j\sin\frac{2(n-1)\pi}{n}\right)=\dot{E}\,\epsilon^{n-1}.$$

The next E.M.F. is again:

$$\dot{E}\,(\cos 2\pi+j\sin 2\pi)=\dot{E}\,\epsilon^n=\dot{E}.$$

Hence, it is

$$\epsilon=\cos\frac{2\pi}{n}+j\sin\frac{2\pi}{n}=\sqrt[n]{1}.$$

Or in other words:

In a symmetrical n-phase system any E.M.F. of the system is expressed by:

$$\epsilon^i\,\dot{E};$$

where:

$$\epsilon=\sqrt[n]{1}.$$

264. Substituting now for n different values, we get the different symmetrical polyphase systems, represented by

$$\epsilon^i\,\dot{E},$$

where,

$$\epsilon=\sqrt[n]{1}=\cos\frac{2\pi}{n}+j\sin\frac{2\pi}{n}.$$

1.) $n=1 \quad \epsilon=1 \quad \epsilon^i\dot{E}=\dot{E},$

the ordinary single-phase system.

2.) $n=2 \quad \epsilon=-1 \quad \epsilon^i\dot{E}=\dot{E}$ and $-\dot{E}$.

Since $-\dot{E}$ is the return of $\dot{E}$, $n=2$ gives again the single-phase system.

3.) $n=3 \quad \epsilon=\cos\frac{2\pi}{3}+j\sin\frac{2\pi}{3}=\frac{-1+j\sqrt{3}}{2}$

$$\epsilon^2=\frac{-1-j\sqrt{3}}{2}$$

The three E.M.Fs. of the three-phase system are:

$$\epsilon^i \dot{E} = \dot{E}, \quad \frac{-1 + j\sqrt{3}}{2}\dot{E}, \quad \frac{-1 - j\sqrt{3}}{2}\dot{E}.$$

Consequently the three-phase system is the lowest symmetrical polyphase system.

4.) $n = 4, \ \epsilon = \cos\frac{2\pi}{4} + j\sin\frac{2\pi}{4} = j, \ \epsilon^2 = -1, \ \epsilon^3 = -j.$

The four E.M.Fs. of the four-phase system are:

$$\epsilon^i \dot{E} = \dot{E}, \quad j\dot{E}, \quad -\dot{E}, \quad -j\dot{E}.$$

They are in pairs opposite to each other:

$$\dot{E} \text{ and } -\dot{E}\,; \ j\dot{E} \text{ and } -j\dot{E}.$$

Hence can be produced by two coils in quadrature with each other, analogous as the two-phase system, or ordinary alternating-current system, can be produced by one coil.

Thus the symmetrical quarter-phase system is a four-phase system.

Higher systems, than the quarter-phase or four-phase system, have not been very extensively used, and are thus of less practical interest. A symmetrical six-phase system, derived by transformation from a three-phase system, has found application in synchronous converters, as offering a higher output from these machines, and a symmetrical eight phase system proposed for the same purpose.

265. A characteristic feature of the symmetrical n phase system is that under certain conditions it can produce a M.M.F. of constant intensity.

If n equal magnetizing coils act upon a point under equal angular displacements in space, and are excited by the n E.M.Fs. of a symmetrical n-phase system, a M.M.F. of constant intensity is produced at this point, whose direction revolves synchronously with uniform velocity.

Let,

n' = number of turns of each magnetizing coil.

E = effective value of impressed E.M.F.
I = effective value of current.

Hence,

$\mathfrak{F} = n'I$ = effective M.M.F. of one of the magnetizing coils.

Then the instantaneous value of the M.M.F. of the coil acting in the direction $2\pi i/n$ is:

$$f_i = \mathfrak{F}\sqrt{2}\sin\left(\beta - \frac{2\pi i}{n}\right)$$
$$= n' I\sqrt{2}\sin\left(\beta - \frac{2\pi i}{n}\right).$$

The two rectangular space components of this M.M.F. are;

$$f_i' = f_i \cos\frac{2\pi i}{n}$$
$$= n' I\sqrt{2}\cos\frac{2\pi i}{n}\sin\left(\beta - \frac{2\pi i}{n}\right)$$

and
$$f_i'' = f_i \sin\frac{2\pi i}{n}$$
$$= n' I\sqrt{2}\sin\frac{2\pi i}{n}\sin\left(\beta - \frac{2\pi i}{n}\right).$$

Hence the M.M.F. of this coil can be expressed by the symbolic formula:

$$f_i = n' I\sqrt{2}\sin\left(\beta - \frac{2\pi i}{n}\right)\left(\cos\frac{2\pi i}{n} + j\sin\frac{2\pi i}{n}\right).$$

Thus the total or resultant M.M.F. of the n coils displaced under the n equal angles is:

$$f = \sum_1^n {}_i f_i = n' I\sqrt{2}\sum_1^n {}_i \sin\left(\beta - \frac{2\pi i}{n}\right)\left(\cos\frac{2\pi i}{n} + j\sin\frac{2\pi i}{n}\right)$$

or, expanded:

$$f = n' I\sqrt{2}\left\{\sin\beta\sum_1^n {}_i\left(\cos^2\frac{2\pi i}{n} + j\sin\frac{2\pi i}{n}\cos\frac{2\pi i}{n}\right) - \cos\beta\sum_1^n {}_i\left(\sin\frac{2\pi i}{n}\cos\frac{2\pi i}{n} + j\sin^2\frac{2\pi i}{n}\right)\right\}.$$

It is, however:

$$\cos^2\frac{2\pi i}{n} + j\sin\frac{2\pi i}{n}\cos\frac{2\pi i}{n} = \tfrac{1}{2}\left(1 + \cos\frac{4\pi i}{n} + j\sin\frac{4\pi i}{n}\right)$$
$$= \tfrac{1}{2}(1 + \epsilon^{2i})$$

$$\sin\frac{2\pi i}{n}\cos\frac{2\pi i}{n} + j\sin^2\frac{2\pi i}{n} = \frac{j}{2}\left(1 - \cos\frac{4\pi i}{n} - j\sin\frac{4\pi i}{n}\right)$$
$$= \frac{j}{2}(1 - \epsilon^{2i}),$$

and, since:

$$\sum_1^n{}_i\, \epsilon^{2i} = 0, \qquad \sum_1^n{}_i\, \epsilon^{-2i} = 0,$$

it is, $$f = \frac{n n' I \sqrt{2}}{2}(\sin\beta - j\cos\beta).$$

or, $$f = \frac{n n' I}{\sqrt{2}}(\sin\beta - j\cos\beta)$$
$$= \frac{n\mathfrak{F}}{\sqrt{2}}(\sin\beta - j\cos\beta);$$

the symbolic expression of the M.M.F. produced by the n circuits of the symmetrical n-phase system, when exciting n equal magnetizing coils displaced in space under equal angles.

The absolute value of this M.M.F. is:

$$F = \frac{n n' I}{\sqrt{2}} = \frac{n\mathfrak{F}}{\sqrt{2}} = \frac{n\mathfrak{F}_{max}}{2}.$$

Hence constant and equal $n/\sqrt{2}$ times the effective M.M.F. of each coil or $n/2$ times the maximum M.M.F. of each coil.

The phase of the resultant M.M.F. at the time represented by the angle β is:

$$\tan\hat{\omega} = -\cot\beta; \text{ hence } \hat{\omega} = \beta - \frac{\pi}{2}$$

That is, the M.M.F. produced by a symmetrical n-phase system revolves with constant intensity:

$$F = \frac{n\mathfrak{F}}{\sqrt{2}};$$

and constant speed, in synchronism with the frequency of the system ; and, if the reluctance of the magnetic circuit is constant, the magnetism revolves with constant intensity and constant speed also, at the point acted upon symmetrically by the n M.M.Fs. of the n-phase system.

This is a characteristic feature of the symmetrical polyphase system.

266. In the three-phase system, $n = 3$, $F = 1.5\ \mathfrak{F}_{max}$ where $\mathfrak{F}_{max}$ is the maximum M.M.F. of each of the magnetizing coils.

In a symmetrical quarter-phase system, $n = 4$, $F = 2\ \mathfrak{F}_{max}$, where $\mathfrak{F}_{max}$ is the maximum M.M.F. of each of the four magnetizing coils, or, if only two coils are used, since the four-phase M.M.Fs. are opposite in phase by two, $F = \mathfrak{F}_{max}$, where $\mathfrak{F}_{max}$ is the maximum M.M.F. of each of the two magnetizing coils of the quarter-phase system.

While the quarter-phase system, consisting of two E.M.Fs. displaced by one-quarter of a period, is by its nature an unsymmetrical system, it shares a number of features — as, for instance, the ability of producing a constant resultant M.M.F. — with the symmetrical system, and may be considered as one-half of a symmetrical four-phase system.

Such systems, consisting of one-half of a symmetrical system, are called *hemisymmetrical systems.*

CHAPTER XXVII.

BALANCED AND UNBALANCED POLYPHASE SYSTEMS.

267. If an alternating E.M.F.:

$$e = E\sqrt{2}\sin\beta,$$

produces a current:

$$i = I\sqrt{2}\sin(\beta - \hat{\omega}),$$

where $\hat{\omega}$ is the angle of lag, the power is:

$$\begin{aligned} p = ei &= 2\,EI\sin\beta\sin(\beta - \hat{\omega}) \\ &= EI\left(\cos\hat{\omega} - \cos(2\,\beta - \hat{\omega})\right), \end{aligned}$$

and the average value of power:

$$P = EI\cos\hat{\omega}.$$

Substituting this, the instantaneous value of power is found as:

$$p = P\left(1 - \frac{\cos(2\,\beta - \hat{\omega})}{\cos\hat{\omega}}\right).$$

Hence the power, or the flow of energy, in an ordinary single-phase alternating-current circuit is fluctuating, and varies with twice the frequency of E.M.F. and current, unlike the power of a continuous-current circuit, which is constant:

$$p = ei.$$

If the angle of lag $\hat{\omega} = 0$ it is:

$$p = P(1 - \cos 2\,\beta);$$

hence the flow of power varies between zero and $2\,P$, where P is the average flow of energy or the effective power of the circuit.

If the current lags or leads the E.M.F. by angle $\hat{\omega}$ the power varies between

$$P\left(1 - \frac{1}{\cos \hat{\omega}}\right) \quad \text{and} \quad P\left(1 + \frac{1}{\cos \hat{\omega}}\right);$$

that is, becomes negative for a certain part of each half-wave. That is, for a time during each half-wave, energy flows back into the generator, while during the other part of the half-wave the generator sends out energy, and the difference between both is the effective power of the circuit.

If $\hat{\omega} = 90°$, it is:

$$p = - EI \sin 2\beta;$$

that is, the effective power: $P = 0$, and the energy flows to and fro between generator and receiving circuit.

Under any circumstances, however, the flow of energy in the single-phase system is fluctuating at least between zero and a maximum value, frequently even reversing.

268. If in a polyphase system

$e_1, e_2, e_3, \ldots . =$ instantaneous values of E.M.F.;
$i_1, i_2, i_3, \ldots . =$ instantaneous values of current produced thereby;

the total flow of power in the system is:

$$p = e_1 i_1 + e_2 i_2 + e_3 i_3 + \ldots$$

The average flow of power is:

$$P = E_1 I_1 \cos \hat{\omega}_1 + E_2 I_2 \cos \hat{\omega}_2 + \ldots .$$

The polyphase system is called a balanced system, if the flow of energy:

$$p = e_1 i_1 + e_2 i_2 + e_3 i_3 + \ldots .$$

is constant, and it is called an unbalanced system if the flow of energy varies periodically, as in the single-phase system; and the ratio of the minimum value to the maximum value of power is called the *balance factor of the system.*

Hence in a single-phase system on non-inductive circuit, that is, at no-phase displacement, the balance factor is zero; and it is negative in a single-phase system with lagging or leading current, and becomes $= -1$, if the phase displacement is 90° — that is, the circuit is wattless.

269. Obviously, in a polyphase system the balance of the system is a function of the distribution of load between the different branch circuits.

A balanced system in particular is called a polyphase system, whose flow of energy is constant, if all the circuits are loaded equally with a load of the same character, that is, the same phase displacement.

270. All the symmetrical systems from the three-phase system upward are balanced systems. Many unsymmetrical systems are balanced systems also.

1.) Three-phase system:

Let

$$e_1 = E\sqrt{2}\sin\beta, \qquad \text{and} \qquad i_1 = I\sqrt{2}\sin(\beta - \hat{\omega});$$
$$e_2 = E\sqrt{2}\sin(\beta - 120), \qquad i_2 = I\sqrt{2}\sin(\beta - \hat{\omega} - 120);$$
$$e_3 = E\sqrt{2}\sin(\beta - 240), \qquad i_3 = I\sqrt{2}\sin(\beta - \hat{\omega} - 240);$$

be the E.M.Fs. of a three-phase system, and the currents produced thereby.

Then the total flow of power is:

$$p = 2\,EI\big(\sin\beta\sin(\beta - \hat{\omega}) + \sin(\beta - 120)\sin(\beta - \hat{\omega} - 120) + \sin(\beta - 240)\sin(\beta - \hat{\omega} - 240)\big)$$
$$= 3\,EI\cos\hat{\omega} = P, \text{ or constant.}$$

Hence the symmetrical three-phase system is a balanced system.

2.) Quarter-phase system:

Let $e_1 = E\sqrt{2}\sin\beta, \qquad i_1 = I\sqrt{2}\sin(\beta - \hat{\omega});$
$e_2 = E\sqrt{2}\cos\beta, \qquad i_2 = I\sqrt{2}\cos(\beta - \hat{\omega});$

be the E.M.Fs. of the quarter-phase system, and the currents produced thereby.

This is an unsymmetrical system, but the instantaneous flow of power is:

$$p = 2\,EI(\sin\beta \sin(\beta - \hat{\omega}) + \cos\beta \cos(\beta - \hat{\omega}))$$
$$= 2\,EI \cos\hat{\omega} = P, \text{ or constant.}$$

Hence the quarter-phase system is an unsymmetrical balanced system.

3.) The symmetrical n-phase system, with equal load and equal phase displacement in all n branches, is a balanced system. For, let:

$$e_i = E\sqrt{2}\sin\left(\beta - \frac{2\pi i}{n}\right) = \text{E.M.F.};$$

$$i_i = I\sqrt{2}\sin\left(\beta - \hat{\omega} - \frac{2\pi i}{n}\right) = \text{current}$$

the instantaneous flow of power is:

$$p = \sum_1^n{}_i\, e_i i_i$$
$$= 2\,EI \sum_1^n{}_i \sin\left(\beta - \frac{2\pi i}{n}\right)\sin\left(\beta - \hat{\omega} - \frac{2\pi i}{n}\right)$$
$$= EI\left\{\sum_1^n{}_i \cos\hat{\omega} - \sum_1^n{}_i \cos\left(2\beta - \hat{\omega} - \frac{4\pi i}{n}\right)\right\};$$

or $$p = n\,EI\cos\hat{\omega} = P, \text{ or constant.}$$

271. An unbalanced polyphase system is the so-called inverted three-phase system,* derived from two branches of a three-phase system by transformation by means of two transformers, whose secondaries are connected in opposite direction with respect to their primaries. Such a system takes an intermediate position between the Edison three-wire system and the three-phase system. It shares with the latter the polyphase feature, and with the Edison three-

* Also called "polyphase monocyclic system," since the E.M.F. triangle is similar to that usual in the single-phase monocyclic system.

wire system the feature that the potential difference between the outside wires is higher than between middle wire and outside wire.

By such a pair of transformers the two primary E.M.Fs. of 120° displacement of phase are transformed into two secondary E.M.Fs. differing from each other by 60°. Thus in the secondary circuit the difference of potential between the outside wires is $\sqrt{3}$ times the difference of potential between middle wire and outside wire. At equal load on the two branches, the three currents are equal, and differ from each other by 120°, that is, have the same relative proportion as in a three-phase system. If the load on one branch is maintained constant, while the load of the other branch is reduced from equality with that in the first branch down to zero, the current in the middle wire first decreases, reaches a minimum value of 87 per cent of its original value, and then increases again, reaching at no load the same value as at full load.

The balance factor of the inverted three-phase system on non-inductive load is .333.

272. In Figs. 185 to 192 are shown the E.M.Fs. as e and currents as i in drawn lines, and the power as p in dotted lines, for :

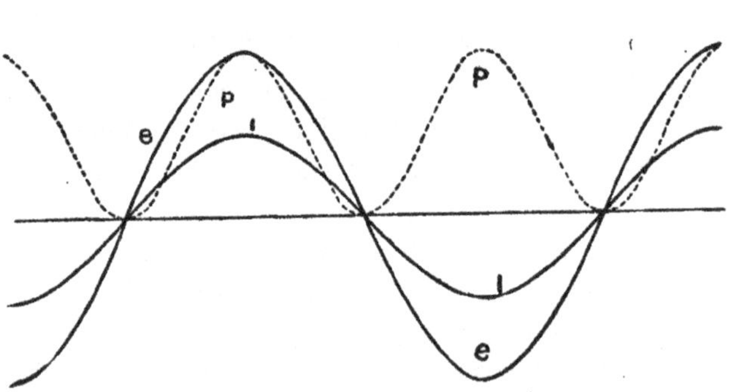

Fig. 185. Single-phase System on Non-inductive Load.
Balance Factor, 0

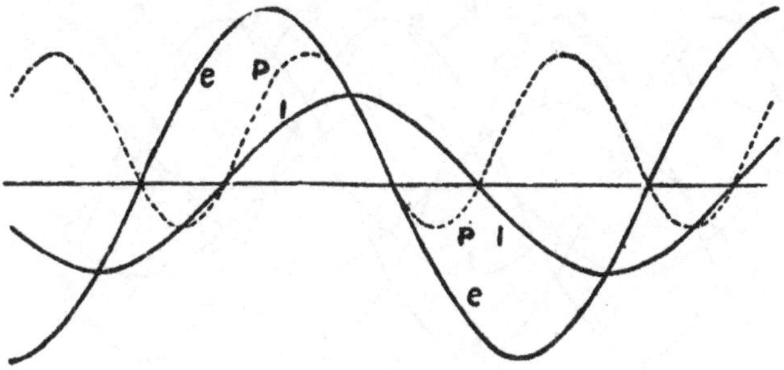

Fig. 186. Single-phase System on Inductive Load of 60° Lag.
Balance Factor, – .333.

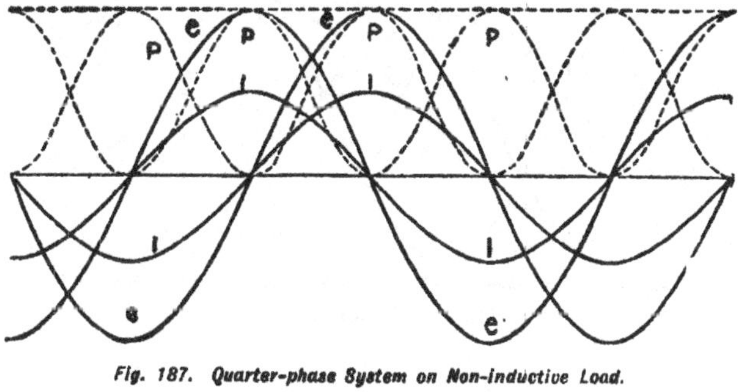

Fig. 187. Quarter-phase System on Non-inductive Load.
Balance Factor, + 1.

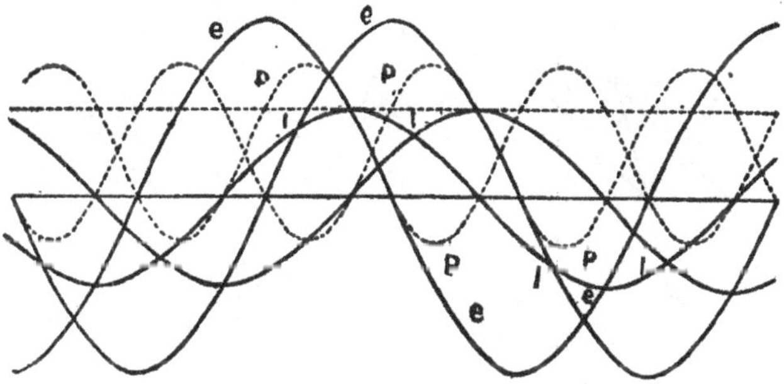

Fig. 188. Quarter-phase System on Inductive Lo d of 60° Lag.
Balance Factor, + 1

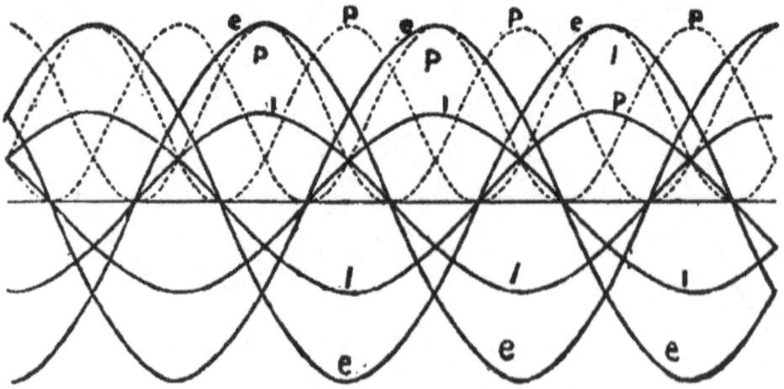

Fig. 189. Three-phase System on Non-inductive Load.

Balance Factor, + 1.

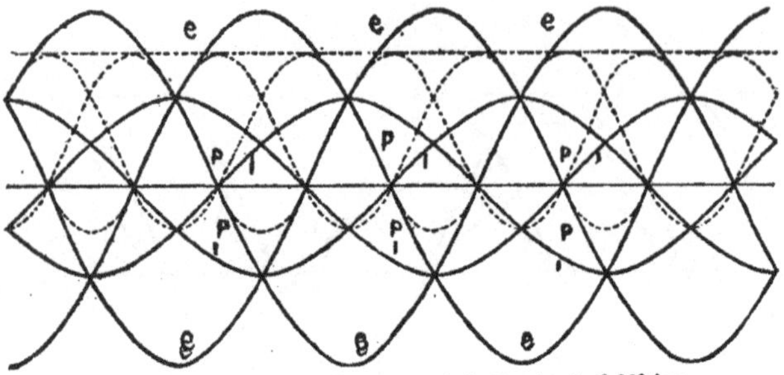

Fig. 190. Three-phase System on Inductive Load of 60° Lag.

Balance Factor, + 1.

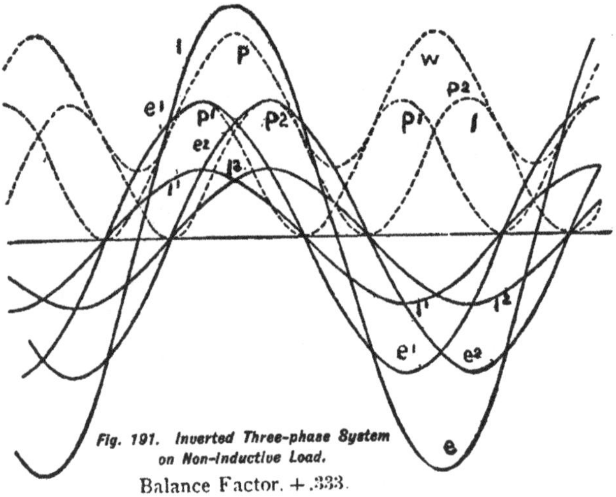

Fig. 191. Inverted Three-phase System on Non-inductive Load.

Balance Factor, + .333.

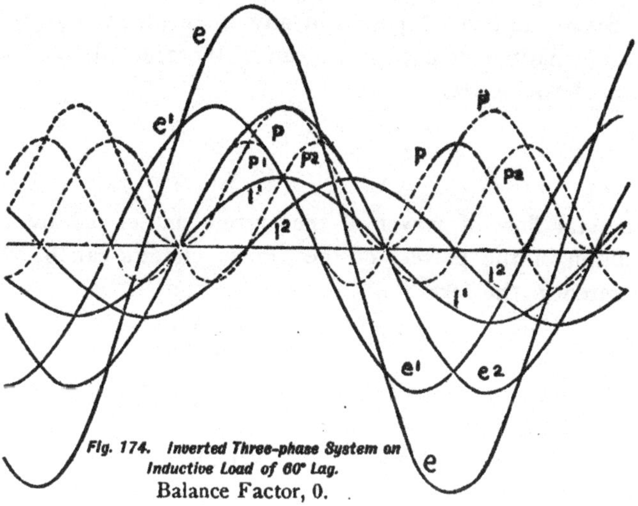

Fig. 174. Inverted Three-phase System on Inductive Load of 60° Lag.
Balance Factor, 0.

273. The flow of power in an alternating-current system is a most important and characteristic feature of the system, and by its nature the systems may be classified into:

Monocyclic systems, or systems with a balance factor zero or negative.

Polycyclic systems, with a positive balance factor.

Balance factor -1 corresponds to a wattless circuit, balance factor zero to a non-inductive single-phase circuit, balance factor $+1$ to a balanced polyphase system.

274. In polar coördinates, the flow of power of an alternating-current system is represented by using the instantaneous flow of power as radius vector, with the angle β corresponding to the time as amplitude, one complete period being represented by one revolution.

In this way the power of an alternating-current system is represented by a closed symmetrical curve, having the zero point as quadruple point. In the monocyclic systems the zero point is quadruple nodal point; in the polycyclic system quadruple isolated point.

Thus these curves are sextics.

Since the flow of power in any single-phase branch of the alternating-current system can be represented by a sine wave of double frequency:

$$p = P\left(1 - \frac{\sin(2\beta - \hat{\omega})}{\cos\hat{\omega}}\right).$$

the total flow of power of the system as derived by the addition of the powers of the branch circuits can be represented in the form:

$$p = P(1 + \epsilon \sin(2\beta - \hat{\omega}_0))$$

This is a wave of double frequency also, with ϵ as amplitude of fluctuation of power.

This is the equation of the power characteristics of the system in polar coördinates.

275. To derive the equation in rectangular coördinates we introduce a substitution which revolves the system of coördinates by an angle $\hat{\omega}_0/2$, so as to make the symmetry axes of the power characteristic the coördinate axes.

$$p = \sqrt{x^2 + y^2},$$

$$\tan\left(\beta - \frac{\hat{\omega}_0}{2}\right) = \frac{y}{x}.$$

hence, $\sin(2\beta - \hat{\omega}_0) = 2\sin\left(\beta - \frac{\hat{\omega}_0}{2}\right)\cos\left(\beta - \frac{\hat{\omega}_0}{2}\right) = \frac{2xy}{x^2 + y^2},$

substituted,

$$\sqrt{x^2 + y^2} = P\left\{1 + \frac{2\epsilon xy}{x^2 + y^2}\right\},$$

or, expanded:

$$(x^2 + y^2)^3 - P^2(x^2 + y^2 + 2\epsilon xy)^2 = 0,$$

the sextic equation of the power characteristic.

Introducing:

$$a = (1 + \epsilon)P = \text{maximum value of power},$$
$$b = (1 - \epsilon)P = \text{minimum value of power};$$

it is
$$P = \frac{a+b}{2},$$
$$\epsilon = \frac{a-b}{a+b};$$

hence, substituted, and expanded:

$$(x^2 + y^2)^3 - \tfrac{1}{4}\{a(x+y)^2 + b(x-y)^2\}^2, = 0$$

the equation of the power characteristic, with the ***main power axes*** a and b, and the balance factor: b/a.

It is thus:

Single-phase non-inductive circuit: $p = P(1 + \sin 2\phi)$, $b = 0$, $a = 2P$

$$(x^2 + y^2)^3 - P^2(x+y)^4 = 0, \quad b/a = 0.$$

Single-phase circuit, 60° lag: $p = P(1 + 2\sin 2\phi)$, $b = -P$, $a = +3P$

$$(x^2 + y^2)^3 - P^2(x^2 + y^2 + 4xy)^2 = 0, \quad b/a = -\tfrac{1}{3}.$$

Single-phase circuit, 90° lag: $p = EI\sin 2\phi$, $b = -EI$, $a = +EI$

$$(x^2 + y^2)^3 - 4(EI)^2 x^2 y^2, \; b/a = -1.$$

Three-phase non-inductive circuit: $p = P$, $b = 1$, $a = 1$

$$x^2 + y^2 - P^2 = 0: \text{circle.} \quad b/a = +1.$$

Three-phase circuit, 60° lag: $p = P$, $b = 1$, $a = 1$

$$x^2 + y^2 - P^2 = 0: \text{circle.} \quad b/a = +1.$$

Quarter-phase non-inductive circuit: $p = P$, $b = 1$, $a = 1$

$$x^2 + y^2 - P^2 = 0: \text{circle.} \quad b/a = +1.$$

Quarter-phase circuit, 60° lag: $p = P$, $b = 1$, $a = 1$

$$x^2 + y^2 - P^2 = 0: \text{circle.} \quad b/a = +1.$$

Inverted three-phase non-inductive circuit:

$$p = P\left(1 + \frac{\sin 2\phi}{2}\right), \quad b = \tfrac{1}{2} P, \quad a = \tfrac{3}{2} P$$

$$(x^2 + y^2)^3 - P^2 (x^2 + y^2 + xy)^2 = 0. \quad b / a = + \tfrac{1}{3}$$

Inverted three-phase circuit 60° lag : $p = P(1 + \sin 2\phi)$, $b = 0$, $a = 2P$

$$(x^2 + y^2)^3 - P^2 (x + y)^4 = 0. \quad b / a = 0.$$

a and b are called the main power axes of the alternating current system, and the ratio b / a is the balance factor of the system.

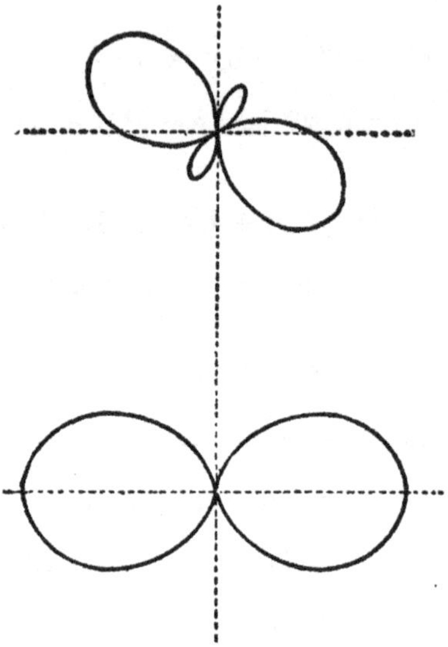

Figs. 193 and 194. Power Characteristic of Single-phase System, at 60° and 0° Lag.

276. As seen, the flow of power of an alternating-current system is completely characterized by its two main power axes a and b.

The power characteristics in polar coördinates, corresponding to the Figs. 185, 186, 191, and 192 are shown in Figs. 193, 194, 195, and 196.

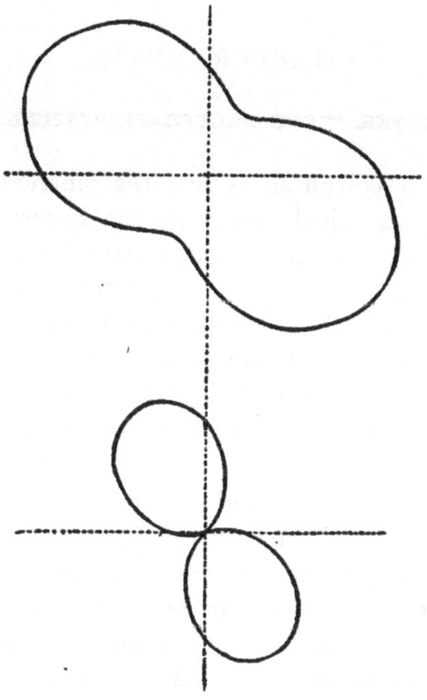

Figs. 195 and 196. Power Characteristic of Inverted Three-phase System, at 0° and 60° Lag.

The balanced quarter-phase and three-phase systems give as polar characteristics concentric circles.

CHAPTER XXVIII.

INTERLINKED POLYPHASE SYSTEMS.

277. In a polyphase system the different circuits of displaced phases, which constitute the system, may either be entirely separate and without electrical connection with each other, or they may be connected with each other electrically, so that a part of the electrical conductors are in common to the different phases, and in this case the system is called an interlinked polyphase system.

Thus, for instance, the quarter-phase system will be called an independent system if the two E.M.Fs. in quadrature with each other are produced by two entirely separate coils of the same, or different but rigidly connected, armatures, and are connected to four wires which energize independent circuits in motors or other receiving devices. If the quarter-phase system is derived by connecting four equidistant points of a closed-circuit drum or ring-wound armature to the four collector rings, the system is an interlinked quarter-phase system.

Similarly in a three-phase system. Since each of the three currents which differ from each other by one-third of a period is equal to the resultant of the other two currents, it can be considered as the return circuit of the other two currents, and an interlinked three-phase system thus consists of three wires conveying currents differing by one-third of a period from each other, so that each of the three currents is a common return of the other two, and inversely.

278. In an interlinked polyphase system two ways exist of connecting apparatus into the system.

1st. The *star connection*, represented diagrammatically in Fig. 197. In this connection the n circuits excited by currents differing from each other by $1/n$ of a period, are connected with their one end together into a neutral point or common connection, which may either be grounded or connected with other corresponding neutral points, or insulated.

In a three-phase system this connection is usually called a Y connection, from a similarity of its diagrammatical representation with the letter Y, as shown in Fig. 181.

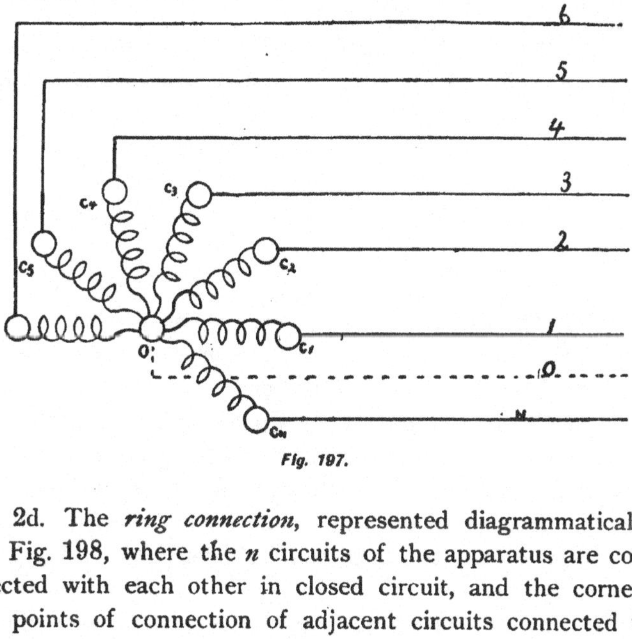

Fig. 197.

2d. The *ring connection*, represented diagrammatically in Fig. 198, where the n circuits of the apparatus are connected with each other in closed circuit, and the corners or points of connection of adjacent circuits connected to the n lines of the polyphase system. In a three-phase system this connection is called the delta connection, from the similarity of its diagrammatic representation with the Greek letter Delta, as shown in Fig. 182.

In consequence hereof we distinguish between star-connected and ring-connected generators, motors, etc., or

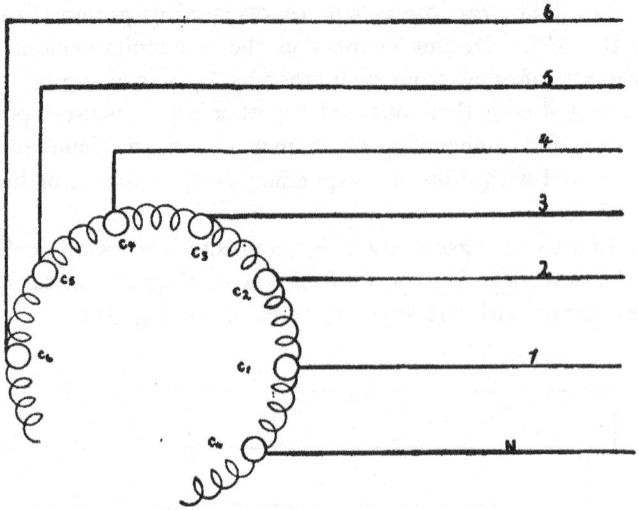

Fig. 198.

in three-phase systems Y-connected and delta-connected apparatus.

279. Obviously, the polyphase system as a whole does not differ, whether star connection or ring connection is used in the generators or other apparatus; and the transmission line of a symmetrical *n*-phase system always consists of *n* wires carrying current of equal strength, when balanced, differing from each other in phase by $1/n$ of a period. Since the line wires radiate from the *n* terminals of the generator, the lines can be considered as being in star connection.

The circuits of all the apparatus, generators, motors, etc., can either be connected in star connection, that is, between one line and a neutral point, or in ring connection, that is, between two adjacent lines.

In general some of the apparatus will be arranged in star connection, some in ring connection, as the occasion may require.

280. In the same way as we speak of star connection and ring connection of the circuits of the apparatus, the term star potential and ring potential, star current and ring current, etc., are used, whereby as star potential or in a three-phase circuit Y potential, the potential difference between one of the lines and the neutral point, that is, a point having the same difference of potential against all the lines, is understood; that is, the potential as measured by a voltmeter connected into star or Y connection. By ring or delta potential is understood the difference of potential between adjacent lines, as measured by a voltmeter connected between adjacent lines, in ring or delta connection.

In the same way the star or Y current is the current flowing from one line to a neutral point; the ring or delta current, the current flowing from one line to the other.

The current in the transmission line is always the star or Y current, and the potential difference between the line wires, the ring or delta potential.

Since the star potential and the ring potential differ from each other, apparatus requiring different voltages can be connected into the same polyphase mains, by using either star or ring connection.

281. If in a generator with star-connected circuits, the E.M.F. per circuit $= E$, and the common connection or neutral point is denoted by zero, the potentials of the n terminals are:

$$\underset{\cdot}{E},\ \epsilon \underset{\cdot}{E},\ \epsilon^2 \underset{\cdot}{E} \ \ldots \ \epsilon^{n-1} \underset{\cdot}{E};$$

or in general: $\epsilon^i \underset{\cdot}{E}$,

at the ith terminal, where:

$$i = 0, 1, 2 \ldots n - 1, \quad \epsilon = \cos\frac{2\pi}{n} + j\sin\frac{2\pi}{n} = \sqrt[n]{1}.$$

Hence the E.M.F. in the circuit from the ith to the kth terminal is:

$$E_{ki} = \epsilon^k E - \epsilon^i E = (\epsilon^k - \epsilon^i) E.$$

The E.M.F. between adjacent terminals i and $i + 1$ is:

$$(\epsilon^{i+1} - \epsilon^i) E = \epsilon^i (\epsilon - 1) E.$$

In a generator with ring-connected circuits, the E.M.F. per circuit:

$$\epsilon^i E$$

is the ring E.M.F., and takes the place of

$$\epsilon^i (\epsilon - 1) E;$$

while the E.M.F. between terminal and neutral point, or the star E.M.F., is:

$$\frac{\epsilon^i}{\epsilon - 1} E.$$

Hence in a star-connected generator with the E.M.F. E per circuit, it is:

Star E.M.F., $\epsilon^i E$.
Ring E.M.F., $\epsilon^i (\epsilon - 1) E$.
E.M.F. between terminal i and terminal k, $(\epsilon^k - \epsilon^i) E$.

In a ring-connected generator with the E.M.F. E per circuit, it is:

Star E.M.F., $\frac{\epsilon^i}{\epsilon - 1} E$.
Ring E.M.F., $\epsilon^i E$.
E.M.F. between terminals i and k, $\frac{\epsilon^k - \epsilon^i}{\epsilon - 1} E$.

In a star-connected apparatus, the E.M.F. and the current per circuit have to be the star E.M.F. and the star current. In a ring-connected apparatus the E.M.F. and current per circuit have to be the ring E.M.F. and ring current.

In the generator of a symmetrical polyphase system, if:

$\epsilon^i E$ are the E.M.Fs. between the n terminals and the neutral point, or star E.M.Fs.,

$\dot{I}_i$ = the currents issuing from terminal i over a line of the impedance Z_i (including generator impedance in star connection), we have:

Potential at end of line i:

$$\epsilon^i \dot{E} - Z_i \dot{I}_i.$$

Difference of potential between terminals k and i:

$$(\epsilon^k - \epsilon^i)\dot{E} - (Z_k \dot{I}_k - Z_i \dot{I}_i),$$

where $\dot{I}_i$ is the star current of the system, Z_i the star impedance.

The ring potential at the end of the line between terminals i and k is $\dot{E}_{ik}$, and it is:

$$\dot{E}_{ik} = -\dot{E}_{ki}.$$

If now $\dot{I}_{ik}$ denotes the current passing from terminal i to terminal k, and Z_{ik} impedance of the circuit between terminal i and terminal k, where:

$$\dot{I}_{ik} = -\dot{I}_{ki},$$
$$Z_{ik} = Z_{ki},$$

it is

$$\dot{E}_{ik} = Z_{ik}\dot{I}_{ik}.$$

If $\dot{I}_{io}$ denotes the current passing from terminal i to a ground or neutral point, and Z_{io} is the impedance of this circuit between terminal i and neutral point, it is:

$$\dot{E}_{io} = \epsilon^i \dot{E} - Z_i \dot{I}_i = Z_{io}\dot{I}_{io}.$$

282. We have thus, by Ohm's law and Kirchhoff's law:

If $\epsilon^i \dot{E}$ is the E.M.F. per circuit of the generator, between the terminal i and the neutral point of the generator, or the star E.M.F.

$\dot{I}_i$ = the current issuing from the terminal i of the generator, or the star current.

Z_i = the impedance of the line connected to a terminal i of the generator, including generator impedance.

$\dot{E}_i$ = the E.M.F. at the end of line connected to a terminal i of the generator.

$\dot{E}_{ik}$ = the difference of potential between the ends of the lines i and k.

$\dot{I}_{ik}$ = the current passing from line i to line k.

Z_{ik} = the impedance of the circuit between lines i and k.

$\dot{I}_{io}, \dot{I}_{ioo} \dots$ = the current passing from line i to neutral points 0, 00, ...

$Z_{io}, Z_{ioo} \dots$ = the impedance of the circuits between line i and neutral points 0, 00, ...

It is then:

1.) $\dot{E}_{ik} = -\dot{E}_{ki}, \quad \dot{I}_{ik} = -\dot{I}_{ki}, \quad Z_{ik} = Z_{ki}, \quad \dot{I}_{io} = -\dot{I}_{oi},$
$Z_{io} = Z_{oi}$, etc.

2.) $\dot{E}_i = \epsilon^i \dot{E} - Z_i \dot{I}_i.$

3.) $\dot{E}_i = Z_{io}\dot{I}_{io} = Z_{ioo}\dot{I}_{ioo} = \dots$

4.) $\dot{E}_{ik} = \dot{E}_k - \dot{E}_i = (\epsilon^k - \epsilon^i)\dot{E} - (Z_k\dot{I}_k - Z_i\dot{I}_i).$

5.) $\dot{E}_{ik} = Z_{ik}\dot{I}_{ik}.$

6.) $\dot{I}_i = \sum_0^n {}^k \dot{I}_{ik}.$

7.) If the neutral point of the generator does not exist, as in ring connection, or is insulated from the other neutral points:

$$\sum_1^n {}^i \dot{I}_i = 0;$$

$$\sum_1^n {}^i \dot{I}_{io} = 0;$$

$$\sum_1^n {}^i \dot{I}_{ioo} = 0, \text{ etc.}$$

Where 0, 00, etc., are the different neutral points which are insulated from each other.

If the neutral point of the generator and all the other neutral points are grounded or connected with each other, it is:

$$\sum_1^n {}^i \dot{I}_i = \sum_1^n {}^i (\dot{I}_{io} + \dot{I}_{ioo} + \dots)$$

$$= \sum_1^n {}^i \dot{I}_{io} + \sum_1^n {}^i \dot{I}_{ioo} + \dots$$

If the neutral point of the generator and all other neutral points are grounded, the system is called a grounded system. If the neutral points are not grounded, the system is an insulated polyphase system, and an insulated polyphase system with equalizing return, if all the neutral points are connected with each other:

8.) The power of the polyphase system is —

$$P = \sum_{1}^{n} {}_{i}\, \epsilon^{i} E I_{i} \cos \phi_{i} \text{ at the generator}$$

$$P = \sum_{0}^{n} {}_{i} \sum_{i}^{n} {}_{k}\, E_{ik} I_{ik} \cos \phi_{ik} \text{ in the receiving circuits.}$$

CHAPTER XXIX.

TRANSFORMATION OF POLYPHASE SYSTEMS.

283. In transforming a polyphase system into another polyphase system, it is obvious that the primary system must have the same flow of power as the secondary system, neglecting losses in transformation, and that consequently a balanced system will be transformed again in a balanced system, and an unbalanced system into an unbalanced system of the same balance factor, since the transformer is an apparatus not able to store energy, and thereby to change the nature of the flow of power. The energy stored as magnetism, amounts in a well-designed transformer only to a very small percentage of the total energy. This shows the futility of producing symmetrical balanced polyphase systems by transformation from the unbalanced single-phase system without additional apparatus able to store energy efficiently, as revolving machinery.

Since any E.M.F. can be resolved into, or produced by, two components of given directions, the E.M.Fs. of any polyphase system can be resolved into components or produced from components of two given directions. This enables the transformation of any polyphase system into any other polyphase system of the same balance factor by two transformers only.

284. Let E_1, E_2, E_3 . . . . be the E.M.Fs. of the primary system which shall be transformed into —

E_1', E_2', E_3' . . . . the E.M.Fs. of the secondary system.

Choosing two magnetic fluxes, $\bar{\phi}$ and $\bar{\bar{\phi}}$, of different phases, as magnetic circuits of the two transformers, which induce the E.M.Fs., $\bar{e}$ and $\bar{\bar{e}}$, per turn, by the law of parallelogram the E.M.Fs., E_1, E_2, ... can be dissolved into two components, $\bar{E}_1$ and $\bar{\bar{E}}_1$, $\bar{E}_2$ and $\bar{\bar{E}}_2$, ... of the phases, $\bar{e}$ and $\bar{\bar{e}}$.

Then, —

$\bar{E}_1$, $\bar{E}_2$, ... are the counter E.M.Fs. which have to be induced in the primary circuits of the first transformer;

$\bar{\bar{E}}_1$, $\bar{\bar{E}}_2$, ... the counter E.M.F.'s which have to be induced in the primary circuits of the second transformer.

hence

$\bar{E}_1 / \bar{e}$, $\bar{E}_2 / \bar{e}$ .... are the numbers of turns of the primary coils of the first transformer.

Analogously

$\bar{\bar{E}}_1 / \bar{\bar{e}}$ $\bar{\bar{E}}_2 / \bar{\bar{e}}$ .... are the number of turns of the primary coils in the second transformer.

In the same manner as the E.M.Fs. of the primary system have been resolved into components in phase with $\bar{e}$ and $\bar{\bar{e}}$, the E.M.Fs. of the secondary system, E_1^1, E_2^1, ... are produced from components, $\bar{E}_1^1$ and $\bar{\bar{E}}_1^1$, $\bar{E}_2^1$ and $\bar{\bar{E}}_2^1$ .... in phase with $\bar{e}$ and $\bar{\bar{e}}$, and give as numbers of secondary turns, —

$\bar{E}_1^1 / \bar{e}$, $\bar{E}_2^1 / \bar{e}$, ... in the first transformer;

$\bar{\bar{E}}_1^1 / \bar{\bar{e}}$, $\bar{\bar{E}}_2^1 / \bar{\bar{e}}$, ... in the second transformer.

That means each of the two transformers $\bar{m}$ and $\bar{\bar{m}}$ contains in general primary turns of each of the primary phases, and secondary turns of each of the secondary phases. Loading now the secondary polyphase system in any desired manner, corresponding to the secondary currents, primary currents will flow in such a manner that the total flow of power in the primary polyphase system is the same as the total flow of power in the secondary system, plus the loss of power in the transformers.

285. As an instance may be considered the transformation of the symmetrical balanced three-phase system

$$E \sin \beta, \quad E \sin (\beta - 120), \quad E \sin (\beta - 240),$$

in an unsymmetrical balanced quarter-phase system:

$$E' \sin \beta, \quad E' \sin (\beta - 90).$$

Let the magnetic flux of the two transformers be

$$\phi \cos \beta \text{ and } \phi \cos (\beta - 90).$$

Then the E.M.Fs. induced per turn in the transformers are

$$e \sin \beta \text{ and } e \sin (\beta - 90);$$

hence, in the primary circuit the first phase, $E \sin \beta$, will give, in the first transformer, E / e primary turns; in the second transformer, 0 primary turns.

The second phase, $E \sin (\beta - 120)$, will give, in the first transformer, $- E / 2\,e$ primary turns; in the second transformer, $\frac{E \times \sqrt{3}}{2\,e}$ primary turns.

The third phase, $E \sin (\beta - 240)$, will give, in the first transformer, $- E / 2\,e$ primary turns; in the second transformer, $\frac{- E \times \sqrt{3}}{2\,e}$ primary turns.

In the secondary circuit the first phase $E' \sin \beta$ will give in the first transformer: E' / e secondary turns; in the second transformer: 0 secondary turns.

The second phase: $E' \sin (\beta - 90)$ will give in the first transformer: 0 secondary turns; in the second transformer, E' / e secondary turns.

Or, if:

$$E = 5{,}000 \quad E' = 100, \quad e = 10.$$

	PRIMARY.			SECONDARY.		
	1st.	2d.	3d.	1st.	2d.	Phase.
first transformer	+ 500	− 250	− 250	10	0	
second transformer	0	+ 433	− 433	0	10	turns.

That means:

Any balanced polyphase system can be transformed by two transformers only, without storage of energy, into any other balanced polyphase system.

286. Some of the more common methods of transformation between polyphase systems are:

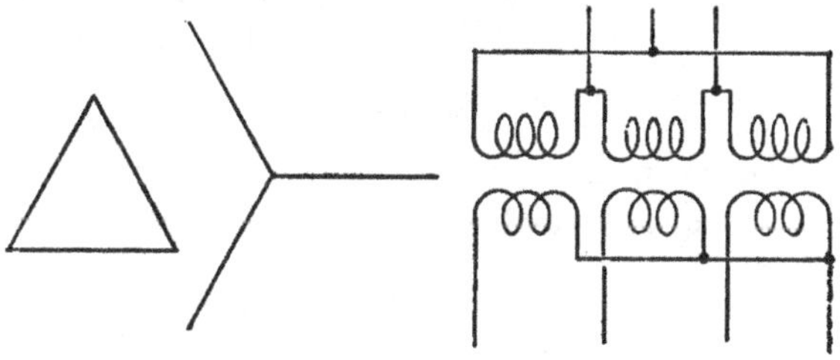

Fig. 199.

1. The delta-*Y* connection of transformers between three-phase systems, shown in Fig. 199. One side of the transformers is connected in delta, the other in *Y*. This arrangement becomes necessary for feeding four wires

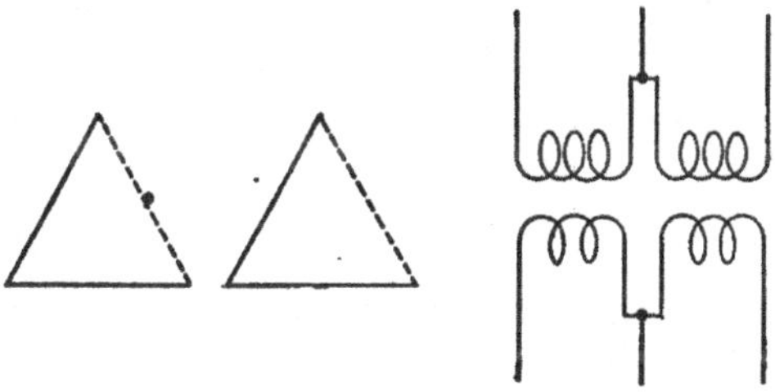

Fig. 200.

three-phase secondary distributions. The *Y* connection of the secondary allows to bring out a neutral wire, while the delta connection of the primary maintains the balance between the phases at unequal distribution of load.

2. The L connection of transformers between three-phase systems, consisting in using two sides of the triangle only, as shown in Fig. 200. This arrangement has the disadvantage of transforming one phase by two transformers in series, hence is less efficient, and is liable to unbalance the system by the internal impedance of the transformers.

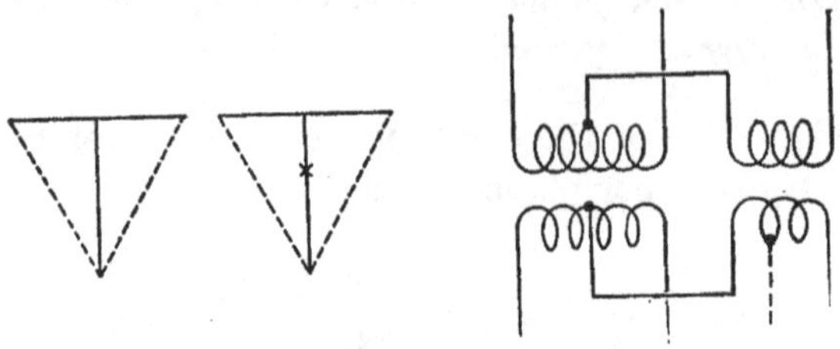

Fig. 201.

3. The main and teaser, or T connection of transformers between three-phase systems, as shown in Fig. 201. One of the two transformers is wound for $\frac{\sqrt{3}}{2}$ times the voltage of the other (the altitude of the equilateral triangle), and connected with one of its ends to the center of the

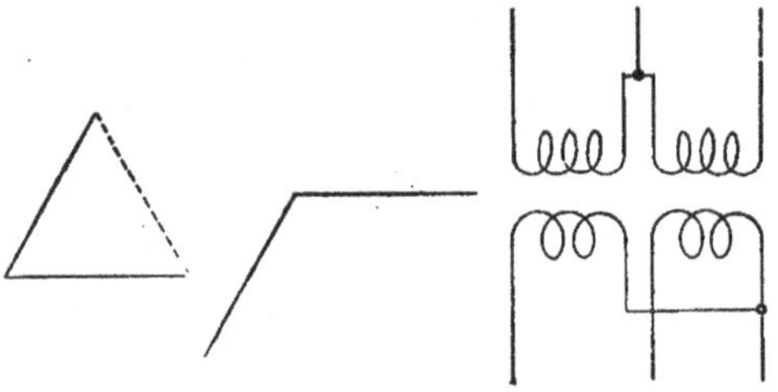

Fig. 202.

other transformer. From the point $\frac{1}{3}$ inside of the teaser transformer, a neutral wire can be brought out in this connection.

4. The monocyclic connection, transforming between three-phase and inverted three-phase or polyphase monocycle, by two transformers, the secondary of one being reversed regarding its primary, as shown in Fig. 202.

5. The *L* connection for transformation between quarter-phase and three-phase as described in the instance, paragraph 257.

6. The *T* connection of transformation between quarter-phase and three-phase, as shown in Fig. 203. The quarter-phase side of the transformers contains two equal and

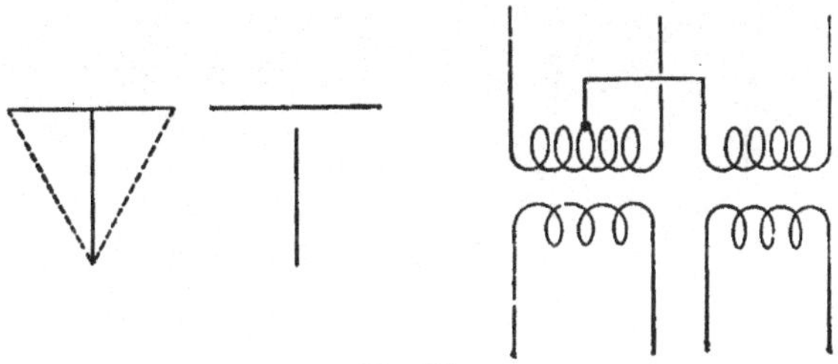

Fig. 203.

independent (or interlinked) coils, the three-phase side two coils with the ratio of turns $1 \div \frac{\sqrt{3}}{2}$ connected in *T*.

7. The double delta connection of transformation from three-phase to six-phase, shown in Fig. 204. Three transformers, with two secondary coils each, are used, one set of

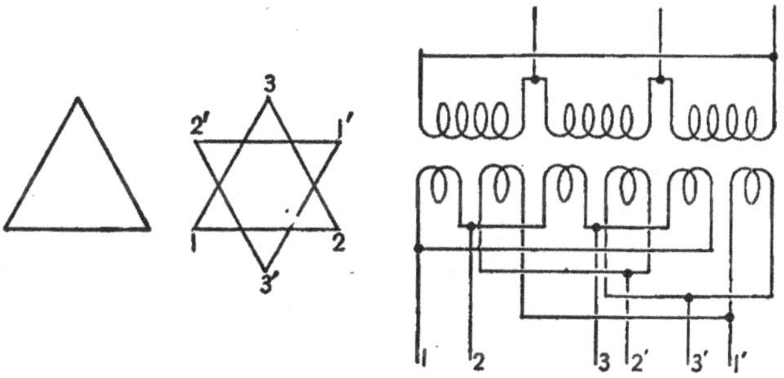

Fig. 204.

secondary coils connected in delta, the other set in delta also, but with reversed terminals, so as to give a reversed E.M.F. triangle. These E.M.F.'s thus give topographically a six-cornered star.

8. The double Y connection of transformation from three-phase to six-phase, shown in Fig. 205. It is analogous to (7), the delta connection merely being replaced by the Y connection. The neutrals of the two Y's may be connected together and to an external neutral if desired.

9. The double T connection of transformation from

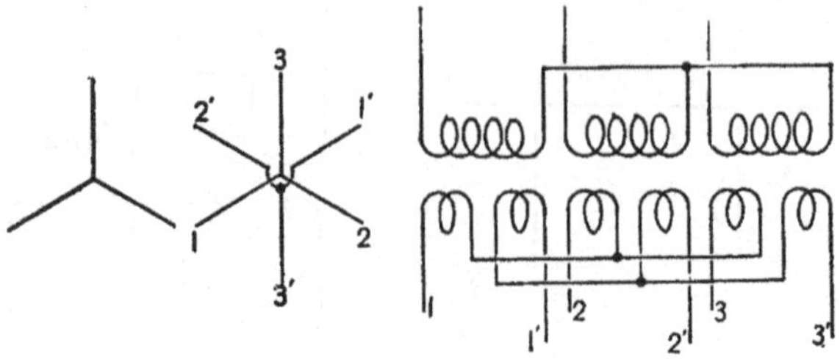

Fig. 205.

three-phase to six-phase, shown in Fig. 206. Two transformers are used with two secondary coils which are T connected, but one with reversed terminals. This method allows a secondary neutral also to be brought out.

287. Transformation with a change of the balance factor of the system is possible only by means of apparatus

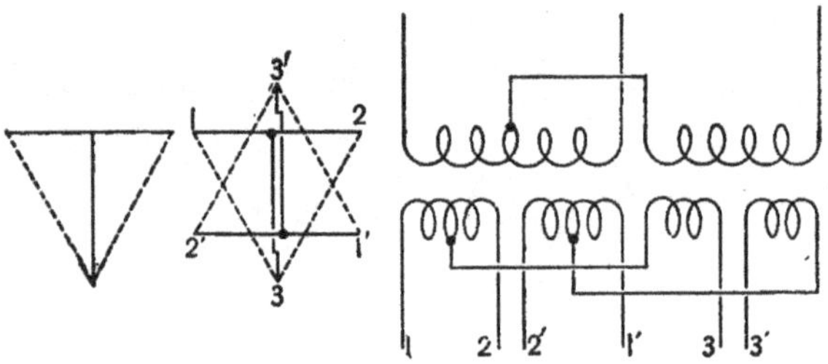

Fig. 206.

able to store energy, since the difference of power between primary and secondary circuit has to be stored at the time when the secondary power is below the primary, and returned during the time when the primary power is below the secondary. The most efficient storing device of electric energy is mechanical momentum in revolving machinery. It has, however, the disadvantage of requiring attendance; fairly efficient also are capacities and inductances, but, as a rule, have the disadvantage not to give constant potential.

CHAPTER XXX.

EFFICIENCY OF SYSTEMS.

288. In electric power transmission and distribution, wherever the place of consumption of the electric energy is distant from the place of production, the conductors which transfer the current are a sufficiently large item to require consideration, when deciding which system and what potential is to be used.

In general, in transmitting a given amount of power at a given loss over a given distance, other things being equal, the amount of copper required in the conductors is inversely proportional to the square of the potential used. Since the total power transmitted is proportional to the product of current and E.M.F., at a given power, the current will vary inversely proportional to the E.M.F., and therefore, since the loss is proportional to the product of current-square and resistance, to give the same loss the resistance must vary inversely proportional to the square of the current, that is, proportional to the square of the E.M.F.; and since the amount of copper is inversely proportional to the resistance, other things being equal, the amount of copper varies inversely proportional to the square of the E.M.F. used.

This holds for any system.

Therefore to compare the different systems, as two-wire single-phase, single-phase three-wire, three-phase and quarter-phase, equality of the potential must be assumed.

Some systems, however, as for instance, the Edison three-wire system, or the inverted three-phase system, have different potentials in the different circuits constituting the system, and thus the comparison can be made either —

1st. On the basis of equality of the maximum potential difference in the system ; or

2d. On the basis of the minimum potential difference in the system, or the potential difference per circuit or phase of the system.

In low potential circuits, as secondary networks, where the potential is not limited by the insulation strain, but by the potential of the apparatus connected into the system, as incandescent lamps, the proper basis of comparison is equality of the potential per branch of the system, or per phase.

On the other hand, in long distance transmissions where the potential is not restricted by any consideration of apparatus suitable for a certain maximum potential only, but where the limitation of potential depends upon the problem of insulating the conductors against disruptive discharge, the proper comparison is on the basis of equality of the maximum difference of potential in the system ; that is, equal maximum dielectric strain on the insulation.

The same consideration holds in moderate potential power circuits, in considering the danger to life from live wires entering human habitations.

Thus the comparison of different systems of long-distance transmission at high potential or power distribution for motors is to be made on the basis of equality of the maximum difference of potential existing in the system. The comparison of low potential distribution circuits for lighting on the basis of equality of the minimum difference of potential between any pair of wires connected to the receiving apparatus.

289. 1st. *Comparison on the basis of equality of the minimum difference of potential, in low potential lighting circuits :*

In the single-phase alternating-current circuit, if $e =$ E.M.F., $i =$ current, $r =$ resistance per line, the total power is $= ei$, the loss of power $2 i^2 r$.

Using, however, a three-wire system, the potential between outside wires and neutral being given $= e$, the potential between the outside wires is $= 2e$, that is, the distribution takes place at twice the potential, or only $\frac{1}{4}$ the copper is needed to transmit the same power at the same loss, if, as it is theoretically possible, the neutral wire has no cross-section. If therefore the neutral wire is made of the same cross-section with each of the outside wires, $\frac{3}{8}$ of the copper of the two-wire system is needed; if the neutral wire is $\frac{1}{2}$ the cross-section of each of the outside wires, $\frac{5}{16}$ of the copper is needed. Obviously, a single-phase five-wire system will be a system of distribution at the potential $4e$, and therefore require only $\frac{1}{16}$ of the copper of the single-phase system in the outside wires; and if each of the three neutral wires is of $\frac{1}{2}$ the cross-section of the outside wires, $\frac{7}{64} = 10.93$ per cent of the copper.

Coming now to the three-phase system with the potential e between the lines as delta potential, if $i =$ the current per line or Y current, the current from line to line or delta current $= i_1 / \sqrt{3}$; and since three branches are used, the total power is $3 e i_1 / \sqrt{3} = e i_1 \sqrt{3}$. Hence if the same power has to be transmitted by the three-phase system as with the single-phase system, the three-phase line current must be $i_1 = i / \sqrt{3}$ where $i =$ single-phase current, $r =$ single-phase resistance per line, at equal power and loss; hence if $r_1 =$ resistance of each of the three wires, the loss per wire is $i_1^2 r_1 = i^2 r_1 / 3$, and the total loss is $i^2 r_1$, while in the single-phase system it is $2 i^2 r$. Hence, to get the same loss, it must be: $r_1 = 2r$, that is, each of the three three-phase lines has twice the resistance — that is, half the copper of each of the two single-phase lines; or in other words, the three-phase system requires three-fourths of the copper of the single-phase system of the same potential.

Introducing, however, a fourth or neutral wire into the three-phase system, and connecting the lamps between the neutral wire and the three outside wires — that is, in Y connection — the potential between the outside wires or delta potential will be $= e \times \sqrt{3}$, since the Y potential $= e$, and the potential of the system is raised thereby from e to $e\sqrt{3}$; that is, only $\frac{1}{3}$ as much copper is required in the outside wires as before — that is $\frac{1}{4}$ as much copper as in the single-phase two-wire system. Making the neutral of the same cross-section as the outside wires, requires $\frac{1}{3}$ more copper, or $\frac{1}{3} = 33.3$ per cent of the copper of the single-phase system; making the neutral of half cross-section, requires $\frac{1}{6}$ more, or $\frac{7}{24} = 29.17$ per cent of the copper of the single-phase system. The system, however, now is a four-wire system.

The independent quarter-phase system with four wires is identical in efficiency to the two-wire single-phase system, since it is nothing but two independent single-phase systems in quadrature.

The four-wire quarter-phase system can be used as two independent Edison three-wire systems also, deriving therefrom the same saving by doubling the potential between the outside wires, and has in this case the advantage, that by interlinkage, the same neutral wire can be used for both phases, and thus one of the neutral wires saved.

In this case the quarter-phase system with common neutral of full cross-section requires $\frac{5}{16} = 31.25$ per cent, the quarter-phase system with common neutral of one-half cross-section requires $\frac{9}{32} = 28.125$ per cent, of the copper of the two-wire single-phase system.

In this case, however, the system is a five-wire system, and as such far inferior to the five-wire single phase system.

Coming now to the quarter-phase system with common return and potential e per branch, denoting the current in the outside wires by i_2, the current in the central wire is $i_2\sqrt{2}$; and if the same current density is chosen for all three wires, as the condition of maximum efficiency, and the resistance of each outside wire denoted by r_2, the resistance of the central wire $= r_2 / \sqrt{2}$, and the loss of power per outside wire is $i_2^2 r_2$, in the central wire $2 i_2^2 r_2 / \sqrt{2} = i_2^2 r_2 \sqrt{2}$; hence the total loss of power is $2 i_2^2 r_2 + i_2^2 r_2 \sqrt{2} = i_2^2 r_2 (2 + \sqrt{2})$. The power transmitted per branch is $i_2 e$, hence the total power $2 i_2 e$. To transmit the same power as by a single-phase system of power, $e i$, it must be $i_2 = i / 2$; hence the loss, $\frac{i^2 r_2 (2 + \sqrt{2})}{4}$. Since this loss shall be the same as the loss $2 i^2 r$ in the single-phase system, it must be $2 r = \frac{(2 + \sqrt{2})}{4} r_2$, or $r_2 = \frac{8 r}{2 + \sqrt{2}}$.

Therefore each of the outside wires must be $\frac{2 + \sqrt{2}}{8}$ times as large as each single-phase wire, the central wire $\sqrt{2}$ times larger; hence the copper required for the quarter-phase system with common return bears to the copper required for the single-phase system the relation:

$$\frac{2 (2 + \sqrt{2})}{8} + \frac{(2 + \sqrt{2}) \sqrt{2}}{8} \div 2, \text{ or, } \frac{3 + 2 \sqrt{2}}{8} \div 1, = 72.9$$

per cent of the copper of the single-phase system.

Hence the quarter-phase system with common return saves 2 per cent more copper than the three-phase system, but is inferior to the single-phase three-wire system.

The inverted three-phase system, consisting of two E.M.Fs. e at 60° displacement, and three equal currents i_3 in the three lines of equal resistance r_3, gives the output $2 e i_3$, that is, compared with the single-phase system, $i_3 = i / 2$. The loss in the three lines is $3 i_3^2 r_3 = \frac{3}{4} i^2 r_3$. Hence, to give the same loss $2 i^2 r$ as the single-phase system, it must be $r_3 = \frac{8}{3} r$, that is, each of the three wires must have $\frac{3}{8}$ of the copper cross-section of the wire in the two-wire single-phase system; or in other words, the inverted three-phase system requires $\frac{9}{16}$ of the copper of the two-wire single-phase system.

We get thus the result,

If a given power has to be transmitted at a given loss, and a given *minimum* potential, as for instance 110 volts for lighting, the amount of copper necessary is:

2 WIRES:	Single-phase system,	100.0
3 WIRES:	Edison three-wire single-phase system, neutral full section,	37.5
	Edison three-wire single-phase system, neutral half-section,	31.25
	Inverted three-phase system,	56.25
	Quarter-phase system with common return,	72.9
	Three-phase system,	75.0
4 WIRES:	Three-phase, with neutral wire full section,	33.3
	Three-phase, with neutral wire half-section,	29.17
	Independent quarter-phase system,	100.0
5 WIRES:	Edison five-wire, single-phase system, full neutral,	15.625
	Edison five-wire, single-phase system, half-neutral,	10.93
	Four-wire, quarter-phase, with common neutral full section,	31.25
	Four-wire, quarter-phase, with common neutral half-section,	28.125

We see herefrom, that in distribution for lighting — that is, with the same minimum potential, and with the same number of wires — the single-phase system is superior to any polyphase system.

The continuous-current system is equivalent in this comparison to the single-phase alternating-current system of the same effective potential, since the comparison is made on the basis of effective potential, and the power depends upon the effective potential also.

290. *Comparison on the Basis of Equality of the Maximum Difference of Potential in the System, in Long-Distance Transmission, Power Distribution, etc.*

Wherever the potential is so high as to bring the question of the strain on the insulation into consideration, or in other cases, to approach the danger limit to life, the proper comparison of different systems is on the basis of equality of maximum potential in the system.

Hence in this case, since the maximum potential is fixed, nothing is gained by three- or five-wire Edison systems. Thus, such systems do not come into consideration.

The comparison of the three-phase system with the single-phase system remains the same, since the three phase system has the same maximum as minimum potential; that is:

The three-phase system requires three-fourths of the copper of the single-phase system to transmit the same power at the same loss over the same distance.

The four-wire quarter-phase system requires the same amount of copper as the single-phase system, since it consists of two single-phase systems.

In a quarter-phase system with common return, the potential between the outside wire is $\sqrt{2}$ times the potential per branch, hence to get the same maximum strain on the insulation — that is, the same potential e between the outside wires as in the single-phase system — the potential per branch will be $e / \sqrt{2}$, hence the current $i_4 = i / \sqrt{2}$, if i equals the current of the single-phase system of equal power, and $i_4 \sqrt{2} = i$ will be the current in the central wire.

Hence, if r_4 = resistance per outside wire, $r_4 / \sqrt{2}$ = resistance of central wire, and the total loss in the system is:

$$2\, i_4^2 r_4 + \frac{i_4^2\, 2\, r_4}{\sqrt{2}} = i_4^2 r_4 (2 + \sqrt{2}) = i^2 r_4 \frac{(2 + \sqrt{2})}{2}$$

Since in the single-phase system, the loss $= 2\, i^2 r$, it is:

$$r_4 = \frac{4\, r}{2 + \sqrt{2}}.$$

That is, each of the outside wires has to contain $\frac{2 + \sqrt{2}}{4}$ times as much copper as each of the single-phase wires. The central wires have to contain $\frac{2 \times \sqrt{2}}{4} \sqrt{2}$ times as much copper; hence the total system contains $\frac{2\,(2 + \sqrt{2})}{4} + \frac{2 + \sqrt{2}}{4} \sqrt{2}$ times as much copper as each of the single phase wires; that is, $\frac{3 + 2\sqrt{2}}{4}$ times the copper of the single-phase system.

Or, in other words,

A quarter-phase system with common return requires $\frac{3 + 2\sqrt{2}}{4} = 1.457$ times as much copper as a single-phase system of the same maximum potential, same power, and same loss.

Since the comparison is made on the basis of equal maximum potential, and the maximum potential of alternating system is $\sqrt{2}$ times that of a continuous-current circuit of equal effective potential, the alternating circuit of effective potential e compares with the continuous-current circuit of potential $e\sqrt{2}$, which latter requires only half the copper of the alternating system.

This comparison of the alternating with the continuous-current system is not proper however, since the continuous-current potential introduces, besides the electrostatic strain, an electrolytic strain on the dielectric which does not exist in the alternating system, and thus makes the action of the continuous-current potential on the insulation more severe than that of an equal alternating potential. Besides, self-induction having no effect on a steady current, continuous current circuits as a rule have a self-induction far in excess of any alternating circuit. During changes of current, as make and break, and changes of load, especially rapid changes, there are consequently induced in these circuits E.M.F.'s far exceeding their normal potentials. At the voltages which came under consideration, the continuous current is excluded to begin with.

Thus we get:

If a given power is to be transmitted at a given loss, and a given *maximum* difference of potential in the system, that is, with the same strain on the insulation, the amount of copper required is:

2 Wires:	Single-phase system,	100.0
	[Continuous-current system,	50.0]
3 Wires:	Three-phase system,	75.0
	Quarter-phase system, with common return,	145.7
4 Wires:	Independent Quarter-phase system,	100.0

Hence the quarter-phase system with common return is practically excluded from long-distance transmission.

291. In a different way the same comparative results between single-phase, three-phase, and quarter-phase systems can be derived by resolving the systems into their single-phase branches.

The three-phase system of E.M.F. e between the lines can be considered as consisting of three single-phase circuits of E.M.F. $e / \sqrt{3}$, and no return. The single-phase system of E.M.F. e between lines as consisting of two single-phase circuits of E.M.F. $e / 2$ and no return. Thus, the relative amount of copper in the two systems being inversely proportional to the square of E.M.F., bears the relation $(\sqrt{3} / e)^2 : (2 / e)^2 = 3 : 4$; that is, the three-phase system requires 75 per cent of the copper of the single-phase system.

The quarter-phase system with four equal wires requires the same copper as the single-phase system, since it consists of two single-phase circuits. Replacing two of the four quarter-phase wires by one wire of the same cross-section as each of the wires replaced thereby, the current in this wire is $\sqrt{2}$ times as large as in the other wires, hence, the loss twice as large — that is, the same as in the two wires replaced by this common wire, or the total loss is not changed — while 25 per cent of the copper is saved, and the system requires only 75 per cent of the copper of the single-phase system, but produces $\sqrt{2}$ times as high a potential between the outside wires. Hence, to give the same maximum potential, the E.M.Fs. of the system have to be reduced by $\sqrt{2}$, that is, the amount of copper doubled, and thus the quarter-phase system with common return of the same cross-section as the outside wires requires 150 per cent of the copper of the single-phase system. In this case, however, the current density in the middle wire is higher, thus the copper not used most economical, and transferring a part of the copper from the outside wire to the middle wire, to bring all three wires to the same current density, reduces the loss, and thereby reduces the amount of copper at a given loss, to 145.7 per cent of that of a single-phase system.

CHAPTER XXXI.

THREE-PHASE SYSTEM.

292. With equal load of the same phase displacement in all three branches, the symmetrical three-phase system offers no special features over those of three equally loaded single-phase systems, and can be treated as such; since the mutual reactions between the three phases balance at equal distribution of load, that is, since each phase is acted upon by the preceding phase in an equal but opposite manner as by the following phase.

With unequal distribution of load between the different branches, the voltages and phase differences become more or less unequal. These unbalancing effects are obviously maximum, if some of the phases are fully loaded, others unloaded,

Let:

$\dot{E}$ = E.M.F. between branches 1 and 2 of a three-phaser.

Then:

$$\epsilon \dot{E} = \text{E.M.F. between 2 and 3,}$$
$$\epsilon^2 \dot{E} = \text{E.M.F. between 3 and 1,}$$

where,
$$\epsilon = \sqrt[3]{1} = \frac{-1 + j\sqrt{3}}{2}.$$

Let

Z_1, Z_2, Z_3 = impedances of the lines issuing from generator terminals 1, 2, 3,

and Y_1, Y_2, Y_3 = admittances of the consumer circuits connected between lines 2 and 3, 3 and 1, 1 and 2.

If then,

$\dot{I}_1, \dot{I}_2, \dot{I}_3$, are the currents issuing from the generator terminals into the lines, it is,

$$\dot{I}_1 + \dot{I}_2 + \dot{I}_3 = 0. \qquad (1)$$

If I_1', I_2', I_3' = **currents flowing through the admittances** Y_1, Y_2, Y_3, **from 2 to 3, 3 to 1, 1 to 2, it is,**

$$\left.\begin{aligned} I_1 &= I_3' - I_2', \text{ or, } I_1 + I_2' - I_3' = 0 \\ I_2 &= I_1' - I_3', \text{ or, } I_2 + I_3' - I_1' = 0 \\ I_3 &= I_2' - I_1', \text{ or, } I_3 + I_1' - I_2' = 0 \end{aligned}\right\} \qquad (2)$$

These three equations (2) added, give (1) as dependent equation.

At the ends of the lines 1, 2, 3, it is:

$$\left.\begin{aligned} E_1' &= E_1 - Z_2 I_2 + Z_3 I_3 \\ E_2' &= E_2 - Z_3 I_3 + Z_1 I_1 \\ E_3' &= E_3 - Z_1 I_1 + Z_2 I_2 \end{aligned}\right\} \qquad (3)$$

the differences of potential, and

$$\left.\begin{aligned} I_1' &= E_1' Y_1 \\ I_2' &= E_2' Y_2 \\ I_3' &= E_3' Y_3 \end{aligned}\right\} \qquad (4)$$

the currents in the receiver circuits.

These nine equations (2), (3), (4), determine the nine quantities: $I_1, I_2, I_3, I_1', I_2', I_3', E_1', E_2', E_3'$.

Equations (4) substituted in (2) give:

$$\left.\begin{aligned} I_1 &= E_3' Y_3 - E_2' Y_2 \\ I_2 &= E_1' Y_1 - E_3' Y_3 \\ I_3 &= E_2' Y_2 - E_1' Y_1 \end{aligned}\right\} \qquad (5)$$

These equations (5) substituted in (3), and transposed, give,

since $\left.\begin{aligned} E_1 &= \epsilon E \\ E_2 &= \epsilon^2 E \\ E_3 &= E \end{aligned}\right\}$ as E.M.Fs. at the generator terminals.

$$\left.\begin{aligned} \epsilon E - E_1'(1 + Y_1 Z_3 + Y_1 Z_2) + E_2' Y_2 Z_3 + E_3' Y_3 Z_2 &= 0 \\ \epsilon^2 E - E_2'(1 + Y_2 Z_3 + Y_2 Z_1) + E_3' Y_3 Z_1 + E_1' Y_1 Z_3 &= 0 \\ E - E_3'(1 + Y_3 Z_1 + Y_3 Z_2) + E_1' Y_1 Z_2 + E_2' Y_2 Z_1 &= 0 \end{aligned}\right\} \qquad (6)$$

as three linear equations with the three quantities $\dot{E}_1'$, $\dot{E}_2'$, $\dot{E}_3'$.

Substituting the abbreviations:

$$\left.\begin{aligned}
D &= \begin{vmatrix} -(1+Y_1Z_2+Y_1Z_3), & Y_2Z_3, & Y_3Z_2 \\ Y_1Z_3, & -(1+Y_2Z_3+Y_2Z_1), & Y_3Z_1 \\ Y_1Z_2, & Y_2Z_1, & -(1+Y_3Z_1+Y_3Z_2) \end{vmatrix} \\
D_1 &= \begin{vmatrix} \epsilon, & Y_2Z_3, & Y_3Z_2 \\ \epsilon^2, & -(1+Y_2Z_3+Y_2Z_1), & Y_3Z_1 \\ 1, & Y_2Z_1, & -(1+Y_3Z_1+Y_3Z_2) \end{vmatrix} \\
D_2 &= \begin{vmatrix} -(1+Y_1Z_2+Y_1Z_3), & \epsilon, & Y_3Z_2 \\ Y_1Z_3, & \epsilon^2, & Y_3Z_1 \\ Y_1Z_2, & 1, & -(1+Y_3Z_1+Y_3Z_2) \end{vmatrix} \\
D_3 &= \begin{vmatrix} -(1+Y_1Z_2+Y_1Z_3), & Y_2Z_3, & \epsilon \\ Y_1Z_3, & -(1+Y_2Z_3+Y_2Z_1), & \epsilon^2 \\ Y_1Z_2, & Y_2Z_1, & 1 \end{vmatrix}
\end{aligned}\right\}. \qquad (7)$$

it is:

$$\left.\begin{aligned}
\dot{E}_1' &= \dot{E}D_1 / D \\
\dot{E}_2' &= \dot{E}D_2 / D \\
\dot{E}_3' &= \dot{E}D_3 / D
\end{aligned}\right\}. \qquad (8)$$

$$\left.\begin{aligned}
\dot{I}_1' &= \frac{Y_1\dot{E}D_1}{D} \\
\dot{I}_2' &= \frac{Y_2\dot{E}D_2}{D} \\
\dot{I}_3' &= \frac{Y_3\dot{E}D_3}{D}
\end{aligned}\right\} \qquad (9)$$

$$\left.\begin{aligned}
\dot{I}_1 &= \frac{Y_3D_3 - Y_2D_2\dot{E}}{D} \\
\dot{I}_2 &= \frac{Y_1D_1 - Y_3D_3\dot{E}}{D} \\
\dot{I}_3 &= \frac{Y_2D_2 - Y_1D_1\dot{E}}{D}
\end{aligned}\right\} \qquad (10)$$

hence,

$$\left.\begin{aligned}
\dot{E}_1' + \dot{E}_2' + \dot{E}_3' &= 0 \\
\dot{I}_1 + \dot{I}_2 + \dot{I}_3 &= 0
\end{aligned}\right\} \qquad (11)$$

293. Special Cases.

A. *Balanced System*

$$Y_1 = Y_2 = Y_3 = Y$$
$$Z_1 = Z_2 = Z_3 = Z.$$

Substituting this in (6), and transposing:

$$\left.\begin{aligned} \dot{E}_1 &= \epsilon \dot{E} \\ \dot{E}_2 &= \epsilon^2 \dot{E} \\ \dot{E}_3 &= \dot{E} \end{aligned}\right\} \qquad \begin{aligned} \dot{E}_1' &= \frac{\epsilon \dot{E}}{1 + 3YZ} \\ \dot{E}_2' &= \frac{\epsilon^2 \dot{E}}{1 + 3YZ} \\ \dot{E}_3' &= \frac{\dot{E}}{1 + 3YZ} \end{aligned}$$

$$\left.\begin{aligned} \dot{I}_1' &= \frac{\epsilon \dot{E} Y}{1 + 3YZ} \\ \dot{I}_2' &= \frac{\epsilon^2 \dot{E} Y}{1 + 3YZ} \\ \dot{I}_3' &= \frac{\dot{E} Y}{1 + 3YZ} \end{aligned}\right\} \qquad \begin{aligned} \dot{I}_1 &= \frac{\epsilon^2 (\epsilon - 1) \dot{E} Y}{1 + 3YZ} \\ \dot{I}_2 &= \frac{(\epsilon - 1) \dot{E} Y}{1 + 3YZ} \\ \dot{I}_3 &= \frac{\epsilon (\epsilon - 1) \dot{E} Y}{1 + 3YZ} \end{aligned} \qquad (12)$$

The equations of the symmetrical balanced three-phase system.

B. *One circuit loaded, two unloaded:*

$$Y_1 = Y_2 = 0, \quad Y_3 = Y$$
$$Z_1 = Z_2 = Z_3 = Z.$$

Substituted in equations (6):

$$\left.\begin{aligned} \epsilon \dot{E} - \dot{E}_1' + \dot{E}_3' YZ &= 0 \\ \epsilon^2 \dot{E} - \dot{E}_2' + \dot{E}_3' YZ &= 0 \end{aligned}\right\} \text{ unloaded branches.}$$

$$\dot{E} - \dot{E}_3'(1 + 2YZ) = 0, \text{ loaded branch.}$$

hence:

$$\left.\begin{aligned} \dot{E}_1' &= \frac{\dot{E}\{\epsilon + (1 + 2\epsilon) YZ\}}{1 + 2YZ} \\ \dot{E}_2' &= \frac{\dot{E}\{\epsilon^2 + (1 + 2\epsilon^2) YZ\}}{1 + 2YZ} \end{aligned}\right\} \text{unloaded;} \qquad \dot{E}_3' = \frac{\dot{E}}{1 + 2YZ} \quad \text{loaded;}$$

all three E.M.F.'s unequal, and of unequal phase angles. (13)

$$\left.\begin{aligned} I_1' &= I_2' = 0 \\ I_3' &= \frac{EY}{1+2YZ} \end{aligned}\right\} \qquad (13)$$

$$\left.\begin{aligned} I_1 &= \frac{EY}{1+2YZ} \\ I_2 &= -\frac{EY}{1+2YZ} \\ I_3 &= 0 \end{aligned}\right\} \qquad (13)$$

C. *Two circuits loaded, one unloaded.*

$$Y_1 = Y_2 = Y, \quad Y_3 = 0,$$
$$Z_1 = Z_2 = Z_3 = Z.$$

Substituting this in equations (6), it is:

$$\left.\begin{aligned} \epsilon E - E_1'(1+2YZ) + E_2' YZ = 0 \\ \epsilon^2 E - E_2'(1+2YZ) + E_1' YZ = 0 \end{aligned}\right\} \text{ loaded branches.}$$

$$E - E_3' + (E_1' + E_2') YZ = 0 \quad \text{unloaded branch.}$$

or, since:

$$(E_1' + E_2') = -E_3';$$
$$E - E_3' - E_3' YZ = 0,$$
$$E_3' = \frac{E}{1+YZ}$$

thus:

$$\left.\begin{aligned} &\left.\begin{aligned} E_1' &= \frac{E\epsilon\{1+(2+\epsilon)YZ\}}{1+4YZ+3Y^2Z^2} \\ E_2' &= \frac{E\epsilon^2\{1+(2+\epsilon^2)YZ\}}{1+4YZ+3Y^2Z^2} \end{aligned}\right\} \text{ loaded branches.} \\ &E_3' = \frac{E}{1+YZ} \qquad\qquad \text{unloaded branch.} \end{aligned}\right\} \qquad (14)$$

As seen, with unsymmetrical distribution of load, all three branches become more or less unequal, and the phase displacement between them unequal also.

CHAPTER XXXII.

QUARTER–PHASE SYSTEM.

294. In a three-wire quarter-phase system, or quarter-phase system with common return wire of both phases, let the two outside terminals and wires be denoted by 1 and 2, the middle wire or common return by 0.

It is then :

$E_1 = E =$ E.M.F. between 0 and 1 in the generator.
$E_2 = jE -$ E.M.F. between 0 and 2 in the generator.

Let :

I_1 and I_2 = currents in 1 and in 2,
I_0 = current in 0,
Z_1 and Z_2 = impedances of lines 1 and 2,
Z_0 = impedance of line 0.
Y_1 and Y_2 = admittances of circuits 0 to 1, and 0 to 2,
I_1' and I_2' = currents in circuits 0 to 1, and 0 to 2,
E_1' and E_2' = potential differences at circuit 0 to 1, and 0 to 2.

it is then, or,

$$\left.\begin{aligned} I_1 + I_2 + I_0 &= 0 \\ I_0 &= -(I_1 + I_2) \end{aligned}\right\} \qquad (1)$$

that is, I_0 is common return of I_1 and I_2.

Further, we have,

$$\left.\begin{aligned} E_1' &= E - I_1 Z_1 + I_0 Z_0 = E - I_1(Z_1 + Z_0) - I_2 Z_0 \\ E_2' &= jE - I_2 Z_0 + I_0 Z_0 = jE - I_2(Z_2 + Z_0) - I_1 Z_1 \end{aligned}\right\} \qquad (2)$$

and

$$\left.\begin{aligned} I_1 &= Y_1 E_1' \\ I_2 &= Y_2 E_2' \\ I_0 &= -(Y_1 E_1' + Y_2 E_2') \end{aligned}\right\} \qquad (3)$$

Substituting (3) in (2); and expanding:

$$\left.\begin{aligned} \dot{E}_1' &= \dot{E}\,\frac{1 + Y_2 Z_2 + Y_2 Z_0 (1 - j)}{(1 + Y_1 Z_0 + Y_1 Z_1)(1 + Y_2 Z_0 + Y_2 Z_2) - Y_1 Y_2 Z_0^2} \\ \dot{E}_2' &= j\dot{E}\,\frac{1 + Y_1 Z_1 + Y_1 Z_0 (1 + j)}{(1 + Y_1 Z_0 + Y_1 Z_1)(1 + Y_2 Z_0 + Y_2 Z_2) - Y_1 Y_2 Z_0^2} \end{aligned}\right\} \quad (4)$$

Hence, the two E.M.Fs. at the end of the line are unequal in magnitude, and not in quadrature any more.

295. SPECIAL CASES:

A. Balanced System.

$$Z_1 = Z_2 = Z;$$
$$Z_0 = Z / \sqrt{2};$$
$$Y_1 = Y_2 = Y.$$

Substituting these values in (4), gives:

$$\left.\begin{aligned} \dot{E}_1' &= \dot{E}\,\frac{1 + \frac{1 + \sqrt{2} - j}{\sqrt{2}} YZ}{1 + \sqrt{2}\,(1 + \sqrt{2})\,YZ + (1 + \sqrt{2})\,Y^2 Z^2} \\ &= \dot{E}\,\frac{1 + (1.707 - .707\,j)\,YZ}{1 + 3.414\,YZ + 2.414\,Y^2 Z^2} \\ \dot{E}_2' &= j\dot{E}\,\frac{1 + \frac{1 + \sqrt{2} + j}{\sqrt{2}} YZ}{1 + \sqrt{2}\,(1 + \sqrt{2})\,YZ + (1 + \sqrt{2})\,Y^2 Z^2} \\ &= j\dot{E}\,\frac{1 + (1.707 + .707\,j)\,YZ}{1 + 3.414\,YZ + 2.414\,Y^2 Z^2} \end{aligned}\right\} \quad (5)$$

Hence, the balanced quarter-phase system with common return is unbalanced with regard to voltage and phase relation, or in other words, even if in a quarter-phase system with common return both branches or phases are loaded equally, with a load of the same phase displacement, nevertheless the system becomes unbalanced, and the two E.M.Fs. at the end of the line are neither equal in magnitude, nor in quadrature with each other.

B. *One branch loaded, one unloaded.*

$$Z_1 = Z_2 = Z;$$
$$Z_0 = Z\sqrt{2}.$$

a.) $Y_1 = 0, \quad Y_2 = Y.$

b.) $Y_1 = Y, \quad Y_2 = 0.$

Substituting these values in (4), gives:

a.)

$$\left.\begin{aligned}
\dot{E}_1' &= \dot{E}\,\frac{1 + YZ\dfrac{1+\sqrt{2}-j}{\sqrt{2}}}{1 + YZ\dfrac{1+\sqrt{2}}{\sqrt{2}}} \\
&= \dot{E}\left\{1 - \frac{j}{1+\sqrt{2}+\dfrac{\sqrt{2}}{YZ}}\right\} \\
&= \dot{E}\left\{1 - \frac{j}{2.414 + \dfrac{1.414}{YZ}}\right\} \\
\dot{E}_2' &= j\dot{E}\,\frac{1}{1 + YZ\dfrac{1+\sqrt{2}}{\sqrt{2}}} \\
&= j\dot{E}\,\frac{1}{1 + 1.707\,YZ}
\end{aligned}\right\} \qquad (6)$$

b.)

$$\left.\begin{aligned}
\dot{E}_1' &= \dot{E}\,\frac{1}{1 + YZ\dfrac{1+\sqrt{2}}{\sqrt{2}}} \\
&= \dot{E}\,\frac{1}{1 + 1.707\,YZ} \\
\dot{E}_2' &= j\dot{E}\,\frac{1 + YZ\dfrac{1+\sqrt{2}+j}{\sqrt{2}}}{1 + YZ\dfrac{1+\sqrt{2}}{\sqrt{2}}} \\
&= j\dot{E}\left\{1 + \frac{j}{1+\sqrt{2}+\dfrac{\sqrt{2}}{YZ}}\right\} \\
&= j\dot{E}\left\{1 + \frac{j}{2.414 + \dfrac{1.414}{YZ}}\right\}
\end{aligned}\right\} \qquad (7)$$

These two E.M.Fs. are unequal, and not in quadrature with each other.

But the values in case *a.*) are different from the values in case *b.*).

That means:

The two phases of a three-wire quarter-phase system are unsymmetrical, and the leading phase 1 reacts upon the lagging phase 2 in a different manner than 2 reacts upon 1.

It is thus undesirable to use a three-wire quarter-phase system, except in cases where the line impedances Z are negligible.

In all other cases, the four-wire quarter-phase system is preferable, which essentially consists of two independent single-phase circuits, and is treated as such.

Obviously, even in such an independent quarter-phase system, at unequal distribution of load, unbalancing effects may take place.

If one of the branches or phases is loaded differently from the other, the drop of voltage and the shift of the phase will be different from that in the other branch; and thus the E.M.Fs. at the end of the lines will be neither equal in magnitude, nor in quadrature with each other.

With both branches however loaded equally, the system remains balanced in voltage and phase, just like the three-phase system under the same conditions.

Thus the four-wire quarter-phase system and the three-phase system are balanced with regard to voltage and phase at equal distribution of load, but are liable to become unbalanced at unequal distribution of load; the three-wire quarter-phase system is unbalanced in voltage and phase, even at equal distribution of load.

APPENDICES.

APPENDIX I.

ALGEBRA OF COMPLEX IMAGINARY QUANTITIES.

INTRODUCTION.

296. The system of numbers, of which the science of algebra treats, finds its ultimate origin in experience. Directly derived from experience, however, are only the absolute integral numbers; fractions, for instance, are not directly derived from experience, but are abstractions expressing relations between different classes of quantities. Thus, for instance, if a quantity is divided in two parts, from one quantity two quantities are derived, and denoting these latter as halves expresses a relation, namely, that two of the new kinds of quantities are derived from, or can be combined to one of the old quantities.

297. Directly derived from experience is the *operation of counting* or of *numeration.*

$$a, \quad a+1, \quad a+2, \quad a+3 \; . \; . \; . \; .$$

Counting by a given number of integers:

$$a + \underbrace{1 + 1 + 1 \; . \; . \; . + 1}_{b \text{ integers}} = c$$

introduces the operation of *addition,* as multiple counting:

$$a + b = c.$$

It is, $$a + b = b + a,$$

that is, the terms of addition, or addenda, are interchange able.

Multiple addition of the same terms:

$$\underbrace{a + a + a + \ldots + a}_{b \text{ equal numbers}} = c$$

introduces the operation of *multiplication:*

$$a \times b = c.$$

It is,
$$a \times b = b \times a,$$

that is, the terms of multiplication, or factors, are interchangeable.

Multiple multiplication of the same factors:

$$\underbrace{a \times a \times a \times \ldots \times a}_{b \text{ equal numbers}} = c$$

introduces the operation of involution:

$$a^b = c$$

Since
$$a^b \text{ is not equal to } b^a,$$

the terms of involution are not interchangeable.

298. The reverse operation of addition introduces the operation of *subtraction:*

If
$$a + b = c,$$
it is
$$c - b = a.$$

This operation cannot be carried out in the system of absolute numbers, if:

$$b > c.$$

Thus, to make it possible to carry out the operation of subtraction under any circumstances, the system of absolute numbers has to be expanded by the introduction of the *negative number:*

$$-a = (-1) \times a,$$

where
$$(-1)$$

is the negative unit.

Thereby the system of numbers is subdivided in the

positive and negative numbers, and the operation of subtraction possible for all values of subtrahend and minuend. From the definition of addition as multiple numeration, and subtraction as its inverse operation, it follows:

$$c - (-b) = c + b,$$

thus:

$$(-1) \times (-1) = 1;$$

that is, the negative unit is defined by, $(-1)^2 = 1$.

299. The reverse operation of multiplication introduces the operation of *division:*

If $$a \times b = c, \text{ then } \frac{c}{b} = a.$$

In the system of integral numbers this operation can only be carried out, if b is a factor of c.

To make it possible to carry out the operation of division under any circumstances, the system of integral numbers has to be expanded by the introduction of the *fraction:*

$$\frac{c}{b} = c \times \left(\frac{1}{b}\right),$$

where $\frac{1}{b}$ is the integer fraction, and is defined by:

$$\left(\frac{1}{b}\right) \times b = 1.$$

300. The reverse operation of involution introduces two new operations, since in the involution:

$$a^b = c,$$

the quantities a and b are not reversible.

Thus $$\sqrt[b]{c} = a, \text{ the } \textit{evolution},$$

$$\log_a c = b, \text{ the } \textit{logarithmation}.$$

The operation of evolution of terms c, which are not complete powers, makes a further expansion of the system of numbers necessary, by the introduction of the ***irrational number*** (endless decimal fraction), as for instance:

$$\sqrt{2} = 1.414213.$$

301. The operation of evolution of negative quantities c with even exponents b, as for instance

$$\sqrt[2]{-a}$$

makes a further expansion of the system of numbers necessary, by the introduction of the ***imaginary unit.***

$$\sqrt{-1}$$

Thus $$\sqrt[2]{-a} = \sqrt[2]{-1} \times \sqrt[2]{a}.$$

where: $\sqrt{-1}$ is denoted by j.

Thus, the imaginary unit j is defined by:

$$j^2 = -1.$$

By addition and subtraction of real and imaginary units, compound numbers are derived of the form:

$$a + jb,$$

which are denoted as ***complex imaginary numbers.***

No further system of numbers is introduced by the operation of evolution.

The operation of logarithmation introduces the irrational and imaginary and complex imaginary numbers also, but no further system of numbers.

302. Thus, starting from the absolute integral numbers of experience, by the two conditions:

1st. Possibility of carrying out the algebraic operations and their reverse operations under all conditions,

2d. Permanence of the laws of calculation,

the expansion of the system of numbers has become necessary, into

Positive and negative numbers,
Integral numbers and fractions,
Rational and irrational numbers,

Real and imaginary numbers and complex imaginary numbers.

Therewith closes the field of algebra, and all the algebraic operations and their reverse operations can be carried out irrespective of the values of terms entering the operation.

Thus within the range of algebra no further extension of the system of numbers is necessary or possible, and the most general number is

$$a + jb.$$

where a and b can be integers or fractions, positive or negative, rational or irrational.

Algebraic Operations with Complex Imaginary Quantities.

303. *Definition of imaginary unit:*

$$j^2 = -1.$$

Complex imaginary number:

$$\dot{A} = a + jb.$$

Substituting:

$$a = r \cos \beta$$
$$b = r \sin \beta,$$

it is
$$\dot{A} = r(\cos \beta + j \sin \beta),$$

where
$$r = \sqrt{a^2 + b^2}.$$

$$\tan \beta = \frac{b}{a},$$

$r =$ vector,
$\beta =$ amplitude of complex imaginary number A.

Substituting:

$$\cos \beta = \frac{\epsilon^{j\beta} + \epsilon^{-j\beta}}{2}$$

$$\sin \beta = \frac{\epsilon^{j\beta} - \epsilon^{-j\beta}}{2j}$$

it is $$A = r\epsilon^{j\beta},$$

where $\epsilon = \lim_{n=\infty}\left(1 + \frac{1}{n}\right)^n = \sum_0^\infty{}_k \frac{1}{1 \times 2 \times 3 \times \ldots \times k}$ •

is the basis of the natural logarithms.

Conjugate numbers:

$$a + jb = r(\cos\beta + j\sin\beta) = r\epsilon^{j\beta}$$

and $$a - jb = r(\cos[-\beta] + j\sin[-\beta]) = r\epsilon^{-j\beta}$$

it is $$(a + jb)(a - jb) = a^2 + b^2 = r^2.$$

Associate numbers:

$$a + jb = r(\cos\beta + j\sin\beta) = r\epsilon^{j\beta}$$

and $$b + ja = r\left(\cos\left[\frac{\pi}{2} - \beta\right] + j\sin\left[\frac{\pi}{2} - \beta\right]\right) = r\epsilon^{j\left(\frac{\pi}{2} - \beta\right)};$$

it is $$(a + jb)(b + ja) = j(a^2 + b^2) = jr^2.$$

If $$a + jb = a' + jb',$$

it is $$a = a'$$
$$b = b'.$$

If $$a + jb = 0;$$

it is $$a = 0,$$
$$b = 0.$$

304. *Addition and Subtraction:*

$$(a + jb) \pm (a' + jb') = (a \pm a') + j(b \pm b').$$

Multiplication:

$$(a + jb)(a' + jb') = (aa' - bb') + j(ab' + ba')$$

or $$r(\cos\beta + j\sin\beta) \times r'(\cos\beta' + j\sin\beta') = rr'(\cos[\beta + \beta'] + j\sin[\beta + \beta']);$$

or $$r\epsilon^{j\beta} \times r'\epsilon^{j\beta'} = rr'\epsilon^{j(\beta + \beta')}.$$

Division:

Expansion of complex imaginary fraction, for rationalization of denominator or numerator, by multiplication with the conjugate quantity:

$$\frac{a+jb}{a'+jb'} = \frac{(a+jb)(a'-jb')}{(a'+jb')(a'-jb')} = \frac{(aa'+bb')+j(ba'-ab')}{a'^2+b'^2}$$

$$= \frac{(a+jb)(a-jb)}{(a'+jb')(a-jb)} = \frac{a^2+b^2}{(aa'+bb')+j(ab'-ba')};$$

or, $$\frac{r(\cos\beta+j\sin\beta)}{r'(\cos\beta'+j\sin\beta')} = \frac{r}{r'}(\cos[\beta-\beta']+j\sin[\beta-\beta']);$$

or, $$\frac{re^{j\beta}}{r'e^{j\beta'}} = \frac{r}{r'}e^{j(\beta-\beta')}.$$

involution:

$$(a+jb)^n = \{r(\cos\beta+j\sin\beta)\}^n = \{re^{j\beta}\}^n$$
$$= r^n(\cos n\beta+j\sin n\beta) = r^n e^{jn\beta}.$$

evolution:

$$\sqrt[n]{a+jb} = \sqrt[n]{r(\cos\beta+j\sin\beta)} = \sqrt[n]{re^{j\beta}}.$$
$$= \sqrt[n]{r}\left(\cos\frac{\beta}{n}+j\sin\frac{\beta}{n}\right) = \sqrt[n]{r}e^{j\frac{\beta}{n}}.$$

305. *Roots of the Unit:*

$\sqrt[2]{1} = +1, \quad -1;$

$\sqrt[3]{1} = +1, \frac{-1+j\sqrt{3}}{2}, \frac{-1-j\sqrt{3}}{2};$

$\sqrt[4]{1} = +1, \quad -1, \quad +j, \quad -j;$

$\sqrt[8]{1} = +1, -1, +j, -j, \frac{+1+j}{\sqrt{2}}, \frac{+1-j}{\sqrt{2}}, \frac{-1+j}{\sqrt{2}}, \frac{-1-j}{\sqrt{2}};$

$\sqrt[n]{1} = \cos\frac{2\pi k}{n}+j\sin\frac{2\pi k}{n} = e^{\frac{2\pi jk}{n}}, \ k = 0, 1, 2 \ldots . n-1.$

306. *Rotation:*

In the complex imaginary plane, multiplication with

$$\sqrt[n]{1} = \cos\frac{2\pi}{n}+j\sin\frac{2\pi}{n} = e^{\frac{2\pi j}{n}}$$

means rotation, in positive direction, by $1/n$ of a revolution,

multiplication with (-1) means reversal, or rotation by 180°,
multiplication with $(+j)$ means positive rotation by 90°,
multiplication with $(-j)$ means negative rotation by 90°.

307. *Complex imaginary plane:*

While the positive and negative numbers can be represented by the points of a line, the complex imaginary numbers are represented by the points of a plane, with the horizontal axis $A' O A$ as real axis, the vertical axis $B' O B$ as imaginary axis. Thus all

the positive real numbers are represented by the points of half axis $\overline{OA}$ towards the right;

the negative real numbers are represented by the points of half axis $\overline{OA'}$ towards the left;

the positive imaginary numbers are represented by the points of half axis $\overline{OB}$ upwards;

the negative imaginary numbers are represented by the points of half axis $\overline{OB'}$ downwards;

the complex imaginary numbers are represented by the points outside of the coördinate axes.

APPENDIX II.

OSCILLATING CURRENTS.

INTRODUCTION.

308. An electric current varying periodically between constant maximum and minimum values, — that is, in equal time intervals repeating the same values, — is called an alternating current if the arithmetic mean value equals zero; and is called a pulsating current if the arithmetic mean value differs from zero.

Assuming the wave as a sine curve, or replacing it by the equivalent sine wave, the alternating current is characterized by the period or the time of one complete cyclic change, and the amplitude or the maximum value of the current. Period and amplitude are constant in the alternating current.

A very important class are the currents of constant period, but geometrically varying amplitude; that is, currents in which the amplitude of each following wave bears to that of the preceding wave a constant ratio. Such currents consist of a series of waves of constant length, decreasing in amplitude, that is in strength, in constant proportion. They are called oscillating currents in analogy with mechanical oscillations, — for instance of the pendulum, — in which the amplitude of the vibration decreases in constant proportion.

Since the amplitude of the oscillating current varies, constantly decreasing, the oscillating current differs from the alternating current in so far that it starts at a definite time, and gradually dies out, reaching zero value theoretically at infinite time, practically in a very short time, short even in comparison with the time of one alternating half-wave. Characteristic constants of the oscillating current are the period T or frequency $N = 1 / T$, the first amplitude and the ratio of any two successive amplitudes, the latter being called the decrement of the wave. The oscillating current will thus be represented by the product of

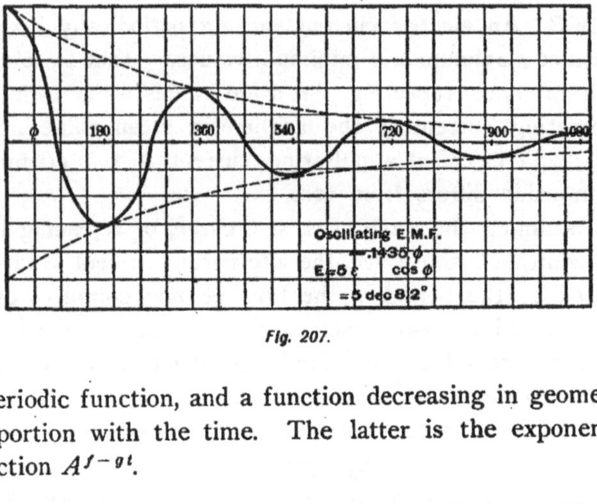

Fig. 207.

a periodic function, and a function decreasing in geometric proportion with the time. The latter is the exponential function A^{f-gt}.

309. Thus, the general expression of the oscillating current is

$$I = A^{f-gt} \cos (2 \pi N t - \hat{\omega}),$$

since

$$A^{f-gt} = A^f A^{-gt} = i \epsilon^{-bt}.$$

Where ϵ = basis of natural logarithms, the current may be expressed

$$I = i \epsilon^{-bt} \cos (2 \pi N t - \hat{\omega}) = i \epsilon^{-a\phi} \cos (\phi - \hat{\omega}),$$

where $\phi = 2 \pi N t$; that is, the period is represented by a complete revolution.

In the same way an oscillating electromotive force will be represented by

$$E = e\epsilon^{-a\phi} \cos (\phi - \hat{\omega}).$$

Such an oscillating electromotive force for the values

$$e = 5, \quad a = .1435 \text{ or } \epsilon^{-2\pi a} = .4, \quad \hat{\omega} = 0,$$

is represented in rectangular coördinates in Fig. 207, and in polar coördinates in Fig. 208. As seen from Fig. 207, the oscillating wave in rectangular coördinates is tangent to the two exponential curves,

$$y = \pm e\epsilon^{-a\phi}.$$

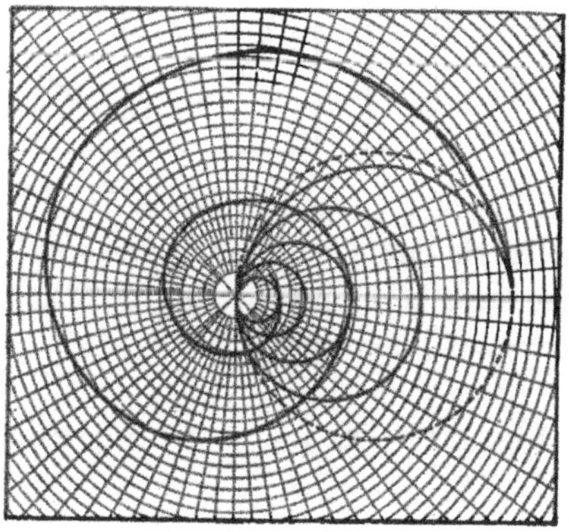

Fig. 208.

310. In polar coördinates, the oscillating wave is represented in Fig. 208 by a spiral curve passing the zero point twice per period, and tangent to the exponential spiral.

$$y = \pm e\epsilon^{-a\phi}.$$

The latter is called the envelope of a system of oscillating waves of which one is shown separately, with the same constants as Figs. 207 and 208, in Fig. 209. Its characteristic feature is: The angle which any concentric circle makes with the curve $y = e\epsilon^{-a\phi}$, is

$$\tan \alpha = \frac{dy}{y\, d\phi} = -a,$$

which is, therefore, constant; or, in other words: "The envelope of the oscillating current is the exponential spiral, which is characterized by a constant angle of intersection

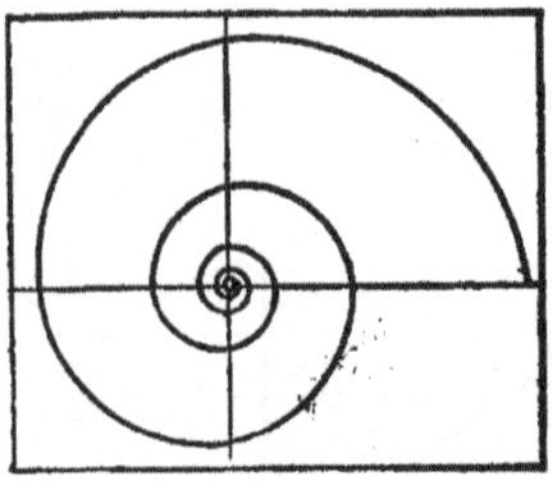

Fig. 209.

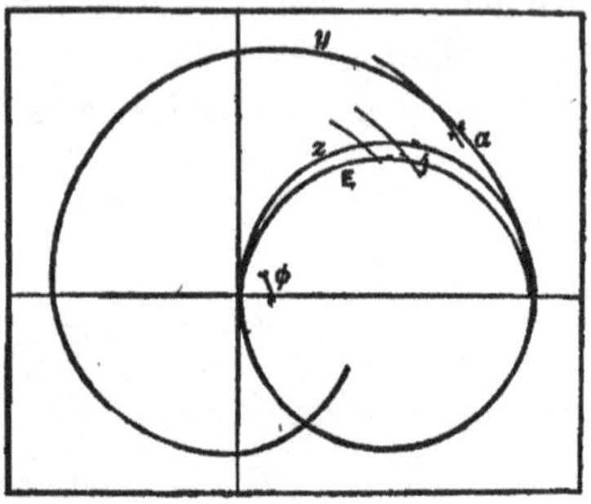

Fig. 210.

with all concentric circles or all radii vectores." The oscillating current wave is the product of the sine wave and the exponential or loxodromic spiral.

311. In Fig. 210 let $y = e\epsilon^{-a\phi}$ represent the exponential spiral;

let $$z = e \cos(\phi - a)$$

represent the sine wave;

and let $$E = e\epsilon^{-a\phi} \cos(\phi - \omega)$$

represent the oscillating wave.

We have then

$$\tan \beta = \frac{dE}{E\, d\phi}$$

$$= \frac{-\sin(\phi - \hat{\omega}) - a\cos(\phi - \hat{\omega})}{\cos(\phi - \hat{\omega})}$$

$$= -\{\tan(\phi - \hat{\omega}) + a\};$$

that is, while the slope of the sine wave, $z = e \cos(\phi - \hat{\omega})$, is represented by

$$\tan \gamma = -\tan(\phi - \hat{\omega}),$$

the slope of the exponential spiral $y = e\epsilon^{-a\phi}$ is

$$\tan \alpha = -a = \text{constant}.$$

That of the oscillating wave $E = e\epsilon^{-a\phi} \cos(\phi - \hat{\omega})$ is

$$\tan \beta = -\{\tan(\phi - \hat{\omega}) + a\}.$$

Hence, it is increased over that of the alternating sine wave by the constant a. The ratio of the amplitudes of two consequent periods is

$$A = \frac{E_{2\pi}}{E_0} = \epsilon^{-2\pi a}.$$

A is called the numerical decrement of the oscillating wave, a the exponential decrement of the oscillating wave, α the angular decrement of the oscillating wave. The oscillating wave can be represented by the equation

$$E = e\epsilon^{-\phi \tan \alpha} \cos(\phi - \hat{\omega}).$$

In the instance represented by Figs. 181 and 182, we have $A = .4$, $a = .1435$, $\alpha = 8.2°$.

Impedance and Admittance.

312. In complex imaginary quantities, the alternating wave

$$z = e \cos(\phi - \hat{\omega})$$

is represented by the symbol

$$\underset{.}{E} = e(\cos \hat{\omega} + j \sin \hat{\omega}) = e_1 + je_2.$$

By an extension of the meaning of this symbolic expression, the oscillating wave $E = e\epsilon^{-a\phi} \cos(\phi - \hat{\omega})$ can be expressed by the symbol

$$\underset{.}{E} = e(\cos \hat{\omega} + j \sin \hat{\omega}) \operatorname{dec} \alpha = (e_1 + je_2) \operatorname{dec} \alpha,$$

where $a = \tan \alpha$ is the exponential decrement, α the angular decrement, $\epsilon^{-2\pi a}$ the numerical decrement.

Inductance.

313. Let r = resistance, L = inductance, and x = $2\pi NL$ = reactance.

In a circuit excited by the oscillating current,

$$I = i\epsilon^{-a\phi}\cos(\phi - \hat{\omega}) = i(\cos\hat{\omega} + j\sin\hat{\omega})\ \text{dec}\ \alpha = (i_1 + ji_2)\ \text{dec}\ \alpha,$$

where $\quad i_1 = i\cos\hat{\omega}, \quad i_2 = i\sin\hat{\omega}, \quad a = \tan\alpha.$

We have then,

The electromotive force consumed by the resistance r of the circuit

$$E_r = rI\ \text{dec}\ \alpha.$$

The electromotive force consumed by the inductance L of the circuit,

$$E_x = L\frac{dI}{dt} = 2\pi NL\frac{dI}{d\phi} = x\frac{dI}{d\phi}.$$

Hence $E_x = -xi\epsilon^{-a\phi}\{\sin(\phi - \hat{\omega}) + a\cos(\phi - \hat{\omega})\}$

$$= -\frac{xi\epsilon^{-a\phi}}{\cos\alpha}\sin(\phi - \hat{\omega} + \alpha).$$

Thus, in symbolic expression,

$$E_x = -\frac{xi}{\cos\alpha}\{-\sin(\hat{\omega} - \alpha) + j\cos(\hat{\omega} - \alpha)\}\ \text{dec}\ \alpha$$

$$= -xi(a + j)(\cos\hat{\omega} + j\sin\hat{\omega})\ \text{dec}\ \alpha;$$

that is, $\quad E_x = -xI(a + j)\ \text{dec}\ \alpha.$

Hence the apparent reactance of the oscillating current circuit is, in symbolic expression,

$$X = x(a + j)\ \text{dec}\ \alpha.$$

Hence it contains an energy component ax, and the impedance is

$$Z = (r - X)\ \text{dec}\ \alpha = \{r - x(a + j)\}\ \text{dec}\ \alpha = (r - ax - jx)\ \text{dec}\ \alpha.$$

Capacity.

314. Let r = resistance, C = capacity, and $x_c = 1/2\pi NC$ = capacity reactance. In a circuit excited by the oscillating current I, the electromotive force consumed by the capacity C is

$$E_x = \frac{1}{C}\int I\,dt = \frac{1}{2\pi NC}\int I\,d\phi = k\int I\,d\phi;$$

or, by substitution,

$$E_x = x\int i\epsilon^{-a\phi}\cos(\phi - \hat{\omega})\,d\phi$$

$$= \frac{x}{1+a^2}\,i\epsilon^{-a\phi}\{\sin(\phi - \hat{\omega}) - a\cos(\phi - \hat{\omega})\}$$

$$= \frac{xi\epsilon^{-a\phi}}{(1+a^2)\cos\alpha}\sin(\phi - \hat{\omega} - \alpha);$$

hence, in symbolic expression,

$$E_x = \frac{xi}{(1+a^2)\cos\alpha}\{-\sin(\hat{\omega}+\alpha) + j\cos(\hat{\omega}+\alpha)\}\text{ dec }a$$

$$= \frac{xi}{1+a^2}(a+j)(\cos\hat{\omega} + j\sin\hat{\omega})\text{ dec }a;$$

hence,

$$E_x = \frac{x}{1+a^2}(-a+j)\,I\text{ dec }a;$$

that is, the apparent capacity reactance of the oscillating circuit is, in symbolic expression,

$$C = \frac{x_c}{1+a^2}(-a+j)\text{ dec }a.$$

315. We have then:

In an oscillating current circuit of resistance r, inductive reactance x, and capacity reactance x_c, with an exponential decrement a, the apparent impedance, in symbolic expression, is:

$$Z = \left\{r - x(a+j) + \frac{x_c}{1+a^2}(-a+j)\right\}\text{ dec }a,$$

$$= \left\{r - a\left(x + \frac{x_c}{1+a^2}\right) - j\left(x - \frac{x_c}{1+a^2}\right)\right\}\text{ dec }a,$$

$$= r_a - jx_a;$$

and, absolute,

$$z_a = \sqrt{r_a^2 + x_a^2}$$

$$= \sqrt{\left[r - a\left(x + \frac{x_c}{1+a^2}\right)\right]^2 + \left[x - \frac{x_c}{1+a^2}\right]^2}.$$

Admittance.

316. Let $I = i\epsilon^{-a\phi} \cos(\phi - \hat{\omega}) =$ current.

Then from the preceding discussion, the electromotive force consumed by resistance r, inductive reactance x, and capacity reactance x_c, is

$$E = ie^{-a\phi}\left\{\cos(\phi - \hat{\omega})\left[r - ax - \frac{a}{1+a^2}x_c\right] - \sin(\phi - \hat{\omega})\left[x - \frac{x_c}{1+a^2}\right]\right\}$$

$$= i z_a \epsilon^{-a\phi} \cos(\phi - \hat{\omega} + \delta),$$

where $\tan \delta = \dfrac{x - \dfrac{x_c}{1+a^2}}{r - ax - \dfrac{a}{1+a^2}x_c}$,

$$z_a = \sqrt{\left(x - \frac{x_c}{1+a^2}\right)^2 + \left(r - ax - \frac{a}{1+a^2}x_c\right)^2};$$

substituting $\hat{\omega} + \delta$ for $\hat{\omega}$, and $e = iz_a$ we have

$$E = e\epsilon^{-a\phi} \cos(\phi - \hat{\omega}),$$

$$I = \frac{e}{z_a}\epsilon^{-a\phi} \cos(\phi - \hat{\omega} - \delta)$$

$$= e\epsilon^{-a\phi}\left\{\frac{\cos\delta}{z_a}\cos(\phi - \hat{\omega}) + \frac{\sin\delta}{z_a}\sin(\phi - \hat{\omega})\right\};$$

hence in complex quantities,

$$\underset{.}{E} = e(\cos\hat{\omega} + j\sin\hat{\omega}) \operatorname{dec} a,$$

$$\underset{.}{I} = \underset{.}{E}\left\{\frac{\cos\delta}{z_a} + j\frac{\sin\delta}{z_a}\right\} \operatorname{dec} a;$$

or, substituting,

$$I = E\left\{\frac{r - ax - \dfrac{a}{1+a^2}x_c}{\left(x - \dfrac{x_c}{1+a^2}\right)^2 + \left(r - ax - \dfrac{a}{1+a^2}x_c\right)^2}\right.$$

$$\left. + j\frac{x - \dfrac{x_c}{1+a^2}}{\left(x - \dfrac{x_c}{1+a^2}\right)^2 + \left(r - ax - \dfrac{a}{1+a^2}x_c\right)^2}\right\} \text{dec } a.$$

317. Thus in complex quantities, for oscillating currents, we have: conductance,

$$g = \frac{r - ax - \dfrac{a}{1+a^2}x_c}{\left(x - \dfrac{x_c}{1+a^2}\right)^2 + \left(r - ax - \dfrac{a}{1+a^2}x_c\right)^2};$$

susceptance,

$$b = \frac{x - \dfrac{x_c}{1+a^2}}{\left(x - \dfrac{x_c}{1+a^2}\right)^2 + \left(r - ax - \dfrac{a}{1+a^2}x_c\right)^2};$$

admittance, in absolute values,

$$y = \sqrt{g^2 + b^2} = \frac{1}{\sqrt{\left(x - \dfrac{x_c}{1+a^2}\right)^2 + \left(r - ax - \dfrac{a}{1+a^2}x_c\right)^2}};$$

in symbolic expression,

$$Y = g + jb = \frac{\left(r - ax - \dfrac{a}{1+a^2}x_c\right) + j\left(x - \dfrac{x_c}{1+a^2}\right)}{\left(x - \dfrac{x_c}{1+a^2}\right)^2 + \left(r - ax - \dfrac{a}{1+a^2}x_c\right)^2}.$$

Since the impedance is

$$Z = \left(r - ax - \frac{a}{1+a^2}x_c\right) - j\left(x - \frac{x_c}{1+a^2}\right) = r_a - jx_a,$$

we have

$$Y = \frac{1}{Z}; \quad y = \frac{1}{z_a}; \quad g = \frac{r_a}{z_a^2}; \quad b = \frac{x_a}{z_a^2};$$

that is, the same relations as in the complex quantities in alternating-current circuits, except that in the present case all the constants r_a, x_a, z_a, g, z, y, depend upon the decrement a.

Circuits of Zero Impedance.

318. In an oscillating-current circuit of decrement a, of resistance r, inductive reactance x, and capacity reactance x_c, the impedance was represented in symbolic expression by

$$Z = r_a - j x_a = \left(r - ax - \frac{a}{1 + a^2} x_c\right) - j\left(x - \frac{x_c}{1 + a^2}\right),$$

or numerically by

$$z = \sqrt{r_a^2 + x_a^2} = \sqrt{\left(r - ax - \frac{a}{1 + a^2} x_c\right)^2 + \left(x - \frac{x_c}{1 + a^2}\right)^2}.$$

Thus the inductive reactance x, as well as the capacity reactance x_c, do not represent wattless electromotive forces as in an alternating-current circuit, but introduce energy components of negative sign

$$- a x - \frac{a}{1 + a^2} x_c;$$

that means,

"In an oscillating-current circuit, the counter electromotive force of self-induction is not in quadrature behind the current, but lags less than 90°, or a quarter period; and the charging current of a condenser is less than 90°, or a quarter period, ahead of the impressed electromotive force."

319. In consequence of the existence of negative energy components of reactance in an oscillating-current circuit, a phenomenon can exist which has no analogy in an alternating-current circuit; that is, under certain conditions the total impedance of the oscillating-current circuit can equal zero:

$$Z = 0.$$

In this case we have

$$r - ax - \frac{a}{1 + a^2} x_c = 0; \quad x - \frac{x_c}{1 + a^2} = 0,$$

substituting in this equation

$$x = 2\pi N L;\ x_c = \frac{1}{2\pi N C}$$

and expanding, we have

$$a = \frac{1}{\sqrt{\frac{4L}{r^2C} - 1}};$$

$$2\pi N = \frac{r}{2L}\sqrt{\frac{4L}{r^2C} - 1} = \frac{r}{2aL}.$$

That is,

"If in an oscillating-current circuit, the decrement

$$a = \frac{1}{\sqrt{\frac{4L}{r^2C} - 1}},$$

and the frequency $N = r/4\pi a L$, the total impedance of the circuit is zero; that is, the oscillating current, when started once, will continue without external energy being impressed upon the circuit."

320. The physical meaning of this is: "If upon an electric circuit a certain amount of energy is impressed and then the circuit left to itself, the current in the circuit will become oscillating, and the oscillations assume the frequency $N = r/4\pi a L$, and the decrement

$$a = \frac{1}{\sqrt{\frac{4L}{r^2C} - 1}}$$

That is, the oscillating currents are the phenomena by which an electric circuit of disturbed equilibrium returns to equilibrium.

This feature shows the origin of the oscillating currents, and the means to produce such currents by disturbing the equilibrium of the electric circuit, for instance, by the discharge of a condenser, by make and break of the circuit, by sudden electrostatic charge, as lightning, etc. Obviously, the most important oscillating currents are those flowing in a circuit of zero impedance, representing oscillating discharges of the circuit. Lightning strokes usually belong to this class.

Oscillating Discharges.

321. The condition of an oscillating discharge is $Z = 0$, that is,

$$a = \frac{1}{\sqrt{\frac{4L}{r^2 C} - 1}}, \quad 2\pi N = \frac{r}{2aL} = \frac{r}{2L}\sqrt{\frac{4L}{r^2 C} - 1}.$$

If $r = 0$, that is, in a circuit without resistance, we have $a = 0$, $N = 1/2\pi\sqrt{LC}$; that is, the currents are alternating with no decrement, and the frequency is that of resonance.

If $4L/r^2 C - 1 < 0$, that is, $r > 2\sqrt{L/C}$, a and N become imaginary; that is, the discharge ceases to be oscillatory. An electrical discharge assumes an oscillating nature only, if $r < 2\sqrt{L/C}$. In the case $r = 2\sqrt{L/C}$ we have $a = \infty$, $N = 0$; that is, the current dies out without oscillation.

From the foregoing we have seen that oscillating discharges, — as for instance the phenomena taking place if a condenser charged to a given potential is discharged through a given circuit, or if lightning strikes the line circuit, — are defined by the equation: $Z = 0$ dec a.

Since

$$\dot{I} = (i_1 + j i_2) \text{ dec } a, \qquad \dot{E}_r = \dot{I} r \text{ dec } a,$$

$$\dot{E}_x = - x \dot{I}(a + j) \text{ dec } a, \quad \dot{E}_{xc} = \frac{x_c}{1 + a^2}\dot{I}(- a + j) \text{ dec } a,$$

we have

$$r - ax - \frac{a}{1 + a^2} x_c = 0,$$

$$- x + \frac{x_c}{1 + a^2} = 0;$$

hence, by substitution,

$$\dot{E}_{xc} = x \dot{I}(- a + j) \text{ dec } a.$$

The two constants, i_1 and i_2, of the discharge, are determined by the initial conditions, that is, the electromotive force and the current at the time $t = 0$.

322. Let a condenser of capacity C be discharged through a circuit of resistance r and inductance L. Let $e =$ electromotive force at the condenser in the moment of closing the circuit, that is, at the time $t = 0$ or $\phi = 0$ At this moment the current is zero; that is,

$$\dot{I} = j i_2, \quad i_1 = 0.$$

Since $\dot{E}_{xc} = x \dot{I} (-a + j) \text{ dec } a = e$ at $\phi = 0$

we have $x i_2 \sqrt{1 + a^2} = e$ or $i_2 = \dfrac{e}{x \sqrt{1 + a^2}}$.

Substituting this, we have,

$$\dot{I} = j \frac{e}{x \sqrt{1 + a^2}} \text{ dec } a, \qquad \dot{E}_r = je \frac{r}{x \sqrt{1 + a^2}} \text{ dec } a,$$

$$\dot{E}_x = \frac{e}{\sqrt{1 + a^2}} (1 - ja) \text{ dec } a, \quad \dot{E}_{xc} = - \frac{e}{\sqrt{1 + a^2}} (1 + ja) \text{ dec } a,$$

the equations of the oscillating discharge of a condenser of initial voltage e.

Since $x = 2 \pi N L$,

$$a = \frac{1}{\sqrt{\frac{4L}{r^2 C} - 1}},$$

$$2 \pi N = \frac{r}{2aL},$$

we have

$$x = \frac{r}{2a} = \frac{r}{2} \sqrt{\frac{4L}{r^2 C} - 1};$$

hence, by substitution,

$$\dot{I} = je \sqrt{\frac{C}{L}} \text{ dec } a, \quad \dot{E}_r = jer \sqrt{\frac{C}{L}} \text{ dec } a;$$

$$\dot{E}_x = \frac{er}{2} \sqrt{\frac{C}{L}} \left(\sqrt{\frac{4L}{r^2 C} - 1} - j \right) \text{ dec } a,$$

$$E_{xc} = -\frac{er}{2}\sqrt{\frac{C}{L}}\left(\sqrt{\frac{4L}{r^2C}-1}+j\right)\text{dec } a,$$

$$a = \frac{1}{\sqrt{\frac{4L}{r^2C}-1}}, \quad N = \frac{r\sqrt{\frac{4L}{r^2C}-1}}{4\pi L}$$

the final equations of the oscillating discharge, in symbolic expression.

Oscillating Current Transformer.

323. As an instance of the application of the symbolic method of analyzing the phenomena caused by oscillating currents, the transformation of such currents may be investigated. If an oscillating current is produced in a circuit including the primary of a transformer, oscillating currents will also flow in the secondary of this transformer. In a transformer let the ratio of secondary to primary turns be p. Let the secondary be closed by a circuit of total resistance, $r_1 = r_1' + r_1''$, where $r_1' =$ external, $r_1'' =$ internal, resistance. The total inductance $L_1 = L_1' + L_1''$, where $L_1' =$ external, $L_1'' =$ internal, inductance; total capacity, C_1. Then the total admittance of the secondary circuit is

$$Y_1 = (g_1 + jb_1)\text{ dec } a = \frac{1}{\left(r_1 - ax_1 - \frac{a}{1+a^2}x_{c1}\right) - j\left(x_1 - \frac{x_c}{1+a^2}\right)},$$

where $x_1 = 2\pi NL_1 =$ inductive reactance: $x_{c1} = 1/2\pi NC =$ capacity reactance. Let $r_0 =$ effecive hysteretic resistance, $L_0 =$ inductance; hence, $x_0 = 2\pi NL_0 =$ reactance; hence,

$$Y_0 = g_0 + jb_0 = \frac{1}{(r_0 - ax_0) - jx_0} = \text{admittance}$$

of the primary exciting circuit of the transformer; that is, the admittance of the primary circuit at open secondary circuit.

As discussed elsewhere, a transformer can be considered as consisting of the secondary circuit supplied by the impressed electromotive force over leads, whose impedance is equal to the sum of primary and secondary transformer impedance, and which are shunted by the exciting circuit, outside of the secondary, but inside of the primary impedance.

Let r = resistance; L = inductance; C = capacity; hence,

$$x = 2\pi NL = \text{inductive reactance},$$

$x_c = 1/2\pi NC$ = capacity reactance of the total primary circuit, including the primary coil of the transformer. If $\dot{E}_1' = E_1'$ dec a denotes the electromotive force induced in the secondary of the transformer by the mutual magnetic flux; that is, by the oscillating magnetism interlinked with the primary and secondary coil, we have $\dot{I}_1 = \dot{E}_1' \dot{Y}_1$ dec a = secondary current.

Hence, $\dot{I}_1' = p \dot{I}_1$ dec $a = p E_1' Y_1$ dec a = primary load current, or component of primary current corresponding to secondary current. Also, $\dot{I}_0 = \frac{1}{p} \dot{E}_1' Y_0$ dec a = primary exciting current; hence, the total primary current is

$$\dot{I} = \dot{I}_1' + \dot{I}_0 = \frac{E_1'}{p}\{Y_0 + p^2 Y_1\} \text{ dec } a.$$

$\dot{E}' = \frac{E_1'}{p}$ dec a = induced primary electromotive force.

Hence the total primary electromotive force is

$$\dot{E} = (\dot{E}' + \dot{I}Z) \text{ dec } a = \frac{E_1'}{p}\{1 + Z Y_0 + p^2 Z Y_1\} \text{ dec } a.$$

In an oscillating discharge the total primary electromotive force $\dot{E} = 0$; that is,

$$1 + Z Y_0 + p^2 Z Y_1 = 0;$$

or, the substitution

$$1 + \frac{\left(r - a x - \frac{a}{1 + a^2} x_c\right) - j\left(x - \frac{x_c}{1 + a^2}\right)}{(r_0 - a x_0) - j x_0} + p^2 \frac{\left(r - a x - \frac{a}{1 + a^2} x_c\right) - j\left(x - \frac{x_c}{1 + a^2}\right)}{\left(r_1 - a x_1 - \frac{a}{1 + a^2} x_{c1}\right) - j\left(x_1 - \frac{x_{c1}}{1 + a^2}\right)} = 0.$$

Substituting in this equation, $x = 2\pi N C$, $x_c = 1/2\pi N C$, etc., we get a complex imaginary equation with the two constants a and N. Separating this equation in the real and the imaginary parts, we derive two equations, from which the two constants a and N of the discharge are calculated.

324. If the exciting current of the transformer is negligible, — that is, if $Y_0 = 0$, the equation becomes essentially simplified, —

$$1 + p^2 \frac{\left(r - a x - \frac{a}{1+a^2} x_c\right) - j\left(x - \frac{x_c}{1+a^2}\right)}{\left(r_1 - a x_1 - \frac{a}{1+a^2} x_{c1}\right) - j\left(x_1 - \frac{x_{c1}}{1+a^2}\right)} = 0;$$

that is,

$$\left(r_1 - a x_1 - \frac{a}{1+a^2} x_{c1}\right) + p^2\left(r - a x - \frac{a}{1+a^2} x_c\right) = 0;$$

$$\left(x_1 - \frac{x_{c1}}{1+a^2}\right) + p^2\left(x - \frac{x_c}{1+a^2}\right) = 0;$$

or, combined, —

$$(r_1 - 2 a x_1) + p^2 (r - 2 a x) = 0,$$
$$r_1 + p^2 r = 2 a (x_1 + p^2 x),$$
$$x_{c1} + p^2 x_c = (1 + a^2)(x_1 + p^2 x).$$

Substituting for x_1, x, x_{c1}, x_c, we have

$$a = \frac{1}{\sqrt{\frac{4(L_1 + p^2 L)}{(r_1 + p^2 r)^2 (C_1 + p^2 C)} - 1}},$$

$$2\pi N = \frac{r_1 + p^2 r}{2a(L_1 + p^2 L)}$$

$$= \frac{r_1 + p^2 r}{2(L_1 + p^2 L)} \sqrt{\frac{4(L_1 + p^2 L)}{(r_1 + p^2 r)^2 (C_1 + p^2 C)} - 1}.$$

$$E = \frac{E_1'}{p}\{1 + p^2 Z Y_1\} \text{ dec } a,$$

$$I = p E_1' Y_1 \text{ dec } a,$$

$$I_1 = E_1' Y_1 \text{ dec } a,$$

the equations of the oscillating-current transformer, with E_1' as parameter.

INDEX.

This page intentionally left blank

Made in the USA
Monee, IL
11 November 2026

68306945R00319